BIRKHÄUSER

Oberwolfach Seminars

Volume3 8

Discrete
Differential Geometry

Alexander I. Bobenko
Peter Schröder
John M. Sullivan
Günter M. Ziegler
Editors

Birkhäuser
Basel · Boston · Berlin

Alexander I. Bobenko
Institut für Mathematik, MA 8-3
Technische Universität Berlin
Strasse des 17. Juni 136
10623 Berlin, Germany
e-mail: bobenko@math.tu-berlin.de

Peter Schröder
Department of Computer Science
Caltech, MS 256-80
1200 E. California Blvd.
Pasadena, CA 91125, USA
e-mail: ps@cs.caltech.edu

John M. Sullivan
Institut für Mathematik, MA 3-2
Technische Universität Berlin
Strasse des 17. Juni 136
10623 Berlin, Germany
e-mail: sullivan@math.tu-berlin.de

Günter M. Ziegler
Institut für Mathematik, MA 6-2
Technische Universität Berlin
Strasse des 17. Juni 136
10623 Berlin, Germany
e-mail: ziegler@math.tu-berlin.de

2000 Mathematics Subject Classification: 53-02 (primary); 52-02, 53-06, 52-06

Library of Congress Control Number: 2007941037

Bibliographic information published by Die Deutsche Bibliothek
Die Deutsche Bibliothek lists this publication in the Deutsche Nationalbibliografie;
detailed bibliographic data is available in the Internet at <http://dnb.ddb.de>.

© 2008 Birkhäuser Verlag AG
Basel · Boston · Berlin
P.O. Box 133, CH-4010 Basel, Switzerland
Part of Springer Science+Business Media
Printed on acid-free paper produced from chlorine-free pulp. TCF ∞

ISBN 978-3-7643-8620-7   ISBN 978-3-7643-8621-4 (eBook)
DOI 10.1007/978-3-7643-8621-4
9 8 7 6 5 4 3 2 1   www.birkhauser.ch

Preface

Discrete differential geometry (DDG) is a new and active mathematical terrain where differential geometry (providing the classical theory of smooth manifolds) interacts with discrete geometry (concerned with polytopes, simplicial complexes, etc.), using tools and ideas from all parts of mathematics. DDG aims to develop discrete equivalents of the geometric notions and methods of classical differential geometry. Current interest in this field derives not only from its importance in pure mathematics but also from its relevance for other fields such as computer graphics.

Discrete differential geometry initially arose from the observation that when a notion from smooth geometry (such as that of a minimal surface) is discretized "properly", the discrete objects are not merely approximations of the smooth ones, but have special properties of their own, which make them form a coherent entity by themselves. One might suggest many different reasonable discretizations with the same smooth limit. Among these, which one is the best? From the theoretical point of view, the best discretization is the one which preserves the fundamental properties of the smooth theory. Often such a discretization clarifies the structures of the smooth theory and possesses important connections to other fields of mathematics, for instance to projective geometry, integrable systems, algebraic geometry, or complex analysis. The discrete theory is in a sense the more fundamental one: the smooth theory can always be recovered as a limit, while it is a nontrivial problem to find which discretization has the desired properties.

The problems considered in discrete differential geometry are numerous and include in particular: discrete notions of curvature, special classes of discrete surfaces (such as those with constant curvature), cubical complexes (including quad-meshes), discrete analogs of special parametrization of surfaces (such as conformal and curvature-line parametrizations), the existence and rigidity of polyhedral surfaces (for example, of a given combinatorial type), discrete analogs of various functionals (such as bending energy), and approximation theory. Since computers work with discrete representations of data, it is no surprise that many of the applications of DDG are found within computer science, particularly in the areas of computational geometry, graphics and geometry processing.

Despite much effort by various individuals with exceptional scientific breadth, large gaps remain between the various mathematical subcommunities working in discrete differential geometry. The scientific opportunities and potential applications here are very substantial. The goal of the Oberwolfach Seminar "Discrete Differential Geometry" held in May–June 2004 was to bring together mathematicians from various subcommunities working in different aspects of DDG to give lecture courses addressed to a general mathematical audience. The seminar was primarily addressed to students and postdocs, but some more senior specialists working in the field also participated.

There were four main lecture courses given by the editors of this volume, corresponding to the four parts of this book:

   I: Discretization of Surfaces: Special Classes and Parametrizations,
  II: Curvatures of Discrete Curves and Surfaces,
 III: Geometric Realizations of Combinatorial Surfaces,
  IV: Geometry Processing and Modeling with Discrete Differential Geometry.

These courses were complemented by related lectures by other participants. The topics were chosen to cover (as much as possible) the whole spectrum of DDG—from differential geometry and discrete geometry to applications in geometry processing.

Part I of this book focuses on special discretizations of surfaces, including those related to integrable systems. Bobenko's "Surfaces from Circles" discusses several ways to discretize surfaces in terms of circles and spheres, in particular a Möbius-invariant discretization of Willmore energy and S-isothermic discrete minimal surfaces. The latter are explored in more detail, with many examples, in Bücking's article. Pinkall constructs discrete surfaces of constant negative curvature, documenting an interactive computer tool that works in real time. The final three articles focus on connections between quad-surfaces and integrable systems: Schief, Bobenko and Hoffmann consider the rigidity of quad-surfaces; Hoffmann constructs discrete versions of the smoke-ring flow and Hashimoto surfaces; and Suris considers discrete holomorphic and harmonic functions on quad-graphs.

Part II considers discretizations of the usual notions of curvature for curves and surfaces in space. Sullivan's "Curves of Finite Total Curvature" gives a unified treatment of curvatures for smooth and polygonal curves in the framework of such FTC curves. The article by Denne and Sullivan considers isotopy and convergence results for FTC graphs, with applications to geometric knot theory. Sullivan's "Curvatures of Smooth and Discrete Surfaces" introduces different discretizations of Gauss and mean curvature for polyhedral surfaces from the point of view of preserving integral curvature relations.

Part III considers the question of realizability: which polyhedral surfaces can be embedded in space with flat faces. Ziegler's "Polyhedral Surfaces of High Genus" describes constructions of triangulated surfaces with n vertices having genus $O(n^2)$ (not known to be realizable) or genus $O(n \log n)$ (realizable). Timmreck gives some new criteria which could be used to show surfaces are not realizable. Lutz discusses automated methods to enumerate triangulated surfaces and to search for realizations. Bokowski discusses heuristic methods for finding realizations, which he has used by hand.

Part IV focuses on applications of discrete differential geometry. Schröder's "What Can We Measure?" gives an overview of intrinsic volumes, Steiner's formula and Hadwiger's theorem. Wardetzky shows that normal convergence of polyhedral surfaces to a smooth limit suffices to get convergence of area and of mean curvature as defined by the cotangent formula. Desbrun, Kanso and Tong discuss the use of a discrete exterior calculus for computational modeling. Grinspun considers a discrete model, based on bending energy, for thin shells.

We wish to express our gratitude to the Mathematisches Forschungsinstitut Oberwolfach for providing the perfect setting for the seminar in 2004. Our work in discrete differential geometry has also been supported by the Deutsche Forschungsgemeinschaft (DFG), as well as other funding agencies. In particular, the DFG Research Unit "Polyhedral Surfaces", based at the Technische Universität Berlin since 2005, has provided direct support to the three of us (Bobenko, Sullivan, Ziegler) based in Berlin, as well as to Bücking and Lutz. Further authors including Hoffmann, Schief, Suris and Timmreck have worked closely with this Research Unit; the DFG also supported Hoffmann through a Heisenberg Fellowship. The DFG Research Center MATHEON in Berlin, through its Application Area F "Visualization", has supported work on the applications of discrete differential geometry. Support from MATHEON went to authors Bücking and Wardetzky as well as to the three of us in Berlin. The National Science Foundation supported the work of Grinspun and Schröder, as detailed in the acknowledgments in their articles.

Our hope is that this book will stimulate the interest of other mathematicians to work in the field of discrete differential geometry, which we find so fascinating.

Alexander I. Bobenko
Peter Schröder
John M. Sullivan
Günter M. Ziegler   Berlin, September 2007

Contents

Part I

Discretization of Surfaces:

Special Classes and Parametrizations

Discrete Differential Geometry, A.I. Bobenko, P. Schröder, J.M. Sullivan and G.M. Ziegler, eds.
Oberwolfach Seminars, Vol. 38, 3–35
© 2008 Birkhäuser Verlag Basel/Switzerland

Surfaces from Circles

Alexander I. Bobenko

Abstract. In the search for appropriate discretizations of surface theory it is crucial to preserve fundamental properties of surfaces such as their invariance with respect to transformation groups. We discuss discretizations based on Möbius-invariant building blocks such as circles and spheres. Concrete problems considered in these lectures include the Willmore energy as well as conformal and curvature-line parametrizations of surfaces. In particular we discuss geometric properties of a recently found discrete Willmore energy. The convergence to the smooth Willmore functional is shown for special refinements of triangulations originating from a curvature-line parametrization of a surface. Further we treat special classes of discrete surfaces such as isothermic, minimal, and constant mean curvature. The construction of these surfaces is based on the theory of circle patterns, in particular on their variational description.

Keywords. Circular nets, discrete Willmore energy, discrete curvature lines, isothermic surfaces, discrete minimal surfaces, circle patterns.

1. Why from circles?

The theory of polyhedral surfaces aims to develop discrete equivalents of the geometric notions and methods of smooth surface theory. The latter appears then as a limit of refinements of the discretization. Current interest in this field derives not only from its importance in pure mathematics but also from its relevance for other fields like computer graphics.

One may suggest many different reasonable discretizations with the same smooth limit. Which one is the best? In the search for appropriate discretizations, it is crucial to preserve the fundamental properties of surfaces. A natural mathematical discretization principle is the invariance with respect to transformation groups. A trivial example of this principle is the invariance of the theory with respect to Euclidean motions. A less trivial but well-known example is the discrete analog for the local Gaussian curvature defined as the angle defect $G(v) = 2\pi - \sum \alpha_i$, at a vertex v of a polyhedral surface. Here the α_i are the angles of all polygonal faces (see Figure 3) of the surface at vertex v. The discrete

FIGURE 1. Discrete surfaces made from circles: general simplicial surface and a discrete minimal Enneper surface.

Gaussian curvature $G(v)$ defined in this way is preserved under isometries, which is a
discrete version of the Theorema Egregium of Gauss.

In these lectures, we focus on surface geometries invariant under Möbius transformations. Recall that Möbius transformations form a finite-dimensional Lie group generated by inversions in spheres; see Figure 2. Möbius transformations can be also thought

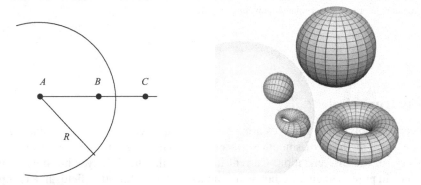

FIGURE 2. Inversion $B \mapsto C$ in a sphere, $|AB||AC| = R^2$. A sphere
and a torus of revolution and their inversions in a sphere: spheres are
mapped to spheres.

as compositions of translations, rotations, homotheties and inversions in spheres. Alternatively, in dimensions $n \geq 3$, Möbius transformations can be characterized as conformal
transformations: Due to Liouville's theorem any conformal mapping $F : U \to V$ between two open subsets $U, V \subset \mathbb{R}^n, n \geq 3$, is a Möbius transformation.

Many important geometric notions and properties are known to be preserved by Möbius transformations. The list includes in particular:

- spheres of any dimension, in particular circles (planes and straight lines are treated as infinite spheres and circles),
- intersection angles between spheres (and circles),
- curvature-line parametrization,
- conformal parametrization,
- isothermic parametrization (conformal curvature-line parametrization),
- the Willmore functional (see Section 2).

For discretization of Möbius-invariant notions it is natural to use Möbius-invariant building blocks. This observation leads us to the conclusion that the discrete conformal or curvature-line parametrizations of surfaces and the discrete Willmore functional should be formulated in terms of circles and spheres.

2. Discrete Willmore energy

The Willmore functional [42] for a smooth surface S in 3-dimensional Euclidean space is

$$\mathcal{W}(S) = \frac{1}{4} \int_S (k_1 - k_2)^2 dA = \int_S H^2 dA - \int_S K dA.$$

Here dA is the area element, k_1 and k_2 the principal curvatures, $H = \frac{1}{2}(k_1 + k_2)$ the mean curvature, and $K = k_1 k_2$ the Gaussian curvature of the surface.

Let us mention two important properties of the Willmore energy:

- $\mathcal{W}(S) \geq 0$ and $\mathcal{W}(S) = 0$ if and only if S is a round sphere.
- $\mathcal{W}(S)$ (and the integrand $(k_1 - k_2)^2 dA$) is Möbius-invariant [1, 42].

Whereas the first claim almost immediately follows from the definition, the second is a nontrivial property. Observe that for closed surfaces $\mathcal{W}(S)$ and $\int_S H^2 dA$ differ by a topo-logical invariant $\int K dA = 2\pi\chi(S)$. We prefer the definition of $\mathcal{W}(S)$ with a Möbius-invariant integrand.

Observe that minimization of the Willmore energy $\mathcal{W}$ seeks to make the surface "as round as possible". This property and the Möbius invariance are two principal goals of the geometric discretization of the Willmore energy suggested in [3]. In this section we present the main results of [3] with complete derivations, some of which were omitted there.

2.1. Discrete Willmore functional for simplicial surfaces

Let S be a simplicial surface in 3-dimensional Euclidean space with vertex set V, edges E and (triangular) faces F. We define the discrete Willmore energy of S using the circum-circles of its faces. Each (internal) edge $e \in E$ is incident to two triangles. A consistent orientation of the triangles naturally induces an orientation of the corresponding circum-circles. Let $\beta(e)$ be the external intersection angle of the circumcircles of the triangles sharing e, meaning the angle between the tangent vectors of the oriented circumcircles (at either intersection point).

Definition 2.1. The *local discrete Willmore energy* at a vertex v is the sum

$$W(v) = \sum_{e \ni v} \beta(e) - 2\pi.$$

over all edges incident to v. The *discrete Willmore energy* of a compact simplicial surface S without boundary is the sum over all vertices

$$W(S) = \frac{1}{2} \sum_{v \in V} W(v) = \sum_{e \in E} \beta(e) - \pi|V|.$$

Here $|V|$ is the number of vertices of S.

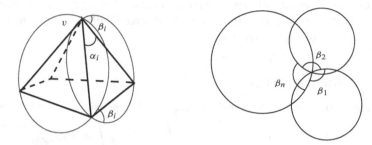

FIGURE 3. Definition of discrete Willmore energy.

Figure 3 presents two neighboring circles with their external intersection angle β_i as well as a view "from the top" at a vertex v showing all n circumcircles passing through v with the corresponding intersection angles $\beta_1, \ldots, \beta_n$. For simplicity we will consider only simplicial surfaces without boundary.

The energy $W(S)$ is obviously invariant with respect to Möbius transformations.

The star $S(v)$ of the vertex v is the subcomplex of S consisting of the triangles incident with v. The vertices of $S(v)$ are v and all its neighbors. We call $S(v)$ convex if for each of its faces $f \in F(S(v))$ the star $S(v)$ lies to one side of the plane of f and strictly convex if the intersection of $S(v)$ with the plane of f is f itself.

Proposition 2.2. *The conformal energy $W(v)$ is non-negative and vanishes if and only if the star $S(v)$ is convex and all its vertices lie on a common sphere.*

The proof of this proposition is based on an elementary lemma.

Lemma 2.3. *Let $\mathcal{P}$ be a (not necessarily planar) n-gon with external angles β_i. Choose a point P and connect it to all vertices of $\mathcal{P}$. Let α_i be the angles of the triangles at the tip P of the pyramid thus obtained (see Figure 4). Then*

$$\sum_{i=1}^{n} \beta_i \geq \sum_{i=1}^{n} \alpha_i,$$

and equality holds if and only if $\mathcal{P}$ is planar and convex and the vertex P lies inside $\mathcal{P}$.

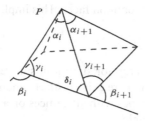

FIGURE 4. Proof of Lemma 2.3

Proof. Denote by γ_i and δ_i the angles of the triangles at the vertices of $\mathcal{P}$, as in Figure 4. The claim of Lemma 2.3 follows from summing over all $i = 1, \ldots, n$ the two obvious relations

$$\beta_{i+1} \geq \pi - (\gamma_{i+1} + \delta_i)$$
$$\alpha_i = \pi - (\gamma_i + \delta_i).$$

All inequalities become equalities only in the case when $\mathcal{P}$ is planar, convex and contains P. □

For P in the convex hull of $\mathcal{P}$ we have $\sum \alpha_i \geq 2\pi$. As a corollary we obtain a polygonal version of Fenchel's theorem [21]:

Corollary 2.4.

$$\sum_{i=1}^{n} \beta_i \geq 2\pi.$$

Proof of Proposition 2.2. The claim of Proposition 2.2 is invariant with respect to Möbius transformations. Applying a Möbius transformation M that maps the vertex v to infinity, $M(v) = \infty$, we make all circles passing through v into straight lines and arrive at the geometry shown in Figure 4 with $P = M(\infty)$. Now the result follows immediately from Corollary 2.4. □

Theorem 2.5. *Let S be a compact simplicial surface without boundary. Then*

$$W(S) \geq 0,$$

and equality holds if and only if S is a convex polyhedron inscribed in a sphere, i.e., a Delaunay triangulation of a sphere.

Proof. Only the second statement needs to be proven. By Proposition 2.2, the equality $W(S) = 0$ implies that the star of each vertex of S is convex (but not necessarily strictly convex). Deleting the edges that separate triangles lying in a common plane, one obtains a polyhedral surface S_P with circular faces and all strictly convex vertices and edges. Proposition 2.2 implies that for every vertex v there exists a sphere S_v with all vertices of the star $S(v)$ lying on it. For any edge (v_1, v_2) of S_P two neighboring spheres S_{v_1} and

S_{v_2} share two different circles of their common faces. This implies $S_{v_1} = S_{v_2}$ and finally the coincidence of all the spheres S_v.   □

2.2. Non-inscribable polyhedra

The minimization of the conformal energy for simplicial spheres is related to a classical result of Steinitz [40], who showed that there exist abstract simplicial 3-polytopes without geometric realizations as convex polytopes with all vertices on a common sphere. We call these combinatorial types non-inscribable.

Let S be a simplicial sphere with vertices colored in black and white. Denote the sets of white and black vertices by V_w and V_b, respectively, $V = V_w \cup V_b$. Assume that there are no edges connecting two white vertices and denote the sets of the edges connecting white and black vertices and two black vertices by E_{wb} and E_{bb}, respectively, $E = E_{wb} \cup E_{bb}$. The sum of the local discrete Willmore energies over all white vertices can be represented as

$$\sum_{v \in V_w} W(v) = \sum_{e \in E_{wb}} \beta(e) - 2\pi |V_w|.$$

Its non-negativity yields $\sum_{e \in E_{wb}} \beta(e) \geq 2\pi |V_w|$. For the discrete Willmore energy of S this implies

$$W(S) = \sum_{e \in E_{wb}} \beta(e) + \sum_{e \in E_{bb}} \beta(e) - \pi(|V_w| + |V_b|) \geq \pi(|V_w| - |V_b|). \qquad (2.1)$$

Equality here holds if and only if $\beta(e) = 0$ for all $e \in E_{bb}$ and the star of any white vertices is convex, with vertices lying on a common sphere. We come to the conclusion that the polyhedra of this combinatorial type with $|V_w| > |V_b|$ have positive Willmore energy and thus cannot be realized as convex polyhedra all of whose vertices belong to a sphere. These are exactly the non-inscribable examples of Steinitz (see [24]).

One such example is presented in Figure 5. Here the centers of the edges of the tetrahedron are black and all other vertices are white, so $|V_w| = 8, |V_b| = 6$. The estimate (2.1) implies that the discrete Willmore energy of any polyhedron of this type is at least 2π. The polyhedra with energy equal to 2π are constructed as follows. Take a tetrahedron, color its vertices white and chose one black vertex per edge. Draw circles through each white vertex and its two black neighbors. We get three circles on each face. Due to Miquel's theorem (see Figure 10) these three circles intersect at one point. Color this new vertex white. Connect it by edges to all black vertices of the triangle and connect pairwise the black vertices of the original faces of the tetrahedron. The constructed polyhedron has $W = 2\pi$.

To construct further polyhedra with $|V_w| > |V_b|$, take a polyhedron $\hat{P}$ whose number of faces is greater than the number of vertices $|\hat{F}| > |\hat{V}|$. Color all the vertices black, add white vertices at the faces and connect them to all black vertices of a face. This yields a polyhedron with $|V_w| = |\hat{F}| > |V_b| = |\hat{V}|$. Hodgson, Rivin and Smith [27] have found a characterization of inscribable combinatorial types, based on a transfer to the Klein model of hyperbolic 3-space. Their method is related to the methods of construction of discrete minimal surfaces in Section 5.

 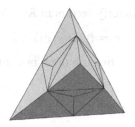

FIGURE 5. Discrete Willmore spheres of inscribable ($W = 0$) and non-inscribable ($W > 0$) types.

The example in Figure 5 (right) is one of the few for which the minimum of the discrete Willmore energy can be found by elementary methods. Generally this is a very appealing (but probably difficult) problem of discrete differential geometry (see the discussion in [3]).

Complete understanding of non-inscribable simplicial spheres is an interesting mathematical problem. However the existence of such spheres might be seen as a problem for using the discrete Willmore functional for applications in computer graphics, such as the fairing of surfaces. Fortunately the problem disappears after just one refinement step: all simplicial spheres become inscribable. Let **S** be an abstract simplicial sphere. Define its refinement **S$_R$** as follows: split every edge of **S** in two by inserting additional vertices, and connect these new vertices sharing a face of **S** by additional edges ($1 \to 4$ refinement, as in Figure 7 (left)).

Proposition 2.6. *The refined simplicial sphere* **S$_R$** *is inscribable, and thus there exists a polyhedron* S_R *with the combinatorics of* **S$_R$** *and* $W(S_R) = 0$.

Proof. Koebe's theorem (see Theorem 5.3, Section 5) states that every abstract simplicial sphere **S** can be realized as a convex polyhedron S all of whose edges touch a common sphere S^2. Starting with this realization S it is easy to construct a geometric realization S_R of the refinement **S$_R$** inscribed in S^2. Indeed, choose the touching points of the edges of S with S^2 as the additional vertices of S_R and project the original vertices of S (which lie outside of the sphere S^2) to S^2. One obtains a convex simplicial polyhedron S_R inscribed in S^2. $\qquad\square$

2.3. Computation of the energy

For derivation of some formulas it will be convenient to use the language of quaternions. Let $\{\mathbf{1}, \mathbf{i}, \mathbf{j}, \mathbf{k}\}$ be the standard basis

$$\mathbf{ij} = \mathbf{k}, \quad \mathbf{jk} = \mathbf{i}, \quad \mathbf{ki} = \mathbf{j}, \quad \mathbf{ii} = \mathbf{jj} = \mathbf{kk} = -1$$

of the quaternion algebra $\mathbb{H}$. A quaternion $q = q_0\mathbf{1} + q_1\mathbf{i} + q_2\mathbf{j} + q_3\mathbf{k}$ is decomposed in its real part $\operatorname{Re} q := q_0 \in \mathbb{R}$ and imaginary part $\operatorname{Im} q := q_1\mathbf{i} + q_2\mathbf{j} + q_3\mathbf{k} \in \operatorname{Im} \mathbb{H}$. The absolute value of q is $|q| := q_0^2 + q_1^2 + q_2^2 + q_3^2$.

We identify vectors in $\mathbb{R}^3$ with imaginary quaternions

$$v = (v_1, v_2, v_3) \in \mathbb{R}^3 \quad \longleftrightarrow \quad v = v_1 \mathbf{i} + v_2 \mathbf{j} + v_3 \mathbf{k} \in \text{Im } \mathbb{H}$$

and do not distinguish them in our notation. For the quaternionic product this implies

$$vw = -\langle v, w \rangle + v \times w, \tag{2.2}$$

where $\langle v, w \rangle$ and $v \times w$ are the scalar and vector products in $\mathbb{R}^3$.

Definition 2.7. Let $x_1, x_2, x_3, x_4 \in \mathbb{R}^3 \cong \text{Im } \mathbb{H}$ be points in 3-dimensional Euclidean space. The quaternion

$$q(x_1, x_2, x_3, x_4) := (x_1 - x_2)(x_2 - x_3)^{-1}(x_3 - x_4)(x_4 - x_1)^{-1}$$

is called the *cross-ratio* of x_1, x_2, x_3, x_4.

The cross-ratio is quite useful due to its Möbius properties:

Lemma 2.8. *The absolute value and real part of the cross-ratio* $q(x_1, x_2, x_3, x_4)$ *are preserved by Möbius transformations. The quadrilateral* x_1, x_2, x_3, x_4 *is circular if and only if* $q(x_1, x_2, x_3, x_4) \in \mathbb{R}$.

Consider two triangles with a common edge. Let $a, b, c, d \in \mathbb{R}^3$ be their other edges, oriented as in Figure 6.

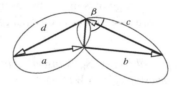

FIGURE 6. Formula for the angle between circumcircles.

Proposition 2.9. *The external angle* $\beta \in [0, \pi]$ *between the circumcircles of the triangles in Figure 6 is given by any of the equivalent formulas:*

$$\cos(\beta) = -\frac{\text{Re } q}{|q|} = -\frac{\text{Re } (abcd)}{|abcd|}$$
$$= \frac{\langle a, c \rangle \langle b, d \rangle - \langle a, b \rangle \langle c, d \rangle - \langle b, c \rangle \langle d, a \rangle}{|a||b||c||d|}. \tag{2.3}$$

Here $q = ab^{-1}cd^{-1}$ *is the cross-ratio of the quadrilateral.*

Proof. Since $\text{Re } q$, $|q|$ and β are Möbius-invariant, it is enough to prove the first formula for the planar case $a, b, c, d \in \mathbb{C}$, mapping all four vertices to a plane by a Möbius transformation. In this case q becomes the classical complex cross-ratio. Considering the arguments $a, b, c, d \in \mathbb{C}$ one easily arrives at $\beta = \pi - \arg q$. The second representation follows from the identity $b^{-1} = -b/|b|$ for imaginary quaternions. Finally applying (2.2) we obtain

$$\text{Re}\,(abcd) = \langle a,b \rangle \langle c,d \rangle - \langle a \times b, c \times d \rangle$$
$$= \langle a,b \rangle \langle c,d \rangle + \langle b,c \rangle \langle d,a \rangle - \langle a,c \rangle \langle b,d \rangle. \qquad \square$$

2.4. Smooth limit

The discrete energy W is not only a discrete analogue of the Willmore energy. In this section we show that it approximates the smooth Willmore energy, although the smooth limit is very sensitive to the refinement method and should be chosen in a special way. We consider a special infinitesimal triangulation which can be obtained in the limit of $1 \to 4$ refinements (see Figure 7 (left)) of a triangulation of a smooth surface. Intuitively it is clear that in the limit one has a regular triangulation such that almost every vertex is of valence 6 and neighboring triangles are congruent up to sufficiently high order in ϵ (ϵ being of the order of the distances between neighboring vertices).

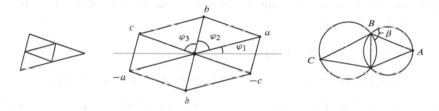

FIGURE 7. Smooth limit of the discrete Willmore energy. *Left:* The $1 \to 4$ refinement. *Middle:* An infinitesimal hexagon in the parameter plane with a (horizontal) curvature line. *Right:* The β-angle corresponding to two neighboring triangles in $\mathbb{R}^3$.

We start with a comparison of the discrete and smooth Willmore energies for an important modeling example. Consider a neighborhood of a vertex $v \in \mathcal{S}$, and represent the smooth surface locally as a graph over the tangent plane at v:

$$\mathbb{R}^2 \ni (x,y) \mapsto f(x,y) = \left(x, y, \frac{1}{2}(k_1 x^2 + k_2 y^2) + o(x^2 + y^2) \right) \in \mathbb{R}^3, \quad (x,y) \to 0.$$

Here x, y are the curvature directions and k_1, k_2 are the principal curvatures at v. Let the vertices $(0,0)$, $a = (a_1, a_2)$ and $b = (b_1, b_2)$ in the parameter plane form an acute triangle. Consider the infinitesimal hexagon with vertices $\epsilon a, \epsilon b, \epsilon c, -\epsilon a, -\epsilon b, -\epsilon c$, (see Figure 7 (middle)), with $b = a + c$. The coordinates of the corresponding points on the smooth surface are

$$f(\pm \epsilon a) = \epsilon(\pm a_1, \pm a_2, \epsilon r_a + o(\epsilon)),$$
$$f(\pm \epsilon c) = \epsilon(\pm c_1, \pm c_2, \epsilon r_c + o(\epsilon)),$$
$$f(\pm \epsilon b) = (f(\pm \epsilon a) + f(\pm \epsilon c)) + \epsilon^2 R, \quad R = (0, 0, r + o(\epsilon)),$$

where

$$r_a = \frac{1}{2}(k_1 a_1^2 + k_2 a_2^2), \quad r_c = \frac{1}{2}(k_1 c_1^2 + k_2 c_2^2), \quad r = (k_1 a_1 c_1 + k_2 a_2 c_2)$$

and $a = (a_1, a_2), c = (c_1, c_2)$.

We will compare the discrete Willmore energy W of the simplicial surface comprised by the vertices $f(\epsilon a), \ldots, f(-\epsilon c)$ of the hexagonal star with the classical Willmore energy $\mathcal{W}$ of the corresponding part of the smooth surface $\mathcal{S}$. Some computations are required for this. Denote by $\epsilon A = f(\epsilon a), \epsilon B = f(\epsilon b), \epsilon C = f(\epsilon c)$ the vertices of two corresponding triangles (as in Figure 7 (right)), and also by $|a|$ the length of a and by $\langle a, c \rangle = a_1 c_1 + a_2 c_2$ the corresponding scalar product.

Lemma 2.10. *The external angle $\beta(\epsilon)$ between the circumcircles of the triangles with the vertices $(0, A, B)$ and $(0, B, C)$ (as in Figure 7 (right)) is given by*

$$\beta(\epsilon) = \beta(0) + w(b) + o(\epsilon^2), \quad \epsilon \to 0, \quad w(b) = \epsilon^2 \frac{g \cos \beta(0) - h}{|a|^2 |c|^2 \sin \beta(0)}. \qquad (2.4)$$

Here $\beta(0)$ is the external angle of the circumcircles of the triangles $(0, a, b)$ and $(0, b, c)$ in the plane, and

$$g = |a|^2 r_c(r + r_c) + |c|^2 r_a(r + r_a) + \frac{r^2}{2}(|a|^2 + |c|^2),$$
$$h = |a|^2 r_c(r + r_c) + |c|^2 r_a(r + r_a) - \langle a, c \rangle(r + 2r_a)(r + 2r_c).$$

Proof. Formula (2.3) with $a = -C, b = A, c = C + \epsilon R, d = -A - \epsilon R$ yields for $\cos \beta$

$$\frac{\langle C, C + \epsilon R \rangle \langle A, A + \epsilon R \rangle - \langle A, C \rangle \langle A + \epsilon R, C + \epsilon R \rangle - \langle A, C + \epsilon R \rangle \langle A + \epsilon R, C \rangle}{|A||C||A + \epsilon R||C + \epsilon R|},$$

where $|A|$ is the length of A. Substituting the expressions for A, C, R we see that the term of order ϵ of the numerator vanishes, and we obtain for the numerator

$$|a|^2 |c|^2 - 2\langle a, c \rangle^2 + \epsilon^2 h + o(\epsilon^2).$$

For the terms in the denominator we get

$$|A| = |a|\left(1 + \frac{r_a^2}{2|a|^2}\epsilon^2 + o(\epsilon^2)\right), \quad |A + \epsilon R| = |a|\left(1 + \frac{(r + r_a)^2}{2|a|^2}\epsilon^2 + o(\epsilon^2)\right)$$

and similar expressions for $|C|$ and $|C + \epsilon R|$. Substituting this to the formula for $\cos \beta$ we obtain

$$\cos \beta = 1 - 2\left(\frac{\langle a, c \rangle}{|a||c|}\right)^2 + \frac{\epsilon^2}{|a|^2 |c|^2}\left(h - g\left(1 - 2\left(\frac{\langle a, c \rangle}{|a||c|}\right)^2\right)\right) + o(\epsilon^2).$$

Observe that this formula can be read as

$$\cos \beta(\epsilon) = \cos \beta(0) + \frac{\epsilon^2}{|a|^2 |c|^2}\left(h - g \cos \beta(0)\right) + o(\epsilon^2),$$

which implies the asymptotics (2.4). $\square$

The term $w(b)$ is in fact the part of the discrete Willmore energy of the vertex v coming from the edge b. Indeed the sum of the angles $\beta(0)$ over all 6 edges meeting at v is 2π. Denote by $w(a)$ and $w(c)$ the parts of the discrete Willmore energy corresponding to the edges a and c. Observe that for the opposite edges (for example a and $-a$) the terms w coincide. Denote by $W_\epsilon(v)$ the discrete Willmore energy of the simplicial hexagon we consider. We have

$$W_\epsilon(v) = (w(a) + w(b) + w(c)) + o(\epsilon^2).$$

On the other hand the part of the classical Willmore functional corresponding to the vertex v is

$$\mathcal{W}_\epsilon(v) = \frac{1}{4}(k_1 - k_2)^2 S + 0(\epsilon^2),$$

where the area S is one third of the area of the hexagon or, equivalently, twice the area of one of the triangles in the parameter domain

$$S = \epsilon^2 |a||c| \sin \gamma.$$

Here γ is the angle between the vectors a and c. An elementary geometric consideration implies

$$\beta(0) = 2\gamma - \pi. \tag{2.5}$$

We are interested in the quotient $W_\epsilon/\mathcal{W}_\epsilon$ which is obviously scale-invariant. Let us normalize $|a| = 1$ and parametrize the triangles by the angles between the edges and by the angle to the curvature line; see Figure 7 (middle).

$$(a_1, a_2) = (\cos\phi_1, \sin\phi_1), \tag{2.6}$$

$$(c_1, c_2) = \left(\frac{\sin\phi_2}{\sin\phi_3} \cos(\phi_1 + \phi_2 + \phi_3), \frac{\sin\phi_2}{\sin\phi_3} \sin(\phi_1 + \phi_2 + \phi_3) \right).$$

The moduli space of the regular lattices of acute triangles is described as follows,

$$\Phi = \{\phi = (\phi_1, \phi_2, \phi_3) \in \mathbb{R}^3 \,|\, 0 \le \phi_1 < \frac{\pi}{2},\ 0 < \phi_2 < \frac{\pi}{2},\ 0 < \phi_3 < \frac{\pi}{2},\ \frac{\pi}{2} < \phi_2 + \phi_3 \}.$$

Proposition 2.11. *The limit of the quotient of the discrete and smooth Willmore energies*

$$Q(\phi) := \lim_{\epsilon \to 0} \frac{W_\epsilon(v)}{\mathcal{W}_\epsilon(v)}$$

is independent of the curvatures of the surface and depends on the geometry of the triangulation only. It is

$$Q(\phi) = 1 - \frac{(\cos 2\phi_1 \cos \phi_3 + \cos(2\phi_1 + 2\phi_2 + \phi_3))^2 + (\sin 2\phi_1 \cos \phi_3)^2}{4\cos\phi_2 \cos\phi_3 \cos(\phi_2 + \phi_3)}, \tag{2.7}$$

and we have $Q > 1$. The infimum $\inf_\Phi Q(\phi) = 1$ corresponds to one of the cases when two of the three lattice vectors a, b, c are in the principal curvature directions:

- $\phi_1 = 0,\ \phi_2 + \phi_3 \to \frac{\pi}{2}$,
- $\phi_1 = 0,\ \phi_2 \to \frac{\pi}{2}$,
- $\phi_1 + \phi_2 = \frac{\pi}{2},\ \phi_3 \to \frac{\pi}{2}$.

Proof. The proof is based on a direct but rather involved computation. We used the *Mathematica* computer algebra system for some of the computations. Introduce

$$\tilde{w} := \frac{4w}{(k_1 - k_2)^2 S}.$$

This gives in particular

$$\tilde{w}(b) = 2\frac{h + g(2\cos^2\gamma - 1)}{(k_1 - k_2)^2|a|^3|c|^3\cos\gamma\sin^2\gamma} = 2\frac{h + g\left(2\frac{\langle a,c\rangle^2}{|a|^2|c|^2} - 1\right)}{(k_1 - k_2)^2\langle a,c\rangle(|a|^2|c|^2 - \langle a,c\rangle^2)}.$$

Here we have used the relation (2.5) between $\beta(0)$ and γ. In the sum over the edges $Q = \tilde{w}(a) + \tilde{w}(b) + \tilde{w}(c)$ the curvatures k_1, k_2 disappear and we get Q in terms of the coordinates of a and c:

$$Q = 2\Big((a_1^2c_2^2 + a_2^2c_1^2)(a_1c_1 + a_2c_2) + a_1^2c_1^2(a_2^2 + c_2^2) + a_2^2c_2^2(a_1^2 + c_1^2)$$

$$+ 2a_1a_2c_1c_2\big((a_1 + c_1)^2 + (a_2 + c_2)^2\big)\Big)\Big/$$

$$\Big((a_1c_1 + a_2c_2)(a_1(a_1 + c_1) + a_2(a_2 + c_2))\big((a_1 + c_1)c_1 + (a_2 + c_2)c_2\big)\Big).$$

Substituting the angle representation (2.6) we obtain

$$Q = \frac{\sin 2\phi_1 \sin 2(\phi_1 + \phi_2) + 2\cos\phi_2 \sin(2\phi_1 + \phi_2)\sin 2(\phi_1 + \phi_2 + \phi_3)}{4\cos\phi_2\cos\phi_3\cos(\phi_2 + \phi_3)}.$$

One can check that this formula is equivalent to (2.7). Since the denominator in (2.7) on the space Φ is always negative we have $Q > 1$. The identity $Q = 1$ holds only if both terms in the nominator of (2.7) vanish. This leads exactly to the cases indicated in the proposition when the lattice vectors are directed along the curvature lines. Indeed the vanishing of the second term in the nominator implies either $\phi_1 = 0$ or $\phi_3 \to \frac{\pi}{2}$. Vanishing of the first term in the nominator with $\phi_1 = 0$ implies $\phi_2 \to \frac{\pi}{2}$ or $\phi_2 + \phi_3 \to \frac{\pi}{2}$. Similarly in the limit $\phi_3 \to \frac{\pi}{2}$ the vanishing of

$$\big(\cos 2\phi_1 \cos\phi_3 + \cos(2\phi_1 + 2\phi_2 + \phi_3)\big)^2\big/\cos\phi_3$$

implies $\phi_1 + \phi_2 = \frac{\pi}{2}$. One can check that in all these cases $Q(\phi) \to 1$. $\square$

Note that for the infinitesimal equilateral triangular lattice $\phi_2 = \phi_3 = \frac{\pi}{3}$ the result is independent of the orientation ϕ_1 with respect to the curvature directions, and the discrete Willmore energy is in the limit $Q = 3/2$ times larger than the smooth one.

Finally, we come to the following conclusion.

Theorem 2.12. *Let S be a smooth surface with Willmore energy $\mathcal{W}(S)$. Consider a simplicial surface S_ϵ such that its vertices lie on S and are of degree 6, the distances between the neighboring vertices are of order ϵ, and the neighboring triangles of S_ϵ meeting at a vertex are congruent up to order ϵ^3 (i.e., the lengths of the corresponding edges differ by terms of order at most ϵ^4), and they build elementary hexagons the lengths of whose* *opposite edges differ by terms of order at most ϵ^4. Then the limit of the discrete Willmore energy is bounded from below by the classical Willmore energy*

$$\lim_{\epsilon \to 0} W(S_\epsilon) \geq \mathcal{W}(\mathcal{S}). \qquad (2.8)$$

Moreover, equality in (2.8) holds if S_ϵ is a regular triangulation of an infinitesimal curvature-line net of S, i.e., the vertices of S_ϵ are at the vertices of a curvature-line net of S.

Proof. Consider an elementary hexagon of S_ϵ. Its projection to the tangent plane of the central vertex is a hexagon which can be obtained from the modeling one considered in Proposition 2.11 by a perturbation of vertices of order $o(\epsilon^3)$. Such perturbations contribute to the terms of order $o(\epsilon^2)$ of the discrete Willmore energy. The latter are irrelevant for the considerations of Proposition 2.11. □

Possibly minimization of the discrete Willmore energy with the vertices constrained to lie on S could be used for computation of a curvature-line net.

2.5. Bending energy for simplicial surfaces

An accurate model for bending of discrete surfaces is important for modeling in computer graphics. The bending energy of smooth thin shells (compare [22]) is given by the integral

$$E = \int (H - H_0)^2 dA,$$

where H_0 and H are the mean curvatures of the original and deformed surface, respectively. For $H_0 = 0$ it reduces to the Willmore energy.

To derive the bending energy for simplicial surfaces let us consider the limit of fine triangulations, where the angles between the normals of neighboring triangles become small. Consider an isometric deformation of two adjacent triangles. Let θ be the external dihedral angle of the edge e, or, equivalently, the angle between the normals of these triangles (see Figure 8) and $\beta(\theta)$ the external intersection angle between the circumcircles of the triangles (see Figure 3) as a function of θ.

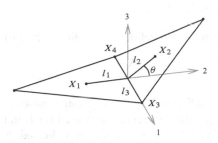

FIGURE 8. Defining the bending energy for simplicial surfaces.

Proposition 2.13. *Assume that the circumcenters of two adjacent triangles do not coincide. Then in the limit of small angles $\theta \to 0$ the angle β between the circles behaves as follows:*

$$\beta(\theta) = \beta(0) + \frac{l}{4L}\theta^2 + o(\theta^3).$$

Here l is the length of the edge and $L \neq 0$ is the distance between the centers of the circles.

Proof. Let us introduce the orthogonal coordinate system with the origin at the middle point of the common edge e, the first basis vector directed along e, and the third basis vector orthogonal to the left triangle. Denote by X_1, X_2 the centers of the circumcircles of the triangles and by X_3, X_4 the end points of the common edge; see Figure 8. The coordinates of these points are $X_1 = (0, -l_1, 0)$, $X_2 = (0, l_2 \cos\theta, l_2 \sin\theta)$, $X_3 = (l_3, 0, 0)$, $X_4 = (-l_3, 0, 0)$. Here $2l_3$ is the length of the edge e, and l_1 and l_2 are the distances from its middle point to the centers of the circumcirlces (for acute triangles). The unit normals to the triangles are $N_1 = (0, 0, 1)$ and $N_2 = (0, -\sin\theta, \cos\theta)$. The angle β between the circumcircles intersecting at the point X_4 is equal to the angle between the vectors $A = N_1 \times (X_4 - X_1)$ and $B = N_2 \times (X_4 - X_2)$. The coordinates of these vectors are $A = (-l_1, -l_3, 0)$, $B = (l_2, -l_3 \cos\theta, -l_3 \sin\theta)$. This implies for the angle

$$\cos\beta(\theta) = \frac{l_3^2 \cos\theta - l_1 l_2}{r_1 r_2}, \tag{2.9}$$

where $r_i = \sqrt{l_i^2 + l_3^2}$, $i = 1, 2$ are the radii of the corresponding circumcircles. Thus $\beta(\theta)$ is an even function, in particular $\beta(\theta) = \beta(0) + B\theta^2 + o(\theta^3)$. Differentiating (2.9) by θ^2 we obtain

$$B = \frac{l_3^2}{2r_1 r_2 \sin\beta(0)}.$$

Also formula (2.9) yields

$$\sin\beta(0) = \frac{l_3 L}{r_1 r_2},$$

where $L = |l_1 + l_2|$ is the distance between the centers of the circles. Finally combining these formulas we obtain $B = l_3/(2L)$. $\qquad\square$

This proposition motivates us to define the bending energy of simplicial surfaces as

$$E = \sum_{e \in E} \frac{l}{L}\theta^2.$$

For discrete thin-shells this bending energy was suggested and analyzed by Grinspun et al. [23, 22]. The distance between the barycenters was used for L in the energy expression, and possible advantages in using circumcenters were indicated. Numerical experiments demonstrate good qualitative simulation of real processes.

Further applications of the discrete Willmore energy in particular for surface restoration, geometry denoising, and smooth filling of a hole can be found in [8].

3. Circular nets as discrete curvature lines

Simplicial surfaces as studied in the previous section are too unstructured for analytical investigation. An important tool in the theory of smooth surfaces is the introduction of (special) parametrizations of a surface. Natural analogues of parametrized surfaces are quadrilateral surfaces, i.e., discrete surfaces made from (not necessarily planar) quadrilaterals. The strips of quadrilaterals obtained by gluing quadrilaterals along opposite edges can be considered as coordinate lines on the quadrilateral surface.

We start with a combinatorial description of the discrete surfaces under consideration.

Definition 3.1. A cellular decomposition $\mathcal{D}$ of a two-dimensional manifold (with boundary) is called a *quad-graph* if the cells have four sides each.

A quadrilateral surface is a mapping f of a quad-graph to $\mathbb{R}^3$. The mapping f is given just by the values at the vertices of $\mathcal{D}$, and vertices, edges and faces of the quad-graph and of the quadrilateral surface correspond. Quadrilateral surfaces with planar faces were suggested by Sauer [35] as discrete analogs of conjugate nets on smooth surfaces. The latter are the mappings $(x, y) \mapsto f(x, y) \in \mathbb{R}^3$ such that the mixed derivative f_{xy} is tangent to the surface.

Definition 3.2. A quadrilateral surface $f : \mathcal{D} \to \mathbb{R}^3$ all faces of which are circular (i.e., the four vertices of each face lie on a common circle) is called a *circular net* (or discrete orthogonal net).

Circular nets as discrete analogues of curvature-line parametrized surfaces were mentioned by Martin, de Pont, Sharrock and Nutbourne [32, 33] . The curvature-lines on smooth surfaces continue through any point. Keeping in mind the analogy to the curvature-line parametrized surfaces one may in addition require that all vertices of a circular net are of even degree.

A smooth conjugate net $f : D \to \mathbb{R}^3$ is a curvature-line parametrization if and only if it is orthogonal. The angle bisectors of the diagonals of a circular quadrilateral intersect orthogonally (see Figure 9) and can be interpreted [14] as discrete principal curvature directions.

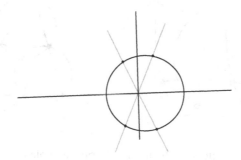

FIGURE 9. Principal curvature directions of a circular quadrilateral.

There are deep reasons to treat circular nets as a discrete curvature-line parametrization.

- The class of circular nets as well as the class of curvature-line parametrized surfaces is invariant under Möbius transformations.
- Take an infinitesimal quadrilateral $(f(x, y), f(x+\epsilon), y), f(x+\epsilon), y+\epsilon), f(x, y+\epsilon))$ of a curvature-line parametrized surface. A direct computation (see [14]) shows that in the limit $\epsilon \to 0$ the imaginary part of its cross-ratio is of order ϵ^3. Note that circular quadrilaterals are characterized by having real cross-ratios.
- For any smooth curvature-line parametrized surface $f : D \to \mathbb{R}^3$ there exists a family of discrete circular nets converging to f. Moreover, the convergence is C^∞, i.e., with all derivatives. The details can be found in [5].

One more argument in favor of Definition 3.2 is that circular nets satisfy the *consistency principle*, which has proven to be one of the organizing principles in discrete differential geometry [10]. The consistency principle singles out fundamental geometries by the requirement that the geometry can be consistently extended to a combinatorial grid one dimension higher. The consistency of circular nets was shown by Cieśliński, Doliwa and Santini [19] based on the following classical theorem.

Theorem 3.3 (Miquel). *Consider a combinatorial cube in $\mathbb{R}^3$ with planar faces. Assume that three neighboring faces of the cube are circular. Then the remaining three faces are also circular.*

Equivalently, provided the four-tuples of black vertices coming from three neighboring faces of the cube lie on circles, the three circles determined by the triples of points corresponding to three remaining faces of the cube all intersect (at the white vertex in Figure 10). It is easy to see that all vertices of Miquel's cube lie on a sphere. Mapping the vertex shared by the three original faces to infinity by a Möbius transformation, we obtain an equivalent planar version of Miquel's theorem. This version, also shown in Figure 10, can be proven by means of elementary geometry.

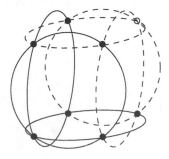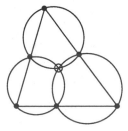

FIGURE 10. Miquel's theorem: spherical and planar versions.

Finally note that circular nets are also treated as a discretization of triply orthogonal coordinate systems. Triply orthogonal coordinate systems in $\mathbb{R}^3$ are maps $(x, y, z) \mapsto f(x, y, z) \in \mathbb{R}^3$ from a subset of $\mathbb{R}^3$ with mutually orthogonal f_x, f_y, f_z. Due to the classical Dupin theorem, the level surfaces of a triply orthogonal coordinate system intersect along their common curvature lines. Accordingly, discrete triply orthogonal systems are defined as maps from $\mathbb{Z}^3$ (or a subset thereof) to $\mathbb{R}^3$ with all elementary hexahedra lying on spheres [2]. Due to Miquel's theorem a discrete orthogonal system is uniquely determined by the circular nets corresponding to its coordinate two-planes (see [19] and [10]).

4. Discrete isothermic surfaces

In this section and in the following one, we investigate discrete analogs of special classes of surfaces obtained by imposing additional conditions in terms of circles and spheres.

We start with minor combinatorial restrictions. Suppose that the vertices of a quad-graph $\mathcal{D}$ are colored black or white so that the two ends of each edge have different colors. Such a coloring is always possible for topological discs. To model the curvature lines, suppose also that the edges of a quad-graph $\mathcal{D}$ may consistently be labelled '$+$' and '$-$', as in Figure 11 (for this it is necessary that each vertex has an even number of edges). Let f_0 be a vertex of a circular net, $f_1, f_3, \ldots, f_{4N-1}$ be its neighbors, and

FIGURE 11. Labelling the edges of a discrete isothermic surface.

$f_2, f_4, \ldots, f_{4N}$ its next-neighbors (see Figure 12 (left)). We call the vertex f_0 generic if it is not co-spherical with all its neighbors and a circular net $f : \mathcal{D} \to \mathbb{R}^3$ generic if all its vertices are generic.

Let $f : \mathcal{D} \to \mathbb{R}^3$ be a generic circular net such that every vertex is co-spherical with all its next-neighbors. We will call the corresponding sphere *central*. For an analytical description of this geometry let us map the vertex f_0 to infinity by a Möbius transformation $\mathcal{M}(f_0) = \infty$, and denote by $F_i = \mathcal{M}(f_i)$, the images of the f_i, for $i = 1, \ldots, 4N$. The points $F_2, F_4, \ldots, F_{4N}$ are obviously coplanar. The circles of the faces are mapped to straight lines. For the cross-ratios we get

$$q(f_0, f_{2k-1}, f_{2k}, f_{2k+1}) = \frac{F_{2k} - F_{2k+1}}{F_{2k} - F_{2k-1}} = \frac{z_{2k+1}}{z_{2k-1}},$$

where z_{2k+1} is the coordinate of F_{2k+1} orthogonal to the plane $\mathcal{P}$ of $F_2, F_4, \ldots, F_{4N}$. (Note that since f_0 is generic none of the z_i vanishes.) As a corollary we get for the product of all cross-ratios:

$$\prod_{k=1}^{n} q(f_0, f_{2k-1}, f_{2k}, f_{2k+1}) = 1. \tag{4.1}$$

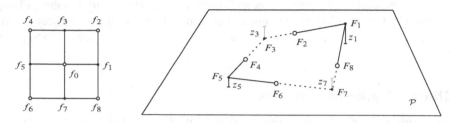

FIGURE 12. Central spheres of a discrete isothermic surface: combinatorics *(left)*, and the Möbius normalized picture for $N = 2$ *(right)*.

Definition 4.1. A circular net $f : \mathcal{D} \to \mathbb{R}^3$ satisfying condition (4.1) at each vertex is called a *discrete isothermic surface*.

This definition was first suggested in [6] for the case of the combinatorial square grid $\mathcal{D} = \mathbb{Z}^2$. In this case if the vertices are labelled by $f_{m,n}$ and the corresponding cross-ratios by $q_{m,n} := q(f_{m,n}, f_{m+1,n}, f_{m+1,n+1}, f_{m,n+1})$, the condition (4.1) reads

$$q_{m,n} q_{m+1,n+1} = q_{m+1,n} q_{m,n+1}.$$

Proposition 4.2. *Each vertex $f_{m,n}$ of a discrete isothermic surface $f : \mathbb{Z}^2 \to \mathbb{R}^3$ has a central sphere, i.e., the points $f_{m,n}$, $f_{m-1,n-1}$, $f_{m+1,n-1}$, $f_{m+1,n+1}$ and $f_{m-1,n+1}$ are co-spherical. Moreover, for generic circular maps $f : \mathbb{Z}^2 \to \mathbb{R}^3$ this property characterizes discrete isothermic surfaces.*

Proof. Use the notation of Figure 12, with $f_0 \equiv f_{m,n}$, and the same argument with the Möbius transformation $\mathcal{M}$ which maps f_0 to ∞. Consider the plane $\mathcal{P}$ determined by the points F_2, F_4 and F_6. Let as above z_k be the coordinates of F_k orthogonal to the plane $\mathcal{P}$. Condition (4.1) yields

$$\frac{F_8 - F_1}{F_8 - F_7} = \frac{z_1}{z_7}.$$

This implies that the z-coordinate of the point F_8 vanishes, thus $F_8 \in \mathcal{P}$. □

The property to be discrete isothermic is also 3D-consistent, i.e., can be consistently imposed on all faces of a cube. This was proven first by Hertrich-Jeromin, Hoffmann and Pinkall [26] (see also [10] for generalizations and modern treatment).

An important subclass of discrete isothermic surfaces is given by the condition that all the faces are conformal squares, i.e., their cross ratio equals -1. All conformal squares are Möbius equivalent, in particular equivalent to the standard square. This is a direct discretization of the definition of smooth isothermic surfaces. The latter are immersions $(x, y) \mapsto f(x, y) \in \mathbb{R}^3$ satisfying

$$\| f_x \| = \| f_y \|, \qquad f_x \perp f_y, \qquad f_{xy} \in \mathrm{span}\{ f_x, f_x \}, \qquad (4.2)$$

i.e., conformal curvature-line parametrizations. Geometrically this definition means that the curvature lines divide the surface into infinitesimal squares.

FIGURE 13. Right-angled kites are conformal squares.

The class of discrete isothermic surfaces is too general and the surfaces are not rigid enough. In particular one can show that the surface can vary preserving all its black vertices. In this case, one white vertex can be chosen arbitrarily [7]. Thus, we introduce a more rigid subclass. To motivate its definition, let us look at the problem of discretizing the class of conformal maps $f : D \to \mathbb{C}$ for $D \subset \mathbb{C} = \mathbb{R}^2$. Conformal maps are characterized by the conditions

$$|f_x| - |f_y|, \qquad f_x \perp f_y. \qquad (4.3)$$

To define discrete conformal maps $f : \mathbb{Z}^2 \supset D \to \mathbb{C}$, it is natural to impose these two conditions on two different sub-lattices (white and black) of $\mathbb{Z}^2$, i.e., to require that the edges meeting at a white vertex have equal length and the edges at a black vertex meet orthogonally. This discretization leads to the circle patterns with the combinatorics of the square grid introduced by Schramm [37]. Each circle intersects four neighboring circles orthogonally and the neighboring circles touch cyclically; see Figure 14 (left).

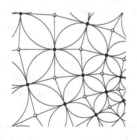 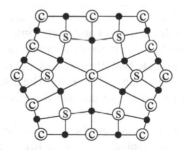

FIGURE 14. Defining discrete S-isothermic surfaces: orthogonal circle patterns as discrete conformal maps *(left)* and combinatorics of S-quad-graphs *(right)*.

The same properties imposed for quadrilateral surfaces with the combinatorics of the square grid $f : \mathbb{Z}^2 \supset \mathcal{D} \to \mathbb{R}^3$ lead to an important subclass of discrete isothermic surfaces. Let us require for a discrete quadrilateral surface that:

- the faces are orthogonal kites, as in Figure 13,
- the edges meet at black vertices orthogonally (black vertices are at orthogonal corners of the kites),
- the kites which do not share a common vertex are not coplanar (locality condition).

Observe that the orthogonality condition (at black vertices) implies that one pair of opposite edges meeting at a black vertex lies on a straight line. Together with the locality condition this implies that there are two kinds of white vertices, which we denote by $\copyright$ and $\text{\textcircled{s}}$. Each kite has white vertices of both types and the kites sharing a white vertex of the first kind $\copyright$ are coplanar.

These conditions imposed on the quad-graphs lead to S-quad-graphs and S-isothermic surfaces (the latter were introduced in [7] for the combinatorics of the square grid).

Definition 4.3. An *S-quad-graph* $\mathcal{D}$ is a quad-graph with black and two kinds of white vertices such that the two ends of each edge have different colors and each quadrilateral has vertices of all kinds; see Figure 14 (right). Let $V_b(\mathcal{D})$ be the set of black vertices. A *discrete S-isothermic surface* is a map

$$f_b : V_b(\mathcal{D}) \to \mathbb{R}^3,$$

with the following properties:

1. If $v_1, \ldots, v_{2n} \in V_b(\mathcal{D})$ are the neighbors of a $\copyright$-vertex in cyclic order, then $f_b(v_1)$, $\ldots, f_b(v_{2n})$ lie on a circle in $\mathbb{R}^3$ in the same cyclic order. This defines a map from the $\copyright$-vertices to the set of circles in $\mathbb{R}^3$.
2. If $v_1, \ldots, v_{2n} \in V_b(\mathcal{D})$ are the neighbors of an $\text{\textcircled{s}}$-vertex, then $f_b(v_1), \ldots, f_b(v_{2n})$ lie on a sphere in $\mathbb{R}^3$. This defines a map from the $\text{\textcircled{s}}$-vertices to the set of spheres in $\mathbb{R}^3$.
3. If v_c and v_s are the $\copyright$- and $\text{\textcircled{s}}$-vertices of a quadrilateral of $\mathcal{D}$, then the circle corresponding to v_c intersects the sphere corresponding to v_s orthogonally.

Discrete S-isothermic surfaces are therefore composed of tangent spheres and tangent circles, with the spheres and circles intersecting orthogonally. The class of discrete S-isothermic surfaces is obviously invariant under Möbius transformations.

Given a discrete S-isothermic surface, one can add the centers of the spheres and circles to it giving a map $V(\mathcal{D}) \to \mathbb{R}^3$. The discrete isothermic surface obtained is called the *central extension* of the discrete S-isothermic surface. All its faces are orthogonal kites.

An important fact of the theory of isothermic surfaces (smooth and discrete) is the existence of a dual isothermic surface [6]. Let $f : \mathbb{R}^2 \supset D \to \mathbb{R}^3$ be an isothermic immersion. Then the formulas

$$f_x^* = \frac{f_x}{\|f_x\|^2}, \quad f_y^* = -\frac{f_y}{\|f_y\|^2}$$

define an isothermic immersion $f^* : \mathbb{R}^2 \supset D \to \mathbb{R}^3$ which is called the *dual isothermic surface*. Indeed, one can easily check that the form df^* is closed and f^* satisfies (4.2). Exactly the same formulas can be applied in the discrete case.

Proposition 4.4. *Suppose* $f : \mathcal{D} \to \mathbb{R}^3$ *is a discrete isothermic surface and suppose the edges have been consistently labeled '$+$' and '$-$', as in Figure* 11. *Then the* dual discrete isothermic surface $f^* : \mathcal{D} \to \mathbb{R}^3$ *is defined by the formula*

$$\Delta f^* = \pm \frac{\Delta f}{\|\Delta f\|^2},$$

where Δf *denotes the difference of values at neighboring vertices and the sign is chosen according to the edge label.*

The closedness of the corresponding discrete form is elementary to check for one kite.

Proposition 4.5. *The dual of the central extension of a discrete S-isothermic surface is the central extension of another discrete S-isothermic surface.*

If we disregard the centers, we obtain the definition of the *dual discrete S-isothermic surface*. The dual discrete S-isothermic surface can be defined also without referring to the central extension [4].

5. Discrete minimal surfaces and circle patterns: geometry from combinatorics

In this section (following [4]) we define *discrete minimal S-isothermic surfaces* (or *discrete minimal surfaces* for short) and present the most important facts from their theory. The main idea of the approach of [4] is the following. Minimal surfaces are characterized among (smooth) isothermic surfaces by the property that they are dual to their Gauss map. The duality transformation and this characterization of minimal surfaces carries over to the discrete domain. The role of the Gauss map is played by discrete S-isothermic surfaces whose spheres all intersect one fixed sphere orthogonally.

5.1. Koebe polyhedra

A *circle packing* in S^2 is a configuration of disjoint discs which may touch but not intersect. The construction of discrete S-isothermic "round spheres" is based on their relation to circle packings in S^2. The following theorem is central in this theory.

Theorem 5.1. *For every polytopal[1] cellular decomposition of the sphere, there exists a pattern of circles in the sphere with the following properties. There is a circle corresponding to each face and to each vertex. The vertex circles form a packing with two circles touching if and only if the corresponding vertices are adjacent. Likewise, the face circles form a packing with circles touching if and only if the corresponding faces are*

---

[1] We call a cellular decomposition of a surface *polytopal* if the closed cells are closed discs, and two closed cells intersect in one closed cell if at all.

FIGURE 15. *Left:* An orthogonal circle pattern corresponding to a cellular decomposition. *Middle:* A circle packing corresponding to a triangulation. *Right:* The orthogonal circles.

adjacent. For each edge, there is a pair of touching vertex circles and a pair of touching face circles. These pairs touch in the same point, intersecting each other orthogonally.
This circle pattern is unique up to Möbius transformations.

The first published statement and proof of this theorem is contained in [16]. For generalizations, see [36, 34, 9], the last also for a variational proof.

Theorem 5.1 is a generalization of the following remarkable statement about circle packings due to Koebe [31].

Theorem 5.2 (Koebe). *For every triangulation of the sphere there is a packing of circles in the sphere such that circles correspond to vertices, and two circles touch if and only if the corresponding vertices are adjacent. This circle packing is unique up to Möbius transformations of the sphere.*

Observe that, for a triangulation, one automatically obtains not one but two orthogonally intersecting circle packings, as shown in Figure 15 (right). Indeed, the circles passing through the points of contact of three mutually touching circles intersect these orthogonally, and thus Theorem 5.2 is a special case of Theorem 5.1.

Consider a circle pattern of Theorem 5.1. Associating white vertices to circles and black vertices to their intersection points, one obtains a quad-graph. Actually we have an S-quad-graph: Since the circle pattern is comprised by two circle packings intersecting orthogonally we have two kinds of white vertices, ⓢ and ⓒ, corresponding to the circles of the two packings.

Now let us construct the spheres intersecting S^2 orthogonally along the circles marked by ⓢ. If we then connect the centers of touching spheres, we obtain a convex polyhedron, all of whose edges are tangent to the sphere S^2. Moreover, the circles marked with ⓒ are inscribed in the faces of the polyhedron. Thus we have a discrete S-isothermic surface.

We arrive at the following theorem, which is equivalent to Theorem 5.1 (compare [43]).

Theorem 5.3. *Every polytopal cell decomposition of the sphere can be realized by a polyhedron with edges tangent to the sphere. This realization is unique up to projective transformations which fix the sphere.*

These polyhedra are called the *Koebe polyhedra*. We interpret the corresponding discrete S-isothermic surfaces as conformal discretizations of the "round sphere".

5.2. Definition of discrete minimal surfaces

Let $f : \mathcal{D} \to \mathbb{R}^3$ be a discrete S-isothermic surface. Suppose $x \in \mathcal{D}$ is a white ⓢ vertex of the quad-graph $\mathcal{D}$, i.e., $f(x)$ is the center of a sphere. Consider all quadrilaterals of $\mathcal{D}$ incident to x and denote by $y_1, \ldots, y_{2n}$ their black vertices and by $x_1, \ldots, x_{2n}$ their white ⓒ vertices. We will call the vertices $x_1, \ldots, x_{2n}$ the neighbors of x in $\mathcal{D}$. (Generically, $n = 2$.) Then $f(y_j)$ are the points of contact with the neighboring spheres and simultaneously points of intersection with the orthogonal circles centered at $f(x_j)$; see Figure 16.

Consider only the white vertices of the quad-graph. Observe that each ⓒ vertex of a discrete S-isothermic surface and all its ⓢ neighbors are coplanar. Indeed the plane of the circle centered at $f(ⓒ)$ contains all its $f(ⓢ)$ neighbors. The same condition imposed at the ⓢ vertices leads to a special class of surfaces.

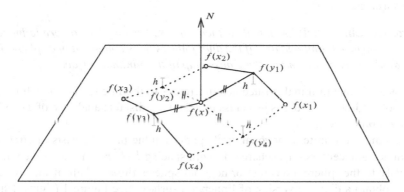

FIGURE 16. Defining discrete minimal surfaces: the tangent plane through the center $f(x)$ of a sphere and the centers $f(x_j)$ of the neighboring circles. The circles and the sphere intersect orthogonally at the black points $f(y_j)$.

Definition 5.4. A discrete S-isothermic surface $f : \mathcal{D} \to \mathbb{R}^3$ is called *discrete minimal* if for each sphere of $f(\mathcal{D})$ there exists a plane which contains the center $f(ⓢ)$ of the sphere as well as the centers $f(ⓒ)$ of all neighboring circles orthogonal to the sphere, i.e., if each white vertex of $f(\mathcal{D})$ is coplanar to all its white neighbors. These planes should be considered as tangent planes to the discrete minimal surface.

Theorem 5.5. *An S-isothermic discrete surface f is a discrete minimal surface if and only if the dual S-isothermic surface f^* corresponds to a Koebe polyhedron. In that case the dual surface $N := f^* : V_w(\mathcal{D}) \to \mathbb{R}^3$ at white vertices $V_w(\mathcal{D})$ can be treated as the Gauss map N of the discrete minimal surface: At ⓢ-vertices, N is orthogonal to the tangent planes, and at ⓒ-vertices, N is orthogonal to the planes of the circles centered at $f(ⓒ)$.*

Proof. That the S-isothermic dual of a Koebe polyhedron is a discrete minimal surface is fairly obvious. On the other hand, let $f : \mathcal{D} \to \mathbb{R}^3$ be a discrete minimal surface and let $x \in \mathcal{D}$ and $y_1, \ldots, y_{2n} \in \mathcal{D}$ be as above. Let $f^* : \mathcal{D} \to \mathbb{R}^3$ be the dual S-isothermic surface. We need to show that all circles of f^* lie in one and the same sphere S and that all the spheres of f^* intersect S orthogonally. Since the quadrilaterals of a discrete S-isothermic surface are kites the minimality condition of Definition 5.4 can be reformulated as follows: There is a plane through $f(x)$ such that the points $\{ f(y_j) \mid j \text{ even} \}$ and the points $\{ f(y_j) \mid j \text{ odd} \}$ lie in planes which are parallel to it at the same distance on opposite sides (see Figure 16). It follows immediately that the points $f^*(y_1), \ldots, f^*(y_{2n})$ lie on a circle c_x in a sphere S_x around $f^*(x)$. The plane of c_x is orthogonal to the normal N to the tangent plane at $f(x)$. Let S be the sphere which intersects S_x orthogonally in c_x. The orthogonal circles through $f^*(y_1), \ldots, f^*(y_{2n})$ also lie in S. Hence, all spheres of f^* intersect S orthogonally and all circles of f^* lie in S. $\qquad\square$

Theorem 5.5 is a complete analogue of Christoffel's characterization [18] of smooth minimal surfaces.

Theorem 5.6 (Christoffel). *Minimal surfaces are isothermic. An isothermic immersion is a minimal surface if and and only if the dual immersion is contained in a sphere. In that case the dual immersion is in fact the Gauss map of the minimal surface.*

Thus a discrete minimal surface is a discrete S-isothermic surface which is dual to a Koebe polyhedron; the latter is its Gauss map and is a discrete analogue of a conformal parametrization of the sphere.

The simplest infinite orthogonal circle pattern in the plane consists of circles with equal radius r and centers on a square grid with spacing $\frac{1}{2}\sqrt{2}\, r$. One can project it stereographically to the sphere, construct orthogonal spheres through half of the circles and dualize to obtain a discrete version of Enneper's surface. See Figure 1 (right). Only the circles are shown.

5.3. Construction of discrete minimal surfaces

A general method to construct discrete minimal surfaces is schematically shown in the following diagram:

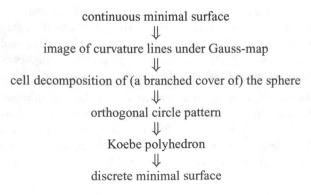

continuous minimal surface

$\Downarrow$

image of curvature lines under Gauss-map

$\Downarrow$

cell decomposition of (a branched cover of) the sphere

$\Downarrow$

orthogonal circle pattern

$\Downarrow$

Koebe polyhedron

$\Downarrow$

discrete minimal surface

As is usual in the theory of minimal surfaces [28], one starts constructing such a surface with a rough idea of how it should look. To use our method, one should understand its Gauss map and the *combinatorics* of the curvature-line pattern. The image of the curvature-line pattern under the Gauss map provides us with a cell decomposition of (a part of) S^2 or a covering. From these data, applying Theorem 5.1, we obtain a Koebe polyhedron with the prescribed combinatorics. Finally, the dualization step yields the desired discrete minimal surface. For the discrete Schwarz P-surface the construction method is demonstrated in Figures 17 and 18.

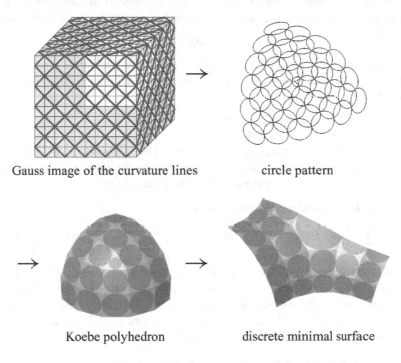

Gauss image of the curvature lines   circle pattern

Koebe polyhedron   discrete minimal surface

FIGURE 17. Construction of the discrete Schwarz P-surface.

Let us emphasize that our data, aside from possible boundary conditions, are purely *combinatorial*—the combinatorics of the curvature-line pattern. All faces are quadrilaterals and typical vertices have four edges. There may exist distinguished vertices (corresponding to the ends or umbilic points of a minimal surface) with a different number of edges.

The most nontrivial step in the above construction is the third one listed in the diagram. It is based on the generalized Koebe Theorem 5.1. It implies the existence and uniqueness for the discrete minimal S-isothermic surface under consideration, but not only this. A constructive proof of the generalized Koebe theorem suggested in [9] is based on a variational principle and also provides a method for the numerical construction of circle patterns. An alternative algorithm by Thurston was implemented in Stephenson's program `circlepack`. (See [41] for an exhaustive presentation of the theory of circle packings and their numerics.) The first step is to transfer the problem from the sphere to the plane by a stereographic projection. Then the radii of the circles are calculated. If the radii are known, it is easy to reconstruct the circle pattern. The above-mentioned variational method is based on the observation that the equations for the radii are the equations for a critical point of a convex function of the radii. The variational method involves minimizing this function to solve the equations.

Let us describe the variational method of [9] for construction of (orthogonal) circle patterns in the plane and demonstrate how it can be applied to construct the discrete Schwarz P-surface. Instead of the radii r of the circles, we use the logarithmic radii

$$\rho = \log r.$$

For each circle j, we need to find a ρ_j such that the corresponding radii solve the circle pattern problem. This leads to the following equations, one for each circle. The equation for circle j is

$$2 \sum_{\text{neighbors } k} \arctan e^{\rho_k - \rho_j} = \Phi_j, \tag{5.1}$$

where the sum is taken over all neighboring circles k. For each circle j, Φ_j is the nominal angle covered by the neighboring circles. It is normally 2π for interior circles, but it differs for circles on the boundary or for circles where the pattern branches.

Theorem 5.7. *The critical points of the functional*

$$S(\rho) = \sum_{(j,k)} \left(\operatorname{Im} \operatorname{Li}_2(i e^{\rho_k - \rho_j}) + \operatorname{Im} \operatorname{Li}_2(i e^{\rho_j - \rho_k}) - \frac{\pi}{2}(\rho_j + \rho_k) \right) + \sum_j \Phi_j \rho_j$$

correspond to orthogonal circle patterns in the plane with cone angles Φ_j at the centers of circles ($\Phi_j = 2\pi$ for internal circles). Here, the first sum is taken over all pairs (j, k) of neighboring circles, the second sum is taken over all circles j, and the dilogarithm function $\operatorname{Li}_2(z)$ is defined by $\operatorname{Li}_2(z) = -\int_0^z \log(1 - \zeta) \, d\zeta / \zeta$. The functional is scale-invariant and, when restricted to the subspace $\sum_j \rho_j = 0$, it is convex.

Proof. The formula for the functional follows from (5.1) and

$$\frac{d}{dx} \operatorname{Im} \operatorname{Li}_2(i e^x) = \frac{1}{2i} \log \frac{1 + i e^x}{1 - i e^x} = \arctan e^x.$$

The second derivative of the functional is the quadratic form

$$D^2 S = \sum_{(j,k)} \frac{1}{\cosh(\rho_k - \rho_j)} (d\rho_k - d\rho_j)^2,$$

where the sum is taken over pairs of neighboring circles, which implies the convexity. □

Now the idea is to minimize $S(\rho)$ restricted to $\sum_j \rho_j = 0$. The convexity of the functional implies the existence and uniqueness of solutions of the classical boundary valued problems: Dirichlet (with prescribed ρ_j on the boundary) and Neumann (with prescribed Φ_j on the boundary).

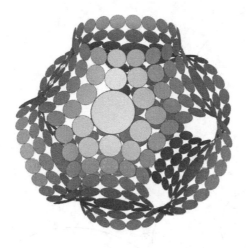

FIGURE 18. A discrete minimal Schwarz P-surface.

The Schwarz P-surface is a triply periodic minimal surface. It is the symmetric case in a 2-parameter family of minimal surfaces with 3 different hole sizes (only the ratios of the hole sizes matter); see [20]. Its Gauss map is a double cover of the sphere with 8 branch points. The image of the curvature-line pattern under the Gauss map is shown schematically in Figure 17 (top left), thin lines. It is a refined cube. More generally, one may consider three different numbers m, n, and k of slices in the three directions. The 8 corners of the cube correspond to the branch points of the Gauss map. Hence, not 3 but 6 edges are incident with each corner vertex. The corner vertices are assigned the label ⓒ. We assume that the numbers m, n, and k are even, so that the vertices of the quad graph may be labelled ⓒ, ⓢ, and • consistently (as in Section 4).

We will take advantage of the symmetry of the surface and construct only a piece of the corresponding circle pattern. Indeed the combinatorics of the quad-graph has the symmetry of a rectangular parallelepiped. We construct an orthogonal circle pattern with the symmetry of the rectangular parallelepiped, eliminating the Möbius ambiguity of Theorem 5.1. Consider one fourth of the sphere bounded by two orthogonal great circles connecting the north and the south poles of the sphere. There are two distinguished vertices (corners of the cube) in this piece. Mapping the north pole to the origin and the south pole to infinity by stereographic projection we obtain a Neumann boundary-value problem for orthogonal circle patterns in the plane. The symmetry great circles become two orthogonal straight lines. The solution of this problem is shown in Figure 19. The Neumann boundary data are $\Phi = \pi/2$ for the lower left and upper right boundary circles and $\Phi = \pi$ for all other boundary circles (along the symmetry lines).

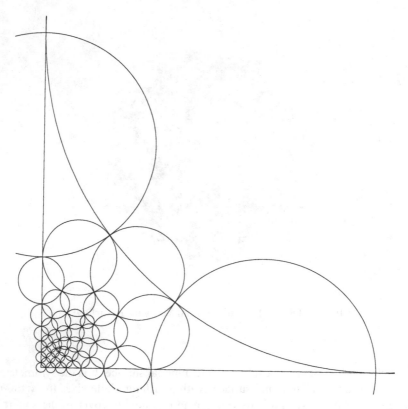

FIGURE 19. A piece of the circle pattern for a Schwarz P-surface after
stereographic projection to the plane.

Now map this circle pattern to the sphere by the same stereographic projection. One
half of the spherical circle pattern obtained (above the equator) is shown in Figure 17 (top
right). This is one eighth of the complete spherical pattern. Now lift the circle pattern to
the branched cover, construct the Koebe polyhedron and dualize it to obtain the Schwarz
P-surface; see Figure 17 (bottom row). A translational fundamental piece of the surface is
shown in Figure 18.

We summarize these results in a theorem.

Theorem 5.8. *Given three even positive integers m, n, k, there exists a corresponding
unique (asymmetric) S-isothermic Schwarz P-surface.*

Surfaces with the same ratios $m : n : k$ are different discretizations of the same
smooth Schwarz P-surface. The cases with $m = n = k$ correspond to the symmetric
Schwarz P-surface.

Further examples of discrete minimal surfaces can be found in [4, 17].

Orthogonal circle patterns on the sphere are treated as the Gauss map of the discrete minimal surface and are central in this theory. Although circle patterns on the plane

and on the sphere differ just by the stereographic projection, some geometric properties of the Gauss map can get lost when represented in the plane. Moreover, to produce branched circle patterns in the sphere it is important to be able to work with circle patterns directly on the sphere. A variational method which works directly on the sphere was suggested in [38, 4]. This variational principle for spherical circle patterns is analogous to the variational principles for Euclidean and hyperbolic patterns presented in [9]. Unlike the Euclidean and hyperbolic cases the spherical functional is not convex, which makes it difficult to use in the theory. However the spherical functional has proved to be amazingly powerful for practical computation of spherical circle patterns (see [4] for details). In particular it can be used to produce branched circle patterns in the sphere.

Numerous examples of discrete minimal surfaces, constructed with the help of this spherical functional, are described in Bücking's contribution [17] to this volume.

6. Discrete conformal surfaces and circle patterns

Conformal immersions

$$f : \mathbb{R}^2 \supset D \to \mathbb{R}^3$$
$$(x, y) \mapsto f(x, y)$$

are characterized by the properties

$$\|f_x\| = \|f_y\|, \quad f_x \perp f_y.$$

Any surface can be conformally parametrized, and a conformal parametrization induces a complex structure on D in which $z = x + iy$ is a local complex coordinate. The development of a theory of discrete conformal meshes and discrete Riemann surfaces is one of the popular topics in discrete differential geometry. Due to their almost square quadrilateral faces and general applicability, discrete conformal parametrizations are important in computer graphics, in particular for texture mapping. Recent progress in this area could be a topic of another paper; it lies beyond the scope of this survey. In this section we mention shortly only the methods based on circle patterns.

Conformal mappings can be characterized as mapping infinitesimal circles to infinitesimal circles. The idea to replace infinitesimal circles with finite circles is quite natural and leads to circle packings and circle patterns (see also Section 4). The corresponding theory for conformal mappings to the plane is well developed (see the recent book of Stephenson [41]). The discrete conformal mappings are constructed using a version of Koebe's theorem or a variational principle which imply the corresponding existence and uniqueness statements as well as a numerical method for construction (see Section 5). It has been shown that a conformal mapping can be approximated by a sequence of increasingly fine, regular circle packings or patterns [37, 25].

Several attempts have been made to generalize this theory for discrete conformal parametrizations of surfaces.

The simplest natural idea is to ignore the geometry and take only the combinatorial data of the simplicial surface [41]. Due to Koebe's theorem there exists an essentially unique circle packing representing this combinatorics. One treats this as a discrete conformal mapping of the surface. This method has been successfully applied by Hurdal et al. [29] for flat mapping of the human cerebellum. However a serious disadvantage is that the results depend only on the combinatorics and not on the geometry of the original mesh.

An extension of Stephenson's circle packing scheme which takes the geometry into account is due to Bowers and Hurdal [15]. They treat circle patterns with non-intersecting circles corresponding to vertices of the original mesh. The geometric data in this case are the so called *inversive distances* of pairs of neighboring circles, which can be treated as imaginary intersection angles of circles. The idea is to get a discrete conformal mapping as a circle pattern in the plane with the inversive distances coinciding with the inversive distances of small spheres in space. The latter are centered at the vertices of the original meshes. The disadvantage of this method is that there are almost no theoretical results regarding the existence and uniqueness of inversive-distance circle patterns.

Kharevich, Springborn and Schröder [30] suggested another way to handle the geometric data. They consider the circumcircles of the faces of a simplicial surface and take their intersection angles θ. The circumcircles are taken intrinsically, i.e., they are round circles with respect to the surface metric. The latter is a flat metric with conical singularities at vertices of the mesh. The intersection angles θ are the geometric data to be respected, i.e., ideally for a discrete conformal mapping one wishes a circle pattern in the plane with the same intersection angles θ. An advantage of this method is that similarly to the circle packing method it is based on a solid theoretical background—the variational principle for patterns of intersecting circles [9]. However to get the circle pattern into the plane one has to change the intersection angles $\theta \mapsto \tilde{\theta}$, and it seems that there is no natural geometric way to do this. (In [30] the angles $\tilde{\theta}$ of the circle pattern in the plane are defined as minimizing the sum of squared differences $(\theta - \tilde{\theta})^2$.) A solution to this problem could possibly be achieved by a method based on Delaunay triangulations of circle patterns with disjoint circles. The corresponding variational principle has been found recently in [39].

It seems that discrete conformal surface parametrizations are at the beginning of a promising development. Although now only some basic ideas about discrete conformal surface parametrizations have been clarified and no approximation results are known, there is good chance for a fundamental theory with practical applications in this field.

Note added in proof. We would like to mention three very recent papers closely related to the topics discussed here: [11, 12, 13]

References

[1] W. Blaschke, *Vorlesungen über Differentialgeometrie III*, Die Grundlehren der mathematischen Wissentschaften, Berlin: Springer-Verlag, 1929.

[2] A.I. Bobenko, *Discrete conformal maps and surfaces*, Symmetries and Integrability of Difference Equations (P.A. Clarkson and F.W. Nijhoff, eds.), London Mathematical Society Lecture Notes Series, vol. 255, Cambridge University Press, 1999, pp. 97–108.

[3] _____, *A conformal functional for simplicial surfaces*, Combinatorial and Computational Geometry (J.E. Goodman, J. Pach, and E. Welzl, eds.), MSRI Publications, vol. 52, Cambridge University Press, 2005, pp. 133–143.

[4] A.I. Bobenko, T. Hoffmann, and B.A. Springborn, *Minimal surfaces from circle patterns: Geometry from combinatorics*, Annals of Math. **164(1)** (2006), 1–34.

[5] A.I. Bobenko, D. Matthes, and Yu.B. Suris, *Discrete and smooth orthogonal systems: C^∞-approximation*, Internat. Math. Research Notices **45** (2003), 2415–2459.

[6] A.I. Bobenko and U.Pinkall, *Discrete isothermic surfaces*, J. reine angew. Math. **475** (1996), 187–208.

[7] _____, *Discretization of surfaces and integrable systems*, Discrete integrable geometry and physics (A.I. Bobenko and R. Seiler, eds.), Oxford Lecture Series in Mathematics and its Applications, vol. 16, Clarendon Press, Oxford, 1999, pp. 3–58.

[8] A.I. Bobenko and P. Schröder, *Discrete Willmore flow*, Eurographics Symposium on Geometry Processing (2005) (M. Desbrun and H. Pottmann, eds.), The Eurographics Association, 2005, pp. 101–110.

[9] A.I. Bobenko and B.A. Springborn, *Variational principles for circle patterns and Koebe's theorem*, Transactions Amer. Math. Soc. **356** (2004), 659–689.

[10] A.I. Bobenko and Yu.B. Suris, *Discrete Differential Geometry. Consistency as Integrability*, 2005, arXiv:math.DG/0504358, preliminary version of a book.

[11] _____, *On organizing principles of discrete differential geometry: geometry of spheres*, Russian Math. Surveys **62**:1 (2007), 1–43.

[12] _____, *Isothermic surfaces in sphere geometries as Moutard nets*, Proc. R. Soc. London A **463** (2007), 3171–3193, arXiv:math.DG/0610434.

[13] _____, *Discrete Koenigs nets and discrete isothermic surfaces*, 2007, arXiv:0709.3408.

[14] A.I. Bobenko and S.P. Tsarev, *Curvature line parametrization from circle patterns*, 2007, arXiv:0706.3221.

[15] P.L. Bowers and M.K. Hurdal, *Planar conformal mappings of piecewise flat surfaces*, Visualization and Mathematics III (H.-C. Hege and K. Polthier, eds.), Springer-Verlag, 2003, pp. 3–34.

[16] G.R. Brightwell and E.R. Scheinerman, *Representations of planar graphs*, SIAM J. Disc. Math. **6(2)** (1993), 214–229.

[17] U. Bücking, *Minimal surfaces from circle patterns: Boundary value problems, examples*, Discrete Differential Geometry (A.I. Bobenko, P. Schröder, J.M. Sullivan, G.M. Ziegler, eds.), Oberwolfach Seminars, vol. 38, Birkhäuser, 2008, this volume, pp. 37–56.

[18] E. Christoffel, *Über einige allgemeine Eigenschaften der Minimumsflächen*, J. reine angew. Math. **67** (1867), 218–228.

[19] J. Cieśliński, A. Doliwa and P.M. Santini, *The integrable discrete analogues of orthogonal coordinate systems are multi-dimensional circular lattices*, Phys. Lett. A **235(5)** (1997), 480–488.

[20] U. Dierkes, S. Hildebrandt, A. Küster, and O. Wohlrab, *Minimal surfaces I.*, Grundlehren der mathematischen Wissenschaften, vol. 295, Berlin: Springer-Verlag, 1992.

[21] W. Fenchel, *Über Krümmung und Windung geschlossener Raumkurven*, Math. Ann. **101** (1929), 238–252.

[22] E. Grinspun, *A discrete model of thin shells*, Discrete Differential Geometry (A.I. Bobenko, P. Schröder, J.M. Sullivan, G.M. Ziegler, eds.), Oberwolfach Seminars, vol. 38, Birkhäuser, 2008, this volume, pp. 325–337.

[23] E. Grinspun, A.N. Hirani, M. Desbrun, and P. Schröder, *Discrete shells*, Eurographics/SIGGRAPH Symposium on Computer Animation (D. Breen and M. Lin, eds.), The Eurographics Association, 2003, pp. 62–67.

[24] B. Grünbaum, *Convex Polytopes*, Graduate Texts in Math., vol. 221, Springer-Verlag, New York, 2003, Second edition prepared by V. Kaibel, V. Klee and G. M. Ziegler (original edition: Interscience, London 1967).

[25] Z.-X. He and O. Schramm, *The C^∞-convergence of hexagonal disc packings to the Riemann map*, Acta Math. **180** (1998), 219–245.

[26] U. Hertrich-Jeromin, T. Hoffmann, and U. Pinkall, *A discrete version of the Darboux transform for isothermic surfaces*, Discrete integrable geometry and physics (A.I. Bobenko and R. Seiler, eds.), Oxford Lecture Series in Mathematics and its Applications, vol. 16, Clarendon Press, Oxford, 1999, pp. 59–81.

[27] C.D. Hodgson, I. Rivin, and W.D. Smith, *A characterization of convex hyperbolic polyhedra and of convex polyhedra inscribed in a sphere*, Bull. Am. Math. Soc., New Ser. **27** (1992), 246–251.

[28] D. Hoffman and H. Karcher, *Complete embedded minimal surfaces of finite total curvature*, Geometry V: Minimal surfaces (R. Osserman, ed.), Encyclopaedia of Mathematical Sciences, vol. 90, Springer-Verlag, Berlin, 1997, pp. 5–93.

[29] M. Hurdal, P.L. Bowers, K. Stephenson, D.W.L. Sumners, K. Rehm, K. Schaper, and D.A. Rottenberg, *Quasi-conformality flat mapping the human cerebellum*, Medical Image Computing and Computer-Assisted Intervention, Springer-Verlag, 1999, pp. 279–286.

[30] L. Kharevich, B. Springborn, and P. Schröder, *Discrete conformal mappings via circle patterns*, ACM Transactions on Graphics **25:2** (2006), 1–27.

[31] P. Koebe, *Kontaktprobleme der konformen Abbildung*, Abh. Sächs. Akad. Wiss. Leipzig Math.-Natur. Kl. **88** (1936), 141–164.

[32] R.R. Martin, J. de Pont, and T.J. Sharrock, *Cyclide surfaces in computer aided design*, The Mathematics of Surfaces (J. Gregory, ed.), The Institute of Mathematics and its Applications Conference Series. New Series, vol. 6, Clarendon Press, Oxford, 1986, pp. 253–268.

[33] A.W. Nutbourne, *The solution of a frame matching equation*, The Mathematics of Surfaces (J. Gregory, ed.), The Institute of Mathematics and its Applications Conference Series. New Series, vol. 6, Clarendon Press, Oxford, 1986, pp. 233–252.

[34] I. Rivin, *A characterization of ideal polyhedra in hyperbolic 3-space*, Ann. of Math. **143** (1996), 51–70.

[35] R. Sauer, *Differenzengeometrie*, Berlin: Springer-Verlag, 1970.

[36] O. Schramm, *How to cage an egg*, Invent. Math. **107(3)** (1992), 543–560.

[37] _____, *Circle patterns with the combinatorics of the square grid*, Duke Math. J. **86** (1997), 347–389.

[38] B.A. Springborn, *Variational principles for circle patterns*, Ph.D. thesis, TU Berlin, 2003, arXiv:math.GT/0312363.

[39] _____, *A variational principle for weighted Delaunay triangulations and hyperideal polyhedra*, 2006, arXiv:math.GT/0603097.

[40] E. Steinitz, *Über isoperimetrische Probleme bei konvexen Polyedern*, J. reine angew. Math. **159** (1928), 133–143.

[41] K. Stephenson, *Introduction to circle packing: The theory of discrete analytic functions*, Cambridge University Press, 2005.

[42] T.J. Willmore, *Riemannian geometry*, Oxford Science Publications, 1993.

[43] G.M. Ziegler, *Lectures on Polytopes*, Graduate Texts in Mathematics, vol. 152, Springer-Verlag, New York, 1995, revised edition, 1998.

Alexander I. Bobenko
Institut für Mathematik, MA 8–3
Technische Universität Berlin
Str. des 17. Juni 136
10623 Berlin
Germany
e-mail: bobenko@math.tu-berlin.de

Discrete Differential Geometry, A.I. Bobenko, P. Schröder, J.M. Sullivan and G.M. Ziegler, eds.
Oberwolfach Seminars, Vol. 38, 37–56

Minimal Surfaces from Circle Patterns: Boundary Value Problems, Examples

Ulrike Bücking

Abstract. We construct discrete solutions to a class of boundary value problems for minimal surfaces without ends, including special classes of Plateau's problem. The boundary consists of finitely many straight line segments lying on the surface and/or planes intersecting the surface orthogonally. The discrete minimal surfaces which satisfy the given boundary conditions are built from a combinatorial parametrization, using an orthogonal circle pattern which approximates the Gauss map and a discrete duality transformation for S-isothermic surfaces.

Keywords. Discrete minimal surfaces, S-isothermic surfaces, combinatorics of curvature lines, construction of examples.

1. Introduction

Discrete minimal surfaces as discrete analogues of smooth minimal surfaces form an important class of surfaces in the field of discrete differential geometry. In this article we focus on some special examples of discrete minimal surfaces. More precisely, we consider minimal surfaces without ends and with a boundary consisting of finitely many parts of straight lines lying on the surface and/or planar curves whose planes intersect the surface orthogonally. Given an example of such a smooth minimal surface, we explain a general algorithm for constructing a discrete minimal analogue. The algorithm is then applied to several examples.

The constructions of discrete minimal surfaces presented in this article rely on an approach designed to conserve many essential properties of smooth minimal surfaces. This approach only considers parametrized surfaces. Since minimal surfaces are isothermic, we use a conformal curvature line parametrization and a suitable discretization which leads to quadrilateral parameter meshes. In Section 2 we briefly introduce the general construction scheme of discrete minimal surfaces which can be found in the contribution of

A.I. Bobenko [1] in this volume and in [2] with more details. Given a smooth minimal surface with boundary properties described above, this construction of the discrete minimal analogue is based on the following two ingredients:

- a combinatorial conformal curvature line parametrization which is a suitable discrete analogue of the corresponding smooth parametrization and
- the boundary data consisting of spherical angles of the corresponding boundary pieces in their image under the Gauss map.

The construction scheme of discrete minimal surfaces with given boundary conditions as above is explained in Section 3. We particularly focus on how to determine the above-mentioned main ingredients. In Section 4 we present several examples of discrete minimal surfaces with pictures, which illustrate the similar appearance of the discrete analogues of known smooth minimal surfaces.

2. General construction of discrete minimal surfaces

The construction used in this article is based on a characterization of minimal surfaces due to E. Christoffel [6]; see standard textbooks like [7] for a proof.

Theorem 2.1 (Christoffel). *A smooth minimal surface is an isothermic surface whose dual is (up to scaling) its image under the Gauss map, i.e., a part of the unit sphere.*

Taking this as a definition for smooth minimal surfaces and discretizing all notions in a convenient way, we are led to the following definition suggested in [2].

Definition 2.2. A *discrete minimal surface* is defined to be a discrete S-isothermic surface (made of touching spheres and orthogonal intersecting circles; see [1, 2] for more details) such that all spheres of the dual discrete S-isothermic surface intersect one fixed additional sphere orthogonally. This fixed sphere is taken to be the unit sphere S^2. Connecting the centers of touching spheres of the dual discrete S-isothermic surface yields (a part of) a Koebe polyhedron with edges tangent to S^2, see [3, 2]. This Koebe polyhedron can be interpreted as a discrete analogue of the Gauss map.

A discrete minimal surface can equivalently be characterized as a discrete S-isothermic surface with the following property. For each sphere, its center and the centers of all (generically four) orthogonally intersecting circles lie in a plane. Also, this property remains true when interchanging the roles of spheres and circles.

From Definition 2.2, a *construction scheme* for discrete minimal analogues of known smooth minimal surfaces can be deduced. We restrict ourselves to the main ideas; see [2] for a detailed presentation. In Section 3, we focus on the first of the following steps which provides the ingredients for the rest of the construction.

Step 1: Consider a given smooth minimal surface together with its conformal curvature line parametrization. Map the curvature lines to the unit sphere by the Gauss map to obtain a qualitative picture. The goal is to understand the combinatorics of the curvature lines. From the combinatorial picture of the curvature lines we choose finitely many curvature lines to obtain a finite cell decomposition of the unit sphere S^2 (or a part or a branched covering of S^2) with quadrilateral cells. Generically, all vertices have degree 4. Exceptional vertices correspond to umbilic or singular points, ends or boundary points of the minimal surface. So the cell decomposition provides a *combinatorial conformal parametrization*. This cell decomposition (together with additional information about the behavior at the boundary or near ends if necessary) is the main ingredient to construct the analogous discrete minimal surface.

Step 2: Given the combinatorics from Step 1, construct an *orthogonal spherical circle pattern* where circles correspond to vertices and faces of the cell decomposition. If two vertices are connected by an edge or if two faces share a common edge, then the corresponding circles touch. Circles corresponding to vertices and adjacent faces intersect orthogonally. At boundary vertices we use information about the smooth minimal surface to prescribe angles. Ends have also to be taken care of, but we do not consider this case in this article.

Step 3: Take the circles corresponding to the vertices (or, equivalently, the circles corresponding to the faces). To these circles determine the spheres which intersect S^2 orthogonally in these circles. Then build the *Koebe polyhedron* by joining the centers of touching spheres by an edge.

Step 4: Dualize the Koebe polyhedron to arrive at the discrete minimal surface.

The cell decomposition constructed in Step 1 (together with additional information about angles and lengths at the boundary if necessary) is the only ingredient in the following construction of the discrete minimal surface. Therefore the main task consists in finding suitable cell decompositions which correspond to the smooth minimal surface of which we are constructing a discrete analogue. Note that in contrast to our discrete minimal surfaces, curvature lines on smooth minimal surfaces are not generally closed modulo the boundary.

Given the combinatorial data (and boundary data if necessary), the goal of Step 2 is to calculate the spherical circle pattern corresponding to the given cell decomposition (and boundary constraints). The rest of the construction does not need additional data.

Existence and uniqueness of such a circle pattern was first proven by Koebe in case of a triangulation of the sphere in [10]. Generalizations of this theorem may be found in [5, 3, 17]. Many of our examples do not rely on cell decompositions of the whole sphere, but only of a part with given boundary data. For this reason we use a method presented in [2]. The solutions of the spherical circle pattern problem with given boundary angles are in one-to-one correspondence with the critical points of a new functional. Since the functional is not convex, a convenient reduced functional is considered instead. In order to compute the circle pattern, we minimize this reduced functional. In our numerical experiments, this method has been used with amazing success to construct circle patterns in the sphere, although existence and uniqueness of a solution are not yet proven.

3. Construction of solutions to special boundary value problems

In this section we explain the construction of discrete minimal surfaces with special boundary conditions. Possible examples include triply periodic minimal surfaces and some examples of Plateau's problem without ends, see also Section 4.

3.1. Boundary conditions

Consider the family of bounded smooth minimal surfaces. We suppose that the boundary curve consists of finitely many pieces and each of these boundary arcs

either (i) *lies within a plane which intersects the surface orthogonally.* Then the boundary curve is a curvature line and the surface may be continued smoothly across the plane by reflection in this plane. Moreover the image of this boundary curve under the Gauss map is (a part of) a great circle on the unit sphere.

or (ii) *it lies on a straight line.* Then the boundary curve is an asymptotic line and the surface may be continued smoothly across this straight line by 180°-rotation about it. Again the image of this boundary curve under the Gauss map is (a part of) a great circle on the unit sphere.

The implications in (i) and (ii) are well-known properties of smooth minimal surfaces and are called *Schwarz's reflection principles*, cf. for example [7].

We assume, that we know all boundary planes and straight lines and also their intersection angles (and lengths if necessary). As indicated in Section 2, to construct a discrete minimal surface, the first step consists in determining the combinatorics of the curvature lines under the Gauss map. To cope with this task, we first try to reduce the problem by symmetry. This is especially helpful when dealing with highly symmetric triply periodic minimal surfaces.

3.2. Reduction of symmetries

To simplify the construction of a discrete minimal surface, we only consider a *fundamental piece* of the smooth minimal surface. This is a piece of the surface which is bounded by planar curvature lines or straight asymptotic lines (like the whole surface itself) with the following two properties.

(i) The surface is obtained from the fundamental piece by successive reflection/rotation in its boundary planes/lines and the new obtained boundary planes/lines.

(ii) There is no smaller piece of the fundamental piece which has property (i).

In general, the fundamental piece is not unique.

Like a smooth minimal surface, a discrete minimal surface may also be obtained from its discrete fundamental piece by reflection in the boundary planes or 180°-rotation about straight boundary lines. This is due to the translation of the boundary conditions for discrete minimal surfaces (see Section 3.4 for more details) which guarantees the same behavior as in the smooth case.

If the symmetry group of a minimal surface contains a reflection in a plane, we can reduce the problem by considering this symmetry plane as a new boundary plane together with one half of the original boundary pieces. An analogous operation works if the symmetry group contains a 180°-rotation about a straight line/edge lying in the

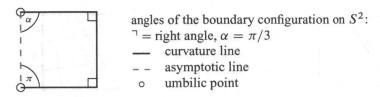

angles of the boundary configuration on S^2:
⌐ = right angle, $\alpha = \pi/3$
—— curvature line
– – asymptotic line
o   umbilic point

FIGURE 1. A combinatorial picture of the boundary conditions on S^2 of a fundamental region of Schwarz's H-surface.

surface. These reductions can be continued until we arrive at a fundamental piece. In the following, we are dealing with a fundamental piece.

3.3. Combinatorics of curvature lines

Knowing the boundary conditions, our next aim is to find the combinatorics of curvature lines. Thus we only draw general combinatorial pictures and merely indicate angles between boundary pieces.

Given a fundamental piece of a smooth minimal surface, we first determine the image of the boundary pieces under the Gauss map. By our assumption, each of these curves is a curvature line or an asymptotic line which is mapped to a part of a great circle on the unit sphere S^2. From the boundary curve, we extract the following ingredients for the construction (see Figure 1 for an example):

- a combinatorial picture of the boundary pieces,
- the angles between different boundary curves on S^2, and
- perhaps the lengths of the image of the boundary pieces under the Gauss map, if the angles do not uniquely determine the image of the boundary curve on S^2 (up to rotations of the sphere). This case is more difficult.

Given the combinatorial picture of the boundary, the remaining step consists in finding a combinatorial parametrization of the bounded domain. This parametrization should correspond to the curvature line parametrization of a smooth minimal surface, so we deduce the following conditions.

(1) Umbilic and singular points are taken from the smooth surface, but only their combinatorial locations matter. The smooth surface also determines the number of curvature lines meeting at these points. For interior umbilic or singular points, we additionally have to take care how many times the interior region is covered by the Gauss map. As this case does not occur in any of the examples presented in Section 4, we assume for simplicity in the following that all umbilic or singular points lie on the boundary.

(2) If a boundary curvature line and a boundary asymptotic line intersect at an angle of 135°, there is another curvature line meeting at this point.

(3) The curvature lines of a combinatorial parametrization divide the domain into combinatorial squares. The only exceptions may occur at asymptotic lines on the boundary, where there are combinatorial triangles built of two curvature lines and the boundary asymptotic line.

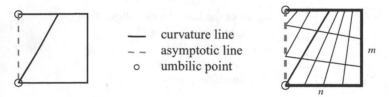

FIGURE 2. A combinatorial conformal parametrization of a fundamental region of Schwarz's H surface.

Hence in order to find the combinatorial curvature line parametrization, first determine all umbilic and singular points and all regular boundary points with 135°-angles. Then continue the additional curvature line(s) meeting at these points. In this way, the combinatorial domain is divided into finitely many subdomains such that conditions (2) and (3) hold; see Figure 2 (left). If there are several combinatorially different possibilities for such a subdivision, we choose the one with the combinatorics of the curvature line pattern of the smooth minimal surface.

Given this subdivision of the combinatorial domain, the parametrization of the subdomains is obvious. The only additional conditions occur at common boundaries of two subdomains, where the number of crossing curvature lines has to be equal on both sides. Hence after this step, we know the maximum number of free integer parameters corresponding to the number of different combinatorial types of curvature lines whose numbers may be chosen independently. These integer parameters correspond to smooth parameters of the continuous minimal surface such as scaling or quotients of lengths (heights, widths, etc.). The number of free integer parameters is greater or equal to the number of smooth parameters.

We restrict our examples presented in Section 4 to cases where the number of integer parameters is equal to the number of smooth parameters.

3.4. Boundary conditions of the discrete minimal surfaces

Given the combinatorics of curvature lines determined in Section 3.3, we can apply the remaining steps of the construction scheme of Section 2. First, the orthogonal spherical circle pattern is determined from the given combinatorial and boundary data. Next, to construct the Koebe polyhedron, half the circles become spheres intersecting S^2 orthogonally and the others remain circles. The choice is free but will affect the appearance of the discrete minimal surface. Corresponding to our smooth boundary conditions, there are three natural types of boundary conditions for the Koebe polyhedron (discrete Gauss map) which are analyzed in the following. The discrete minimal surface is finally obtained by dualization.

3.4.1. Touching spheres on the boundary.
First consider the case as in Figure 3, when the boundary circles of the spherical pattern touch. Their centers lie on a given great circle $\mathcal{C}$, and these boundary circles are labeled as spheres.

—	part of the great circle C
O	circle
⟨⟩	sphere
□	planar face

FIGURE 3. Touching spheres on the boundary.

Then a vector v which joins the centers of two touching spheres on the boundary lies in the plane $\mathcal{P}$ of the great circle C and is tangent to C. The discrete dualization transformation only changes the length of v and the orientation if necessary. So after dualization, all boundary vectors tangent to C form a discrete planar curvature line.

3.4.2. Touching circles on the boundary. Next consider the case of touching boundary circles as above, but which are labeled as circles as in Figure 4. The circle pattern can be continued symmetrically by reflection in the circle C.

—	part of the great circle C
O	circle
⟨⟩	sphere
⟦⟧	symmetric planar face

FIGURE 4. Touching circles on the boundary.

Consider a vector v joining the centers of two touching spheres which lie on different sides of C. Then by symmetry, v is orthogonal to the plane $\mathcal{P}$ of C. Hence, all such vectors are parallel. After dualization, the circles are orthogonal to a translation of the plane $\mathcal{P}$ and therefore build a discrete planar curvature line.

3.4.3. Orthogonally intersecting circles and spheres on the boundary. The case of orthogonally intersecting circles on the boundary is illustrated in Figure 5. The centers of these circles lie on the great circle C. As in the previous case, the circle pattern can be continued symmetrically by reflection in the circle C.

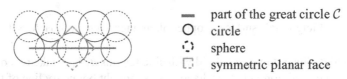

—	part of the great circle C
O	circle
⟨⟩	sphere
⟦⟧	symmetric planar face

FIGURE 5. Orthogonally intersecting circles and spheres on the boundary.

All boundary faces are orthogonal to the plane $\mathcal{P}$ of C and are symmetric with respect to this plane. After dualization, the boundary faces are still orthogonal to $\mathcal{P}$. Furthermore, the different signs accorded by the dualization transformation and the symmetry of the faces imply that the diagonals of all dualized boundary lie on the same straight line. This is the case of an asymptotic boundary curve.

4. Examples

In this section, the construction of discrete minimal surfaces explained in Section 3 is applied to some examples. To present discrete minimal surfaces resembling their smooth analogues, only the circles filled up with disks are shown in the pictures. In each case we present the (reduced) boundary conditions, a combinatorial picture of the curvature lines, and the image of the reduced boundary conditions under the Gauss map. Note that this image is uniquely determined (up to rotation of the sphere) by the intersection angles between the arcs of great circles for all our examples. Due to the simple combinatorics, the corresponding spherical circle patterns are also unique (up to rotations of the sphere). Pictures of the corresponding smooth minimal surfaces can, for example, be found in textbooks like [7, 12] or in some of the original treatises cited below. Triply periodic minimal surfaces can also be found in the article [9] by H. Karcher or on K. Brakke's webpage [4].

4.1. Gergonne's surface

Gergonne's surface traces back to J.D. Gergonne [8], who posed the first geometric problem leading to minimal surfaces with free boundaries in 1816. A correct solution was found by H.A. Schwarz in 1872; see [16, pp. 126–148].

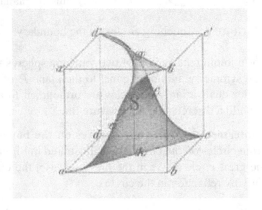

FIGURE 6. Gergonne's surface. Copper plate engraving from H.A. Schwarz [16].

Boundary conditions: Given a cuboid with side lengths a, b, c, take two opposite faces as boundary faces and non-collinear diagonals as straight boundary line of two other opposite faces, as in Figure 8 (left). The cuboid with these boundary conditions has two axes of 180°-rotation symmetry. These axes will lie on the minimal surface and cut the surface into four congruent parts; see Figures 6 and 8 (right). Therefore a fundamental piece of this surface is bounded by three orthogonal straight lines and one planar boundary face or equivalently three straight asymptotic lines and one planar curvature line, as in Figure 9.

The image of the reduced boundary conditions under the Gauss map is a spherical triangle with angles $\frac{\pi}{2}$, $\frac{\pi}{2}$ and $(\frac{\pi}{2} - \alpha)$.

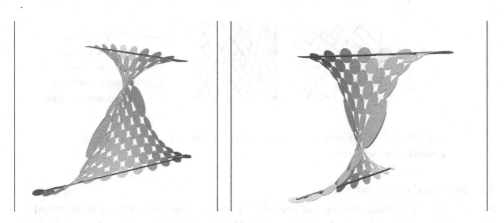

FIGURE 7. Discrete Gergonne's surface with $\alpha = \frac{\pi}{6}$ (left) and $\alpha = \frac{\pi}{4}$ (right).

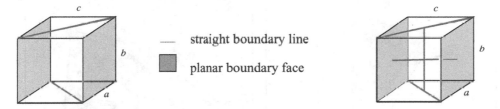

—— straight boundary line

▪ planar boundary face

FIGURE 8. Gergonne's surface: boundary conditions (left) and symmetry axes (right).

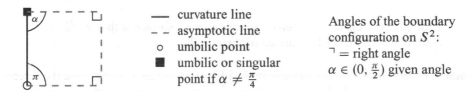

—— curvature line
– – asymptotic line
o umbilic point
■ umbilic or singular
  point if $\alpha \neq \frac{\pi}{4}$

Angles of the boundary
configuration on S^2:
⌐ = right angle
$\alpha \in (0, \frac{\pi}{2})$ given angle

FIGURE 9. Gergonne's surface: reduced boundary conditions (image under the Gauss map).

Now a combinatorial picture of the asymtotic lines or of the curvature lines can be determined; see Figure 10. All cases are parametrized by two integer numbers. This corresponds to the free choice of two length parameters of the cuboid with a given angle α between the diagonal and the planar boundary face.

In the case $\alpha = \frac{\pi}{4}$, the minimal surface can be continued by reflection and rotation in the boundary faces/edges to result in a triply periodic minimal surface.

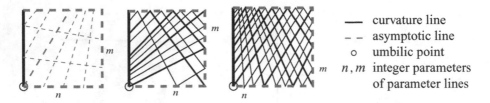

—— curvature line
- - asymptotic line
o   umbilic point
n, m integer parameters
     of parameter lines

FIGURE 10. Gergonne's surface: Different types of combinatorial conformal parametrizations.

4.2. Schwarz's CLP-surface

Schwarz's CLP-surface was one of the triply periodic minimal surfaces constructed by H.A. Schwarz; see [16, vol. 1, pp. 92–125]. The name CLP is due to A. Schoen [15]. If we consider the labyrinth formed by the periodic surface, the name is associated to properties of the underlying spatial lattice.

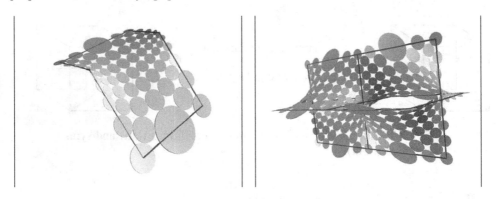

FIGURE 11. A generalization of Schwarz's CLP-surface with $\alpha = \frac{2\pi}{3}$ (left) and Schwarz's CLP-surface (right, $\alpha = \frac{\pi}{2}$).

Boundary conditions: Take a rectangle with side lengths a and b and choose the three boundary edges with lengths b, a and b. Then take another rectangle with side lengths a and c and choose the three boundary edges with lengths c, a and c. Paste these two frames at the open ends of the edges. The pasted edges will enclose an angle $\alpha \in (0, \pi)$, as in Figure 12 (left). By construction, there is a plane of reflection symmetry orthogonal to the sides with length a. Thus we get one planar curvature line. If $b = c$, there is another plane of reflection symmetry through both corners with angle α and yet another curvature line. This curvature line persists if $b \neq c$. To determine the general combinatorics of curvature lines, it therefore suffices to consider one fourth of the whole combinatorial picture bounded by two asymptotic and two curvature lines.

These combinatorics are shown in Figure 12 (right). There are two integer parameters which correspond to the two boundary lengths of the first rectangle. Analogously, there are two other integer parameters corresponding to the two boundary lengths of the

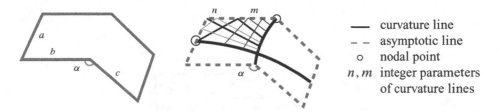

FIGURE 12. A generalization of Schwarz's CLP-surface: boundary frame (left) and combinatorics of curvature lines (right).

second rectangle. The pasting procedure translates into the condition that the number of curvature lines on the curvature line joining the pasted corners has to be equal on both sides. Therefore there are three parameters left corresponding to the free choice of the three boundary lengths.

The image of the reduced boundary conditions (plane of reflection and half of the boundary frame) under the Gauss map is a spherical triangle with angles $\frac{\pi}{2}$, $\frac{\pi}{2}$ and $(\pi - \alpha)$. Note that the angle at the umbilic points in the middle of the straight boundary lines of length a is π in the image under the normal map.

If $\alpha = \frac{\pi}{2}$ and $b = c$, the surface can be continued across its boundaries to a triply periodic discrete minimal surface. Figure 11 (right) shows a cubical unit cell.

4.3. Schwarz's D-surface

Schwarz's D-surface is another triply periodic minimal surface constructed by H.A. Schwarz [16, vol. 1, pp. 92–125]. The surface was named D by A. Schoen because its labyrinth graphs are 4-connected "diamond" networks.

Boundary conditions: Take a cuboid and a closed boundary frame consisting of two edges of each side; see Figure 14 (left). By construction, there are three straight lines of 180°-rotation symmetry lying within the minimal surface (see Figure 14 (right)) and three planes of reflection symmetry orthogonal to the boundary frame which give planar curvature lines (see Figure 14 (right), where only one of the three planes of reflection symmetry is shown). Therefore a fundamental piece is a triangle bounded by two straight asymptotic lines and one planar curvature line. The combinatorics of the curvature lines of this fundamental piece is shown in Figure 15.

The image of the reduced boundary conditions under the Gauss map is a spherical triangle with angles $\frac{\pi}{2}$ (between the two asymptotic lines) and $\frac{\pi}{4}$ and $\frac{\pi}{3}$ (between an asymptotic line and the curvature line).

The minimal surface constructed by the boundary frame in Figure 14 can be continued across the boundary to result in a triply periodic minimal surface. Figure 13 shows one cubical unit cell of the periodic lattice.

4.4. Neovius's surface

H.A. Schwarz [16] began to consider minimal surfaces bounded by two straight lines and an orthogonal plane. His student E.R. Neovius continued and deepened this study [11]

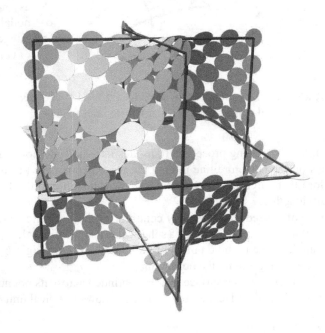

FIGURE 13. Schwarz's D-surface.

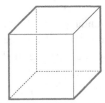 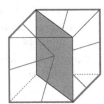

FIGURE 14. Schwarz's D-surface: boundary frame (left) and symmetries (right).

and found another triply periodic surface. This surface has the same symmetry group as
Schwarz's P-surface and was named C(P) by A. Schoen.

Boundary conditions: One unit cell of the lattice of Neovius's surface is basically a
cubical cell with one central chamber and necks out of the middle of each edge of the cube;
see Figure 16. By symmetry, it is sufficient to consider a cuboid which is one eighth of the
unit cell. All the planes of this cuboid are boundary planes. Furthermore, the surface piece
has three planes of reflection symmetry and three lines of 180° rotation symmetry (see
Figure 14 (right), where only one of the three planes of reflection symmetry is shown).

— curvature line

-- asymptotic line

○ nodal point

n integer parameter of curvature lines

FIGURE 15. Schwarz's D-surface: combinatorics of curvature lines of a fundamental piece.

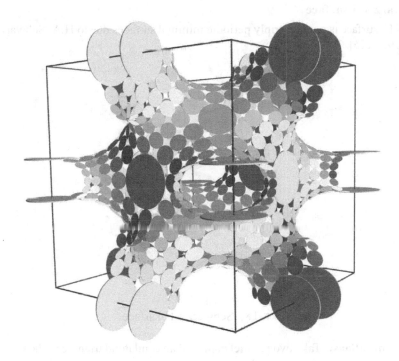

FIGURE 16. Neovius's surface.

Therefore the fundamental piece of this minimal surface consists of a triangle bounded by one straight asymptotic line and two planar curvature lines. The combinatorics of the curvature lines of this fundamental piece is shown in Figure 17.

The image of the reduced boundary conditions under the Gauss map is a spherical triangle with angles $\frac{\pi}{4}$ and $\frac{\pi}{3}$ (between a curvature line and the asymptotic line) and $\frac{3\pi}{4}$ (between the two curvature lines).

The minimal surface constructed by the above boundary conditions can be continued across the boundary to result in a triply periodic minimal surface. Figure 16 shows one cubical unit cell of the periodic lattice.

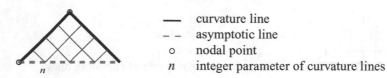

FIGURE 17. Neovius's surface: combinatorics of curvature lines of a fundamental piece.

4.5. Schwarz's H-surface

Schwarz's H-surface is another triply periodic minimal surfaces due to H.A. Schwarz [16, vol. 1, pp. 92–125].

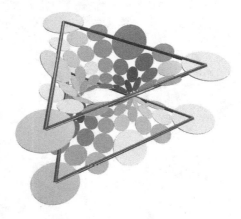

FIGURE 18. Schwarz's H-surface.

Boundary conditions: Take two parallel copies of an equilateral triangle as the boundary frame for a minimal surface spanned in between them, as in Figure 19. By construction this minimal surface has one plane of reflection symmetry parallel to the planes of the triangles and three other orthogonal planes as symmetry group of the equilateral triangle. Thus a fundamental piece is bounded by three planar curvature lines and one straight asymptotic line, as in Figure 19.

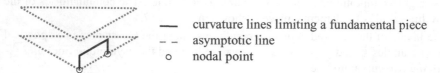

FIGURE 19. The boundary frame of Schwarz's H-surface with a fundamental piece.

The combinatorics of the curvature lines are shown in Figure 2. The two integer parameters correspond to the side length of the equilateral triangle and the distance between the two triangles.

The image of the reduced boundary conditions under the Gauss map is a spherical triangle with angles $\frac{\pi}{2}$ (between two of the curvature lines, the angle between the other two curvature lines is π) and $\frac{\pi}{2}$ and $\frac{\pi}{3}$ (between the asymptotic line and a curvature line).

The minimal surface constructed by the boundary conditions shown in Figure 18 can be continued across the boundary by 180°-rotation about the straight lines to result in a triply periodic minimal surface.

4.6. Schoen's I-6-surface and generalizations

About 1970, the physicist and crystallographer A. Schoen discovered many triply periodic minimal surfaces. His reports [15] were a bit sketchy, but H. Karcher [9] established the existence of all of Schoen's surfaces.

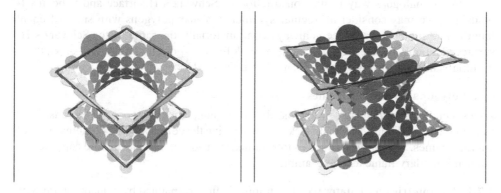

FIGURE 20. Schoen's I-6-surface (left) and a generalization (right).

Boundary conditions: They are similar to Schwarz's H-surface. Take two parallel copies of a rectangle as the boundary frame for a minimal surface between them. By construction this minimal surface has one plane of reflection symmetry parallel to the planes of the rectangles and two other orthogonal planes as symmetry group of the rectangle (three in case of a square). Thus we arrive at a fundamental piece bounded by three planar curvature lines and one straight asymptotic line; see Figure 21. There is one additional curvature line splitting the piece into two parts which can easily be found by reflection symmetry in the case of squares.

For both parts of the fundamental piece, the combinatorics of the curvature lines can be derived from the fundamental piece of Schwarz's H-surface; see Figures 2 and 21. Then we have to choose the four integer parameters such that the number of curvature lines meeting at the middle curvature line is the same on both sides. Thus three integer parameters remain corresponding to the lengths of the two sides of the rectangles and the distance between them.

The image of the reduced boundary conditions under the Gauss map is a spherical triangle with angles $\frac{\pi}{2}$, $\frac{\pi}{2}$ and $\frac{\pi}{2}$.

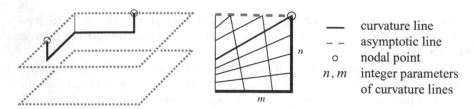

FIGURE 21. The boundary frame of (a generalization of) Schoen's I-6-surface with a fundamental piece and the combinatorics of curvature lines of one half of a fundamental piece.

The minimal surfaces shown in Figure 20 can be continued across the boundary to result in triply periodic minimal surfaces.

In an analogous way to the construction of Schwarz's H-surface and Schoen's I-6-surface, we may consider all regular symmetric planar polygons with sides of equal length. The fundamental piece is always combinatorially the same as for Schwarz's H-surface. The only difference is the angle $\alpha = \frac{\pi}{n}$ for an n-gon. In the limit $n \to \infty$, these minimal surfaces converge to the catenoid; see Figure 22.

4.7. Polygonal boundary frames

As a special class of Plateau's problems, all sorts of nonplanar polygons can be considered as boundary conditions. The main task is to determine the corresponding combinatorics of curvature lines. We confine ourselves to symmetric quadrilaterals and a more complicated cubical boundary frame as two examples.

4.7.1. Symmetric quadrilaterals. The minimal surface spanned by a quadrilateral with equal side lengths and equal angles of $\frac{\pi}{3}$ was the first solution in the class of Plateau's problems. In 1865, H.A. Schwarz found the explicit solution [16, p. 6–125]. About the same time, B. Riemann independently solved this problem [14, pp. 326–329]. His paper [13] appeared posthumously in 1867. At the same time H. A. Schwarz sent his prize essay to the Berlin academy. Later on, Plateau's problem was tackled for other polygonal boundaries; see, for example, [7, 12].

For a nonplanar quadrilateral boundary frame with a mirror symmetry plane the combinatorics of curvature lines is obvious; see Figure 24.

4.7.2. Cubical frame. Take a cuboid with edge lengths a, b, c and select eight of its boundary edges as illustrated in Figure 25 (left).

By construction, the minimal surface spanned by this frame has two planes of reflection symmetry and also two planar curvature lines; see Figure 25 (right). A fundamental piece is therefore bounded by three straight asymptotic lines and two curvature lines. The possible combinatorics of curvature lines are depicted in Figure 26. The three integer parameters correspond to the three edge lengths a, b, c.

The image of the reduced boundary conditions under the Gauss map is a spherical triangle with angles $\frac{\pi}{2}$, $\frac{\pi}{2}$ and $\frac{\pi}{2}$.

FIGURE 22. An approximation of a catenoid by parallel polygonal boundary frames.

FIGURE 23. A symmetric quadrilateral boundary frame.

The minimal surface with this cubical boundary conditions is shown in Figure 27 (left) and can be continued across the boundary to result in a triply periodic minimal

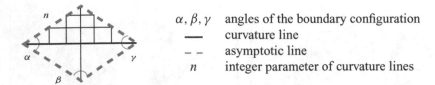

α, β, γ angles of the boundary configuration
—   curvature line
– –   asymptotic line
n integer parameter of curvature lines

FIGURE 24. Symmetric quadrilateral: boundary conditions and combinatorics of curvature lines of a fundamental piece.

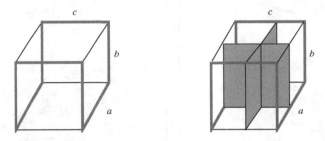

FIGURE 25. Cubic boundary frame and symmetry planes

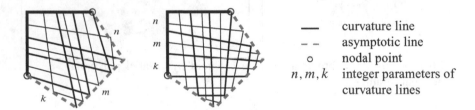

—   curvature line
– –   asymptotic line
o   nodal point
n, m, k integer parameters of curvature lines

FIGURE 26. General cubic frame: Combinatorics of curvature lines of a fundamental piece

surface. This example may be generalized to less symmetric cases. In Figure 27 (right) such an example is shown.

References

[1] Alexander I. Bobenko, *Surfaces from circles*, Discrete Differential Geometry (A.I. Bobenko, P. Schröder, J.M. Sullivan, G.M. Ziegler, eds.), Oberwolfach Seminars, vol. 38, Birkhäuser, 2008, this volume, pp. 3–35.

[2] Alexander I. Bobenko, Tim Hoffmann, and Boris A. Springborn, *Minimal surfaces from circle patterns: Geometry from combinatorics*, Ann. of Math. **164** (2006), no. 1, 231–264.

[3] Alexander I. Bobenko and Boris A. Springborn, *Variational principles for circle patterns and Koebe's theorem*, Trans. Amer. Math. Soc **356** (2004), 659–689.

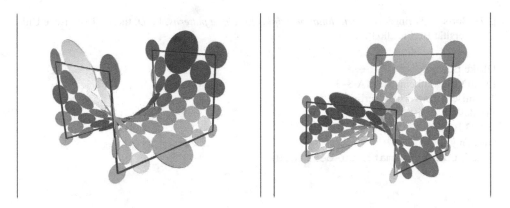

FIGURE 27. A cubical boundary frame (left) and a generalization (right).

[4] Ken Brakke, *Triply periodic minimal surfaces*, www.susqu.edu/facstaff/b/brakke/evolver/examples/periodic/periodic.html.

[5] Graham R. Brightwell and Edward R. Schreinerman, *Representations of planar graphs*, SIAM J. Disc. Math. **6** (1993), no. 2, 214–229.

[6] Elwin Christoffel, *Ueber einige allgemeine Eigenschaften der Minimumsflächen*, J. Reine Angew. Math. **67** (1867), 218–228.

[7] Ullrich Dierkes, Stefan Hildebrandt, Albrecht Küster, and Ortwin Wohlrab, *Minimal Surfaces I*, Springer-Verlag, Berlin, 1992.

[8] Joseph D. Gergonne, *Questions proposées/résolues*, Ann. Math. Pure Appl. **7** (1816), 68, 99–100, 156, 143–147.

[9] Hermann Karcher, *The triply periodic minimal surfaces of Alan Schoen and their constant mean curvature companions*, Manuscr. Math. **64** (1989), 291–357.

[10] Paul Kocbc, *Kontaktprobleme der konformen Abbildung*, Abh. Sächs. Akad. Wiss. Leipzig Math.-Natur. Kl. **88** (1936), 141–164.

[11] Edvard R. Neovius, *Bestimmung zweier spezieller periodischer Minimalflächen, auf welchen unendlich viele gerade Linien und unendlich viele ebene geodätische Linien liegen*, J. C. Frenckell & Sohn, Helsingfors, 1883.

[12] Johannes C. C. Nitsche, *Vorlesungen über Minimalfächen*, Grundlehren math. Wiss., vol. 199, Springer Verlag, 1975.

[13] Bernhard Riemann, *Über die Fläche vom kleinsten Inhalt bei gegebener Begrenzung*, Abh. Königl. Ges. d. Wiss. Göttingen, Mathem. Cl. **13** (1867), 3–52, K. Hattendorff, ed.

[14] ———, *Gesammelte Mathematische Werke*, B.G. Teubner, 1876 (1st ed.), 1892 (2nd ed.), and additions 1902.

[15] Alan H. Schoen, *Infinite periodic minimal surfaces without self-intersections*, NASA Technical Note (1970), no. D–5541.

[16] Hermann A. Schwarz, *Gesammelte Mathematische Abhandlungen*, vol. I and II, Springer-Verlag, Berlin, 1890.

[17] Boris A. Springborn, *Variational principles for circle patterns*, Ph.D. thesis, Technische Universität Berlin, 2003.

Ulrike Bücking
Institut für Mathematik, MA 8–3
Technische Universität Berlin
Str. des 17. Juni 136
10623 Berlin
Germany
e-mail: buecking@math.tu-berlin.de

Discrete Differential Geometry, A.I. Bobenko, P. Schröder, J.M. Sullivan and G.M. Ziegler, eds.
Oberwolfach Seminars, Vol. 38, 57–66

Designing Cylinders with Constant Negative Curvature

Ulrich Pinkall

Abstract. We describe algorithms that can be used to interactively construct ("design") surfaces with constant negative curvature, in particularly those that touch a plane along a closed curve and those exhibiting a cone point. Both smooth and discrete versions of the algorithms are given.

Keywords. Constant Gauss curvature, discrete K-surfaces.

1. Smooth K-surfaces

Here we give a brief introduction to the differential geometry of surfaces with constant negative Gaussian curvature.

1.1. Overview

Surfaces with constant negative Gaussian curvature, known as K-surfaces, are a classical topic in differential geometry. One reason is that their intrinsic geometry provides a model for the hyperbolic plane. The oldest example known is the so-called pseudosphere, a certain surface of revolution with Gaussian curvature -1:

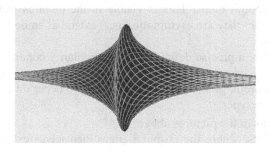

FIGURE 1. The pseudosphere.

Other surfaces of revolution with $K = -1$ come in two types as shown in the pictures below. The first type looks like a series of barrels joined along cuspidal edges:

FIGURE 2. K-surface of revolution.

The other type in addition exhibits singular points where the surface behaves like a cone:

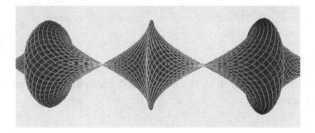

FIGURE 3. K-surface of revolution with cone points.

By a theorem of Hilbert there is no complete immersed surface with constant negative Gaussian curvature in 3-space. Nevertheless, all the above surfaces can be parametrized by perfectly smooth maps. This is visible in the pictures from the fact that all parameter curves (in fact they are asymptotic lines) extend as smooth curves through the apparent singularities.

Later we will give a precise definition of the regularity conditions we impose on a K-surface.

By Hilbert's theorem, globally, singularities have to occur on a K-surface. Generically, they come in two types:

- cuspidal edges as in the pictures above,
- swallowtail points, where the cuspidal edges themselves exhibit cusps, as in the picture below.

Cone points do not occur generically.

FIGURE 4. Swallowtail singularity.

1.2. The Gauss map of a K-surface

The Gauss map of a K-surface has the characterizing property that the images of the asymptotic lines form a Chebyshev net on the 2-sphere. This means that in suitable asymptotic coordinates the Gauss images of the parameter lines are parametrized with constant speed. Visually this implies that the parameter lines form "infinitesimally small parallelograms" on the sphere.

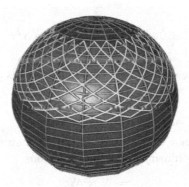

FIGURE 5. Chebyshev net on S^2.

One way to state this property is that the Gauss map of a K-surface is a harmonic map from the plane $\mathbb{R}^2$ (endowed with a Lorentz metric where the coordinate lines are lightlike) to the 2-sphere S^2. Using arbitrary asymptotic coordinates u, v this is equivalent to the following partial differential equation (subscripts indicate partial derivatives):

$$N \times N_{uv} = 0.$$

Switching the coordinates u, v to the coordinates

$$x = u + v,$$
$$t = u - v,$$

one obtains another physical interpretation of the Gauss map of a K-surface. The above PDE for N becomes

$$N \times (N_{xx} - N_{tt}) = 0$$

and the images of the lines $t = $ const therefore model the evolution of an elastic string on the sphere S^2. Think of the continuum limit of a sequence of massive balls coupled by elastic rubber bands.

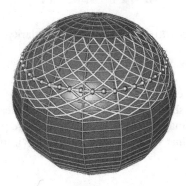

FIGURE 6. Elastic string on S^2.

In the physics literature such a string evolution on S^2 is referred to as a solution of the "nonlinear σ-model".

1.3. Reconstruction of a K-surface from its Gauss map

Using asymptotic coordinates u, v one can reconstruct a K-surface f from its Gauss map N by solving the following ordinary differential equations:

$$f_u = N \times N_u,$$
$$f_v = -N \times N_v.$$

Using the x, t coordinates introduced above these equations become:

$$f_x = N \times N_t,$$
$$f_t = N \times N_x.$$

1.4. Precise definition of a K-surface

The most convenient way to give a precise definition of a (not necessarily immersed) K-surface is by using its Gauss map: A map $f : \mathbb{R}^2 \mapsto \mathbb{R}^3$ is called a K-surface if there is a smooth map $N : \mathbb{R}^2 \mapsto S^2$ such that the formulas of the previous section apply and

- the partial derivatives N_u and N_v are nowhere vanishing.
- $N \times N_{uv} = 0$.

Since the second condition implies $\langle N_u, N_u \rangle_v = 0$ and $\langle N_v, N_v \rangle_u = 0$, it is enough to ensure the first condition on some line $u - v = $ const.

It is easy to see that the second condition is equivalent to the existence of a map $f : \mathbb{R}^2 \mapsto \mathbb{R}^3$ satisfying (3) and (4), and the first condition is then equivalent to the map $(f, N) : \mathbb{R}^2 \mapsto \mathbb{R}^3 \times S^2$ being an immersion.

By the formulas of the last section, all parameter lines of f in the (u, v)-coordinates will be curves with non-vanishing derivative (wherever f is an immersion, these curves will be asymptotic lines of f).

2. Discrete K-surfaces

Here we explain the discrete differential geometry used to implement a polyhedral version of K-surfaces [1, 2]. This should not just be considered as a numerical approximation to the smooth constructions, but as geometrically interesting in its own right.

2.1. Definition

A map $f : \mathbb{Z}^2 \mapsto \mathbb{R}^3$ from the integer lattice into 3-space is called a *discrete K-surface* if

- the length $|f_{n+1,m} - f_{n,m}|$ of a horizontal edge is independent of m;
- the length $|f_{n,m+1} - f_{n,m}|$ of a vertical edge is independent of n;
- each vertex $f_{n,m}$ together with its four neighbors $f_{n-1,m}, f_{n+1,m}, f_{n,m-1}, f_{n,m+1}$ lies in some plane $E_{n,m}$.

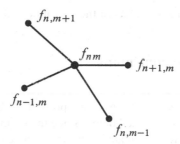

FIGURE 7. To the definition of a discrete K-surface.

The first two properties mean that a discrete K-surface is composed out of "skew parallelograms", i.e., quadrilaterals in space where opposite sides have the same length (without being planar). The third property ensures that we have a well-defined normal vector $N_{n,m}$ at each vertex $f_{n,m}$, given by the normal vector to the plane $E_{n,m}$. See the pictures in the section on K-surfaces. The surfaces depicted there were, in reality, discrete K-surfaces.

FIGURE 8. To the initial zig-zag.

2.2. Properties of discrete K-surfaces

The Gauss map of a discrete K-surface is always a *discrete Chebyshev net* in the sense that all quadrilaterals $N_{n,m}$, $N_{n+1,m}$, $N_{n+1,m+1}$, $N_{n,m+1}$ are spherical parallelograms (they admit a 180°-rotation that interchanges opposite points).

A discrete Chebyshev net is completely determined by giving initial data $N_{n,n}$ (time 0) and $N_{n+1,n}$ (time 1) for all integers n. We will refer to such initial data as an "initial zig-zag". By successively adding the missing fourth point of a spherical parallelogram (three points of which already have been determined) one can reconstruct a discrete Chebyshev net from an initial zig-zag uniquely.

2.3. Reconstruction of a discrete K-surface from its Gauss map

One can reconstruct a discrete K-surface f from its Gauss map N by solving the following difference equation:

$$f_{n+1,m} - f_{n,m} = N_{n,m} \times N_{n+1,m},$$
$$f_{n,m+1} - f_{n,m} = N_{n,m+1} \times N_{n,m}.$$

Here $v \times w$ denotes the cross-product of the vectors v and w.

3. K-surfaces with a cone point

3.1. Smooth case

A smooth K-surface f is said to have a cone point if a whole regular curve γ in the parameter domain is mapped to a fixed point. Since the asymptotic lines are always regular by definition, in asymptotic coordinates u, v such a curve can never be tangent to a coordinate line $u = $ const or $v = $ const. Hence, by passing to different asymptotic coordinates we locally may assume that γ is equal to the x-axis in the coordinates

$$x = u + v,$$
$$t = u - v.$$

From the formulas describing the reconstruction of a K-surface from its Gauss map we see that $f_x = 0$ implies $N_t = 0$. This means that at time 0 the moving string on the 2-sphere is at rest, i.e., all points have velocity zero. This is then the algorithm to construct cylinders with constant negative Gaussian curvature that exhibit a cone point:

- Place a closed elastic string $x \mapsto N(x, 0)$ on the 2-sphere and use it as initial data with zero velocity for a string evolution $N : \mathbb{R}^2 \mapsto S^2$.
- Reconstruct the surface f from N as described in the section on the reconstruction of a K-surface from its Gauss map.

3.2. Discrete case

It is now clear how to construct discrete K-surfaces with a cone point: According to the section on Gauss maps of discrete K-surfaces one has to start with an initial zig-zag that is stationary, i.e., a fixed point of its discrete evolution. This will occur if and only if every point of the zig-zag at time 1 is situated exactly at the center of the spherical great circle arc connecting the neighboring points at time 0.

In this case, the points at time -1 obtained by completing the parallelograms will coincide with the corresponding points at time 1. Furthermore the whole Gauss map for negative values of the time will be an exact copy of the Gauss map for positive times.

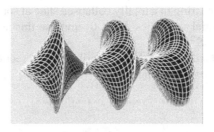

FIGURE 9. A discrete K-surface with a cone point.

FIGURE 10. Gauss map of a discrete K-surface with a cone point.

4. K-surfaces with a planar strip

4.1. Smooth case

A smooth K-surface f is said to have a planar strip if a whole regular curve γ in the parameter domain is mapped to a fixed point on the 2-sphere by its Gauss map N. Since the asymptotic lines are always regular curves by definition, in asymptotic coordinates u, v such a curve γ can never be tangent to a coordinate line $u = $ const or $v = $ const. Hence, by passing to different asymptotic coordinates we locally may assume that γ is the x-axis in the coordinates

$$
x = u + v,
$$
$$
t = u - v.
$$

From the formulas describing the reconstruction of a K-surface from its Gauss map we see that $N(x, 0)$ is a fixed point on S^2. This is then the algorithm to construct cylinders with constant negative Gaussian curvature with a planar strip:

- Place a closed elastic string in a totally collapsed state at some point on the 2-sphere, provide arbitrary initial velocities $N_t(x, 0)$ and use this as initial data for a string evolution $N : \mathbb{R}^2 \mapsto S^2$.
- Reconstruct the surface f from N as described in the section on the reconstruction of a K-surface from its Gauss map.

4.2. Discrete case

It now clear how to construct discrete K-surfaces with a planar strip: One has to start with an initial zig-zag with all $N_{n,n}$ equal to a fixed point a on S^2.

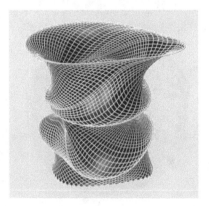

FIGURE 11. A discrete K-surface with a planar strip.

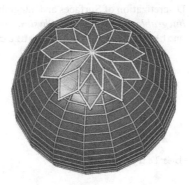

FIGURE 12. Gauss map of a discrete K-surface with a planar strip.

According to the section on the reconstruction of a discrete K-surface from its Gauss map the corresponding points $f_{n,n}$ will then satisfy

$$f_{n+1,n} - f_{n,n} = a \times N_{n+1,n},$$
$$f_{n+1,m+1} - f_{n+1,n} = a \times N_{n+1,n},$$

and therefore

$$f_{n+1,n+1} - f_{n,n} = 2a \times N_{n+1,n}.$$

To make a discrete K-surface that touches a plane with normal vector a along a prescribed closed polygon $\gamma_1, \ldots, \gamma_m$ one therefore has to proceed as follows:

- Rotate the polygon by 90° around a and scale it so that all edge vectors $\gamma_{n+1} - \gamma_n$ have length smaller than 2 (this leaves a free parameter for the construction).
- Project the scaled edge vectors $(\gamma_{n+1} - \gamma_n)/2$ to the unit 2-sphere to obtain the normal vectors $N_{2,1}, \ldots, N_{m+1,m}$.
- Set $N_{n,n} = a$ for all n and use the defined normal vectors as an initial zig-zag for a discrete Chebyshev net.
- Reconstruct the discrete K-surface f from its Gauss map N.

5. Software

Interactive Java webstart applications that implement the algorithms in this paper are available at:

www.math.tu-berlin.de/geometrie/lab

The material in this paper is also included with these applications, under the help menu.

References

[1] A.I. Bobenko, U. Pinkall, Discrete surfaces with constant negative Gaussian curvature and the Hirota equation. *J. Differential Geom.*, 43(3):527–611, 1996

[2] A.I. Bobenko, U. Pinkall, Discretization of Surfaces and Integrable Systems. In: A.I. Bobenko, R. Seiler (eds.) *Discrete Integrable Geometry and Physics*, Oxford University Press (1999) pp. 3–58, www-sfb288.math.tu-berlin.de/abstractNew/296.

Ulrich Pinkall
Institut für Mathematik, MA 3–2
Technische Universität Berlin
Str. des 17. Juni 136
10623 Berlin
Germany
e-mail: pinkall@math.tu-berlin.de

Discrete Differential Geometry, A.I. Bobenko, P. Schröder, J.M. Sullivan and G.M. Ziegler, eds.
Oberwolfach Seminars, Vol. 38, 67–93
© 2008 Birkhäuser Verlag Basel/Switzerland

On the Integrability
of Infinitesimal and Finite Deformations
# of Polyhedral Surfaces

Wolfgang K. Schief, Alexander I. Bobenko and Tim Hoffmann

Abstract. It is established that there exists an intimate connection between isometric deformations of polyhedral surfaces and discrete integrable systems. In particular, Sauer's kinematic approach is adopted to show that second-order infinitesimal isometric deformations of discrete surfaces composed of planar quadrilaterals (discrete conjugate nets) are determined by the solutions of an integrable discrete version of Bianchi's classical equation governing finite isometric deformations of conjugate nets. Moreover, it is demonstrated that finite isometric deformations of discrete conjugate nets are completely encapsulated in the standard integrable discretization of a particular nonlinear σ-model subject to a constraint. The deformability of discrete Voss surfaces is thereby retrieved in a natural manner.

Keywords. Isometric deformations, polyhedral surfaces, integrable systems.

1. Introduction

The study of infinitesimal and finite deformations of both polyhedral and smooth surfaces has a long history. Various proofs of the fact that closed convex polyhedra are infinitesimally rigid are due to Cauchy [10] (1813), Dehn [13] (1916), Weyl [28] (1917) and Alexandrov [2] (1958). Analogous results for smooth surfaces were obtained by Liebmann [21] (1899) and Cohn-Vossen [12] (1936). Apart from their significance in differential geometry, isometric deformations of smooth surfaces also find diverse application in physics. For instance, it was observed by Blaschke [5] that the standard theory of shell membranes which are in equilibrium and not subjected to external forces may be set in correspondence with infinitesimal isometric deformations of surfaces. Thus, the geometric determination of such deformations corresponds to finding solutions of the equilibrium equations in membrane theory as set down and discussed by such luminaries as Beltrami, Clapeyron, Kirchhoff, Lagally, Lamé, Lecornu, Love and Rayleigh [22].

The main aim of the present paper is to show that there exists an intrinsic connection between isometric deformations of (open) quadrilateral surfaces and discrete integrable systems. Thus, the study of isometrically deformable quadrilateral surfaces is canonically embedded in the emerging field of *integrable discrete differential geometry* [8]. In particular, it is demonstrated that the deformation parameter may be identified as the 'spectral parameter' which constitutes the key ingredient in the theory of integrable systems [1].

Here, in the main, we focus on isometric deformations of *discrete conjugate nets* which constitute discrete surfaces composed of planar quadrilaterals. We adopt the kinematic approach due to Sauer [24] and retrieve his fundamental notion of *reciprocal-parallel* discrete surfaces associated with the existence of *infinitesimal* isometric deformations. We show that, remarkably, discrete conjugate nets admit infinitesimal isometric deformations of *second-order* if and only if the reciprocal-parallel surfaces constitute *discrete Bianchi surfaces*. These constitute natural discrete analogues of an integrable class of classical surfaces which was analysed by Bianchi [4] in connection with isometric deformations of conjugate nets. The nonlinear equation underlying discrete Bianchi surfaces has been shown to be integrable [26] in a different geometric context, namely discrete isothermic surfaces [8]. By construction, discrete Bianchi surfaces constitute particular *discrete asymptotic nets*. Integrable reductions of discrete asymptotic nets including discrete Bianchi surfaces have been the subject of [14].

If a discrete conjugate net admits a *finite* isometric deformation, then any quadrilateral undergoes a rigid motion which may be decomposed into a translation and a rotation. We demonstrate that the rotational component interpreted as an $SU(2)$-valued lattice function obeys a *pair* of *linear* equations which bears the hallmarks of a 'Lax pair' [1] for a discrete integrable system in that it depends parametrically on the deformation parameter and gives rise to a *discrete zero-curvature condition*. As an illustration, it is shown that the discrete zero-curvature condition contains as a special case the 'Gauss map' of *discrete K-surfaces*. These are integrable [6] and have been proposed as natural discrete analogues of surfaces of constant negative Gaussian curvature by Sauer [23] and Wunderlich [29]. As observed by Sauer, discrete K-surfaces are reciprocal-parallel to *discrete Voss surfaces* which have indeed been shown by Sauer and Graf [25] to admit *finite* isometric deformations. Finally, it is established that the discrete zero-curvature condition may be formulated in terms of a standard integrable *discrete nonlinear σ-model* [8, 26] subject to a constraint involving the deformation parameter.

2. Infinitesimal deformations of discrete surfaces

In the following, a *discrete surface* F is defined as (the image of) a mapping

$$F : V(\mathcal{G}) \to \mathbb{R}^3,$$

where $V(\mathcal{G})$ denotes the set of vertices of a cellular decomposition $\mathcal{G}$ of the plane. The edges and (combinatorial) faces of the discrete surface are those induced naturally by the mapping F. A *dual* discrete surface F^* is a mapping which is defined on the vertices of

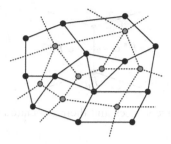

FIGURE 1. A cellular decomposition $\mathcal{G}$ (black vertices) and its dual $\mathcal{G}^*$ (grey vertices).

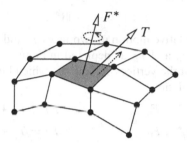

FIGURE 2. Infinitesimal rigid motion of a quadrilateral.

the dual cellular decomposition $\mathcal{G}^*$ (cf. Figure 1), that is,

$$F^* : V(\mathcal{G}^*) \to \mathbb{R}^3.$$

In the present paper, we are mainly concerned with *quadrilateral surfaces* corresponding to the choice $\mathcal{G} = \mathbb{Z}^2$ as illustrated in Figure 2. However, we begin with a generic discrete surface $F : V(\mathcal{G}) \to \mathbb{R}^3$ and consider a 'deformed surface'

$$F^\epsilon = F + \epsilon \bar{F},$$

where the constant ϵ constitutes a 'small' deformation parameter, that is, $|\epsilon| \ll 1$, and $\bar{F} : V(\mathcal{G}) \to \mathbb{R}^3$ defines the displacement of the vertices of F. It is convenient to imagine the edges of any face of the discrete surface as the boundary of a small piece of a surface. Accordingly, it is meaningful to define an *infinitesimal isometric deformation* F^ϵ of a discrete surface F as a deformation which does not change the shapes of the faces to order $O(\epsilon)$. In kinematic terms, when the discrete surface is being deformed, any face undergoes an infinitesimal rigid motion, that is, a combination of an infinitesimal translation and an infinitesimal rotation. Accordingly, if P is a point on a face f, then its displacement $P^\epsilon - P = \epsilon \bar{P}$ is given by

$$\bar{P} = T + F^* \times P, \qquad (2.1)$$

where the translation T and the oriented axis of rotation F^* are *independent* of P. This is illustrated in Figure 2 for a quadrilateral surface. Since each face is associated with a vector of rotation F^*, we may regard F^* as another discrete surface which is dual to F.

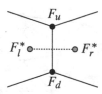

FIGURE 3. Two adjacent faces and their associated vectors of rotation F_r^* and F_l^*.

Thus, the vertices, edges and faces of F are in one-to-one correspondence with the faces, edges and vertices of F^*, respectively. Similarly, T is defined on the faces of $\mathcal{G}$, that is,

$$T : V(\mathcal{G}^*) \to \mathbb{R}^3.$$

We now focus on the relative motion of two faces f_r and f_l which are joined by an edge e linking the vertices F_u and F_d as depicted in Figure 3. Thus, if we evaluate the displacement relation (2.1) at the vertices F_u and F_d which belong to both faces f_r and f_l, then we obtain the four relations

$$\bar{F}_u = T_r + F_r^* \times F_u, \qquad\qquad \bar{F}_d = T_r + F_r^* \times F_d, \qquad (2.2a)$$
$$\bar{F}_u = T_l + F_l^* \times F_u, \qquad\qquad \bar{F}_d = T_l + F_l^* \times F_d. \qquad (2.2b)$$

The latter imply that the *dual* edges $[F_l^*, F_r^*]$ and $[F_d, F_u]$ are *parallel*, that is,

$$(F_r^* - F_l^*) \times (F_u - F_d) = 0.$$

This merely expresses the fact that the relative motion of the two adjacent faces f_r and f_l represents an infinitesimal rotation about the common edge $[F_d, F_u]$. Thus, the following definition is natural.

Definition 2.1. Two combinatorially dual discrete surfaces F and F^* are *reciprocal-parallel* if dual edges are parallel.

The above reasoning may now be inverted and hence we are led to a result which is due to Sauer [24] in the case of quadrilateral surfaces.

Theorem 2.2. *A discrete surface F admits an infinitesimal isometric deformation if and only if there exists a reciprocal-parallel discrete surface F^*.*

Proof. In the preceding, it has been established that an infinitesimal isometric deformation of a discrete surface F gives rise to a reciprocal-parallel surface generated by the vectors of rotation F^*. Conversely, if F^* constitutes a reciprocal-parallel discrete surface, then the relations (2.2) imply that

$$T_r - T_l = -(F_r^* - F_l^*) \times F_u, \quad T_r - T_l = -(F_r^* - F_l^*) \times F_d \qquad (2.3)$$

constitute necessary conditions on the vectors of translation associated with two adjacent faces. In fact, the above pair uniquely defines a dual discrete surface $T : V(\mathcal{G}^*) \to \mathbb{R}^3$ up to a single vector T_0 defined on a face f_0. In kinematic terms, the latter corresponds to a uniform translation of the discrete surface F. In order to make good the assertion that T

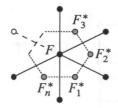

FIGURE 4. The vertices F_k^* of a face dual to a vertex F.

is well defined, it is required to verify two properties. Firstly, the two relations (2.3) are equivalent since F^* is reciprocal-parallel to F and hence $(F_r^* - F_l^*) \times (F_u - F_d) = 0$. Secondly, the 'closing condition' associated with any closed polygon composed of edges of F^* is satisfied. Indeed, it is sufficient to consider the boundary of a dual face as depicted in Figure 4. If we denote the vertices of the face which is dual to the vertex F by $F_1^*, \ldots, F_n^*$, then the n relations

$$T_{k+1} - T_k = -(F_{k+1}^* - F_k^*) \times F, \quad k = 1, \ldots, n,$$

hold (with the identification $T_{n+1} - T_1$) and the corresponding closing condition

$$\sum_{k=1}^{n} (T_{k+1} - T_k) = -\sum_{k=1}^{n} (F_{k+1}^* - F_k^*) \times F = 0$$

is satisfied. The 'displacement' $\bar{F}$ of the vertex F may now be defined by

$$\bar{F} = T_k + F_k^* \times F$$

since the latter is independent of k. In this manner, one may construct a discrete surface

$$F^\epsilon = F + \epsilon \bar{F}$$

which represents an infinitesimal isometric deformation of F. □

2.1. Quadrilateral surfaces

We are now concerned with (portions of) quadrilateral surfaces

$$F : \mathbb{Z}^2 \to \mathbb{R}^3.$$

In this case, the relation between two reciprocal-parallel surfaces F and F^* is illustrated in Figure 5. Here and throughout the remainder of the paper, we indicate unit increments of the discrete variables n_1 and n_2 which label the lattice $\mathbb{Z}^2$ by subscripts, that is, for instance,

$$F = F(n_1, n_2), \quad F_1 = F(n_1 + 1, n_2), \quad F_{12} = F(n_1 + 1, n_2 + 1),$$

so that a quadrilateral $\diamond$ is represented by $[F, F_1, F_{12}, F_2]$. Similarly, overbars on subscripts designate unit decrements. Thus, a vertex F is linked to the quadrilaterals $\diamond$, $\diamond_{\bar{1}}$, $\diamond_{\bar{1}\bar{2}}$ and $\diamond_{\bar{2}}$. Furthermore, we adopt the notation

$$\Delta_i F = F_i - F, \quad \Delta_{12} F = F_{12} - F_1 - F_2 + F$$

for the first-order and mixed second-order difference operators, respectively.

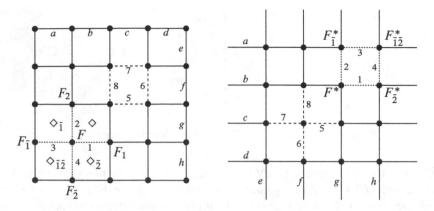

FIGURE 5. Schematic depiction of reciprocal-parallel quadrilateral surfaces F and F^*.

In the following, we are concerned with the deformation of discrete surfaces which are composed of *planar* quadrilaterals.

Definition 2.3. A quadrilateral surface is termed a *discrete conjugate net* if all quadrilaterals are planar.

Discrete conjugate nets constitute natural analogues of conjugate nets in classical differential geometry (see [8] and references therein). Their reciprocal-parallel counterparts (if they exist) represent discrete versions of classical asymptotic nets.

Definition 2.4. A quadrilateral surface is termed a *discrete asymptotic net* if all stars are planar.

The following observation [25] which provides an important connection between discrete conjugate and asymptotic nets is a direct consequence of the analysis undertaken in the previous subsection. It is illustrated in Figure 6.

Theorem 2.5. *A discrete conjugate net with nonplanar stars is infinitesimally isometrically deformable if and only if there exists a reciprocal-parallel discrete asymptotic net. The latter is uniquely determined up to a scaling and hence the deformation is unique.*

The uniqueness of the reciprocal-parallel discrete asymptotic net is due to the fact that any star of the discrete conjugate net determines the directions of the edges of the corresponding dual quadrilateral of the discrete asymptotic net. By assumption, this quadrilateral is nonplanar and hence it is known up to a scaling. However, if the length of one edge of the discrete asymptotic net is arbitrarily prescribed, the scalings of all quadrilaterals are uniquely determined. In this connection, it proves useful to adopt the following definition [19].

Definition 2.6. Two discrete conjugate nets are *Combescure transforms* of each other if corresponding edges are parallel.

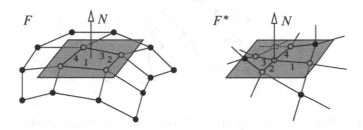

FIGURE 6. Reciprocal-parallel discrete conjugate and asymptotic nets F and F^*.

It is evident that any discrete conjugate net admits an infinite number of Combescure transforms. Their relevance in the context of discrete asymptotic nets is the content of the following theorem.

Theorem 2.7. *Any discrete asymptotic net with nonplanar quadrilaterals possesses an infinity of reciprocal-parallel discrete conjugate nets. These are related by Combescure transformations and admit an infinitesimal isometric deformation.*

Proof. Since the stars of an asymptotic net are planar, each star may be associated with a unit normal N as illustrated in Figure 6. The reciprocal-parallel conjugate nets are constructed by successively drawing planes which are orthogonal to the vectors N and have the property that the four planes associated with any four 'neighbouring' normals meet at a point. $\qquad\square$

3. Finite deformations

The subject of *finite isometric deformations* of polyhedral surfaces is classical and is highlighted by Cauchy's non-existence theorem [10] for deformations of convex polyhedra. Here, we confine ourselves to isometric deformations of a discrete conjugate net $F : \mathbb{Z}^2 \to \mathbb{R}^3$, that is a one-parameter family of discrete surfaces

$$F^\epsilon : \mathbb{Z}^2 \to \mathbb{R}^3$$

which depends continuously on the parameter ϵ in such a manner that the planar quadrilaterals of F^ϵ are congruent to those of the undeformed discrete conjugate net $F^0 = F$. In the following, it is understood that the term deformation implies its finite character while infinitesimal deformations are explicitly referred to as such.

In the previous section, it has been concluded that for any given discrete conjugate net with nonplanar stars there exists at most a one-parameter family of infinitesimal isometric deformations. Accordingly, finite deformations of such conjugate nets are unique if they exist. In fact, this statement is still correct if one replaces the condition of nonplanar stars by the assumption that the discrete conjugate net is *non-degenerate* in the sense that opposite edges of any star are not collinear. This may be deduced by considering a non-degenerate complex of 2×2 planar quadrilaterals meeting at a vertex (cf. Figure 7). Indeed, if we vary the angle between two adjacent quadrilaterals, then the remaining two

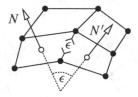

FIGURE 7. The deformability of a 2×2 complex of four planar quadrilaterals.

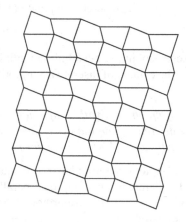

FIGURE 8. A deformable 'tessellation of the plane'.

quadrilaterals move in such a way that the complex is not torn. Thus, deformations of 2×2 complexes always exist. If we hold that angle constant, then the complex is rigid and if the complex forms part of a discrete conjugate net, then the entire discrete surface is rigid.

In conclusion, if a discrete conjugate net is isometrically deformable, then any Combescure transform is likewise isometrically deformable. In fact, the deformability of a discrete conjugate net is a property of its *normals only* and may therefore be dealt with in the realm of spherical geometry. Accordingly, we may state the following:

Theorem 3.1. *A non-degenerate discrete conjugate net admits at most a one-parameter family of isometric deformations. It is isometrically deformable if and only if all its Combescure transforms are isometrically deformable.*

3.1. General considerations

The complete classification of deformable discrete conjugate nets has not been achieved yet. Particular classes of such discrete surfaces were constructed by Sauer and Graf [25] and Kokotsakis [17] in the 1930s. An interesting example is displayed in Figure 8 and represents the 'tessellation of the plane' by means of copies of an arbitrary quadrilateral which are glued at corresponding edges. Some of the known deformable discrete conjugate nets do not possess smooth counterparts. This is a first indication that the class of

FIGURE 9. A complex of eight planar quadrilaterals is always deformable while a 3×3 complex is generically rigid.

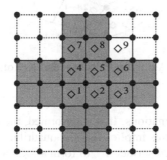

FIGURE 10. Illustration of the fact that a discrete conjugate net is deformable if and only if its 3×3 complexes are deformable.

deformable discrete conjugate nets is richer than in the smooth setting. In this connection, it is interesting to note that the conditions for the existence of isometric deformations which preserve conjugate nets on smooth surfaces were given by Bianchi [4] in as early as 1892 and related to an integrable class of surfaces which have been termed *Bianchi surfaces*. This will be discussed in more detail in Section 4.4.

The problem of isometric deformations becomes nontrivial as soon as the discrete conjugate net consists of at least 3×3 quadrilaterals. Indeed, if we consider a 3×3 complex and remove a corner quadrilateral, then deformation of the 'opposite' 2×2 complex determines the new position of the remaining four adjacent quadrilaterals (cf. Figure 9). However, in general, it will be impossible to re-insert the ninth quadrilateral and form a 3×3 complex. Thus, a 3×3 complex is generically rigid but admits a unique one-parameter family of isometric deformations if certain constraints on the normals to the quadrilaterals are satisfied. In fact, it is sufficient to focus on 3×3 complexes in the following sense:

Theorem 3.2. *A non-degenerate discrete conjugate net is isometrically deformable if and only if its 3×3 complexes are isometrically deformable.*

Proof. Let F be a non-degenerate discrete conjugate net which is such that all its 3×3 complexes are isometrically deformable. We remove all quadrilaterals up to two neighbouring 'horizontal' and two neighbouring 'vertical' strips which intersect at four quadrilaterals $\diamond^1, \diamond^2, \diamond^4, \diamond^5$ as indicated in Figure 10. This 'cross' of quadrilaterals admits a

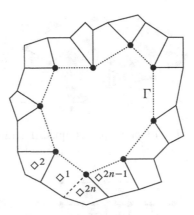

FIGURE 11. A polyhedral surface of Kokotsakis type.

one-parameter family of deformations. If it is deformed, then the assumption of the deformability of the 3×3 complexes implies that the quadrilateral $\diamond^9$ may be inserted into the existing part of the deformed surface F^ϵ to form a 3×3 complex $[\diamond^{1\epsilon}, \ldots, \diamond^{9\epsilon}]$. The complete deformed surface F^ϵ is constructed by repeating this procedure iteratively. $\quad\square$

Kokotsakis [17] investigated infinitesimal and finite deformations of a special class of (open) polyhedral surfaces which contains, for instance, the above-mentioned 3×3 complexes and closed octahedra. His investigations naturally led to Cauchy's theorem in the case of convex octahedra and the infinitesimal deformability of the octahedra of Bricard type [9]. Specifically, his polyhedral surfaces consist of a closed strip of $2n$ planar quadrilaterals $\diamond^1, \ldots, \diamond^{2n}$ which are alternately attached to the n edges and n vertices of a rigid and closed (not necessarily planar) polygon Γ as indicated in Figure 11. In particular, if Γ constitutes a planar quadrilateral, then it may be regarded as the central quadrilateral of a 3×3 complex. In order to investigate the (infinitesimal) deformability of this polyhedral surface, one now cuts the surface at the edge between the quadrilaterals $\diamond^1$ and $\diamond^{2n}$ and (infinitesimally) rotates the quadrilateral $\diamond^1$ about the corresponding edge of Γ. The remaining quadrilaterals then move accordingly and, in general, there now exists a gap between the quadrilaterals $\diamond^1$ and $\diamond^{2n}$. If the motion is infinitesimal, then the condition of a closed deformed strip imposes one constraint on the polyhedral surface. However, if the motion is finite and one demands that the strip is closed for *all* possible positions of the quadrilateral $\diamond^1$, then this constraint constitutes a one-parameter family of constraints the solution of which is, in general, unknown. In the case of a 3×3 complex, it may be shown that the determination of solutions of these constraints amounts to finding common zeros of certain polynomials. Once again, in general, the solution to this problem is unknown. Nevertheless, as noted by Kokotsakis, direct inspection of these constraints gives rise to a class of isometrically deformable 3×3 complexes which was first recorded by Sauer and Graf [25]. Its extension to discrete surfaces is discussed below.

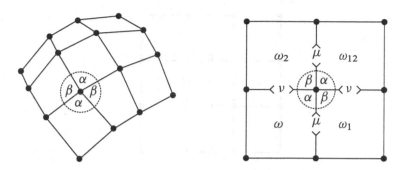

FIGURE 12. Definition and properties of discrete Voss surfaces.

3.2. The deformability of discrete Voss surfaces

As stated in the preceding, it is not known under which circumstances a discrete conjugate net is isometrically deformable. However, there exists an important class of discrete conjugate nets which admit isometric deformations. It consists of natural discrete analogues of classical Voss surfaces [27]. The latter terminology is due to the fact that, in the formal continuum limit, the coordinate polygons of any discrete Voss surface become two families of *geodesic* conjugate lines. The definition of discrete Voss surfaces is depicted in Figure 12.

Definition 3.3. A non-degenerate discrete conjugate net which is such that opposite angles made by the edges of any star are equal is termed a *discrete Voss surface*.

The existence of discrete Voss surfaces is readily shown by considering well-posed Cauchy problems. For instance, it is not difficult to verify that any arbitrarily prescribed spatial stairway gives rise to a unique discrete Voss surface. Moreover, it may be shown that the angle made by two adjacent quadrilaterals (dihedral angle) is constant along the coordinate polygon containing the common edge. This property may be taken as an alternative definition of discrete Voss surfaces and is illustrated in Figure 12. The connection between the 'interior' angles α, β and the dihedral angles μ, ν is given by [25]

$$\tan \frac{\mu}{2} \tan \frac{\nu}{2} = \frac{\sin(\alpha + \beta)}{\sin \alpha + \sin \beta}. \tag{3.1}$$

Thus, the one-parameter family of deformations of 2×2 complexes of Voss type may be described algebraically by the transformation

$$\tan \frac{\mu}{2} \to \lambda \tan \frac{\mu}{2}, \quad \tan \frac{\nu}{2} \to \frac{1}{\lambda} \tan \frac{\nu}{2}.$$

The deformability of discrete Voss surfaces is now an immediate consequence of the above invariance.

Theorem 3.4. *Discrete Voss surfaces admit a one-parameter family of isometric deformations.*

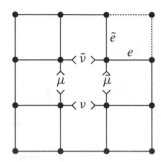

FIGURE 13. Illustration of the deformability of discrete Voss surfaces.

Proof. It is sufficient to consider a discrete Voss surface composed of nine quadrilaterals as displayed in Figure 13. If the top-right quadrilateral is removed and the discrete surface is isometrically deformed, then the dihedral angles change according to

$$\tan\frac{\mu}{2} \to \lambda \tan\frac{\mu}{2}, \qquad\qquad \tan\frac{\nu}{2} \to \frac{1}{\lambda}\tan\frac{\nu}{2}$$

$$\tan\frac{\tilde{\mu}}{2} \to \lambda \tan\frac{\tilde{\mu}}{2}, \qquad\qquad \tan\frac{\tilde{\nu}}{2} \to \frac{1}{\lambda}\tan\frac{\tilde{\nu}}{2}$$

so that the quantity

$$\tan\frac{\tilde{\mu}}{2}\tan\frac{\tilde{\nu}}{2}$$

is preserved. Relation (3.1) now implies that the angle made by the edges e and $\tilde{e}$ is unchanged. Thus, the missing quadrilateral may be inserted back into the discrete surface and the 3×3 complex is indeed deformable.   □

Since discrete Voss surfaces are isometrically deformable, any discrete Voss surface F admits a discrete asymptotic reciprocal-parallel net F^* which is unique up to an overall scaling. Moreover, the defining property of discrete Voss surfaces implies that opposite angles in the quadrilaterals of F^* are equal. This is equivalent to stating that the lengths of opposite edges of the quadrilaterals are equal. The latter is the defining property of *discrete (generalized) Chebyshev nets*. Thus, based on the definition below, we have come to the following conclusion [24]:

Definition 3.5. A discrete surface is called a *discrete K-surface* or *discrete pseudospherical* if the coordinate polygons form both a discrete asymptotic net and a discrete (generalized) Chebyshev net.

Corollary 3.6. *The reciprocal-parallel counterparts of discrete Voss surfaces are discrete K-surfaces.*

Discrete K-surfaces have been proposed independently by Sauer [23] and Wunderlich [29] as natural discrete versions of classical pseudospherical surfaces, that is, surfaces of constant negative Gaussian curvature. The significance of discrete K-surfaces in *integrable* discrete differential geometry has been revealed in [6]. At the level of discrete

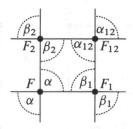

FIGURE 14. Definition of the lattice functions α and β.

Voss surfaces, the connection with the theory of integrable systems [1] is readily established. Thus, relation (3.1) may be exploited to characterize Voss surfaces in the following manner. If we arbitrarily prescribe two families of dihedral angles which are taken to be constant along the respective coordinate polygons, then the angles of a discrete Voss surface are constrained by the relation (3.1) which holds on each star and by the condition that the four angles in any quadrilateral must add up to 2π. If we regard the angles α and β as introduced in Figure 14 as functions defined on the vertices of the discrete Voss surfaces, then the latter condition may be formulated as

$$\alpha + \beta_1 + \alpha_{12} + \beta_2 = 2\pi.$$

It may be satisfied identically by introducing a lattice function ω which is defined on the quadrilaterals as illustrated in Figure 12. Thus, on each star of Voss type with angles α, β and dihedral angles μ, ν, the angles α and β are parametrized by

$$\alpha = \frac{\omega_1 + \omega_2}{2}, \quad \beta = \pi - \frac{\omega_{12} + \omega}{2}$$

and the relation (3.1) becomes

$$\sin\left(\frac{\omega_{12} - \omega_2 - \omega_1 + \omega}{4}\right) = \tan\frac{\mu}{2}\tan\frac{\nu}{2}\sin\left(\frac{\omega_{12} + \omega_2 + \omega_1 + \omega}{4}\right).$$

For given dihedral angles and modulo the Combescure transformation, any solution of the above lattice equation gives rise to a unique discrete Voss surface. Remarkably, this lattice equation is but another avatar of Hirota's integrable discrete version of the classical sine-Gordon equation [16]. Moreover, as demonstrated in Section 5.2, the deformation parameter λ turns out to play the role of a 'spectral parameter' in the corresponding Lax pair [6].

4. Infinitesimal deformations of second order

It has been demonstrated that the existence of an infinitesimal isometric deformation of a discrete surface F corresponds to the existence of a reciprocal-parallel discrete surface F^*. Accordingly, if there exists an infinitesimal isometric deformation of F of *second order*, then the reciprocal-parallel surface is infinitesimally deformable and the deformation is such that the stars are unchanged except for the length of their edges. In particular, the

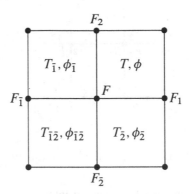

FIGURE 15. The translation T and rotation ϕ associated with finite deformations.

angles between the edges of the coordinate polygons are preserved and hence the deformation is infinitesimally *conformal*. If the original discrete surface constitutes a discrete conjugate net F, then the second-order isometric deformation corresponds to an infinitesimal conformal deformation of its discrete asymptotic reciprocal-parallel counterpart F^*.

4.1. Finite deformations

In order to analyse second-order isometric deformations, it is convenient to make some general statements about finite isometric deformations of arbitrary quadrilateral surfaces. Thus, if a discrete surface F is isometrically deformable, then its (nonplanar) quadrilaterals are subjected to rigid motions. Since any rigid motion may be decomposed into a translation and a rotation, the motion of any point P on the quadrilateral $\diamond = [F, F_1, F_{12}, F_2]$ may be described by

$$P^\epsilon = T(\epsilon) + \phi^{-1}(\epsilon) P \phi(\epsilon),$$

where we have made the standard identification of the ambient space $\mathbb{R}^3$ with the Lie algebra $su(2)$, that is,

$$P_{su(2)} = \langle P_{\mathbb{R}^3}, e \rangle, \quad e = (e_1, e_2, e_3),$$

$$e_1 = \frac{1}{i} \begin{pmatrix} 0 & 1 \\ 1 & 0 \end{pmatrix}, \quad e_2 = \frac{1}{i} \begin{pmatrix} 0 & -i \\ i & 0 \end{pmatrix}, \quad e_3 = \frac{1}{i} \begin{pmatrix} 1 & 0 \\ 0 & -1 \end{pmatrix},$$

so that $T \in su(2)$ constitutes the translation and $\phi \in SU(2)$ encodes the rotation. Here, T and ϕ are independent of P and $T(0) = 0$, $\phi(0) = \mathbb{1}$ so that $P^0 = P$ as required. As in the case of infinitesimal deformations, each quadrilateral is associated with a translation T and a rotation ϕ. Hence, T and ϕ are defined on the dual lattice (cf. Figure 15).

If we now consider the motion of the vertices F and F_2 which are common to the quadrilaterals $\diamond$ and $\diamond_{\bar{1}}$, then

$$F^\epsilon = T + \phi^{-1} F \phi, \qquad\qquad F_2^\epsilon = T + \phi^{-1} F_2 \phi,$$

$$F^\epsilon = T_{\bar{1}} + \phi_{\bar{1}}^{-1} F \phi_{\bar{1}}, \qquad\qquad F_2^\epsilon = T_{\bar{1}} + \phi_{\bar{1}}^{-1} F_2 \phi_{\bar{1}},$$

and elimination of F^ϵ, F_2^ϵ, T, $T_{\bar{1}}$ produces

$$\phi^{-1}(F_2 - F)\phi = \phi_{\bar{1}}^{-1}(F_2 - F)\phi_{\bar{1}}. \tag{4.1}$$

Similarly, consideration of the quadrilaterals $\diamond$ and $\diamond_{\bar{2}}$ leads to

$$\phi^{-1}(F_1 - F)\phi = \phi_{\bar{2}}^{-1}(F_1 - F)\phi_{\bar{2}}. \tag{4.2}$$

Thus, if the discrete surface F is isometrically deformable, then the latter two edge relations hold. Conversely, if for a given discrete surface F there exists a one-parameter family of matrices $\phi(\epsilon) \in SU(2)$ with $\phi(0) = \mathbb{1}$ which satisfies the edge relations (4.1), (4.2), then the compatible relations

$$T - T_{\bar{1}} = \phi_{\bar{1}}^{-1} F \phi_{\bar{1}} - \phi^{-1} F \phi, \tag{4.3a}$$

$$T - T_{\bar{2}} = \phi_{\bar{2}}^{-1} F \phi_{\bar{2}} - \phi^{-1} F \phi, \tag{4.3b}$$

and $T(0) = 0$ uniquely[1] define $T \in su(2)$ and the discrete surface F admits a one-parameter family of isometric deformations given by

$$F^\epsilon = T + \phi^{-1} F \phi.$$

It is noted that the existence of T is guaranteed since, as in the case of infinitesimal deformations, the associated closing condition is satisfied modulo the edge relations (4.1) and (4.2).

4.2. Second-order deformations

Instead of demanding that the edge relations (4.1) and (4.2) be satisfied for all values of ϵ, we may now require that only the first n nontrivial orders in ϵ vanish. Here, we are interested in the case $n = 2$ so that it is sufficient to deal with Taylor series about $\epsilon = 0$ which are truncated at the second level. Thus, we consider deformations of the form

$$P^\epsilon = P + \epsilon(T^* + [P, F^*]) + \frac{\epsilon^2}{2}(S^* + [P, G^*] + [[P, F^*], F^*]),$$

where P is any point on the quadrilateral $\diamond$ and

$$T^* = T_\epsilon, \quad S^* = T_{\epsilon\epsilon}, \quad F^* = \phi_\epsilon \phi^{-1}, \quad G^* = (\phi_\epsilon \phi^{-1})_\epsilon,$$

evaluated at $\epsilon = 0$. Once again, T^*, S^*, F^* and G^* are $su(2)$-valued objects defined on the quadrilaterals of F. Deformations of this type do not change the quadrilaterals to order $O(\epsilon^2)$ since it is readily seen that

$$\left\langle \hat{P}^\epsilon - P^\epsilon, \tilde{P}^\epsilon - P^\epsilon \right\rangle = \left\langle \hat{P} - P, \tilde{P} - P \right\rangle + O(\epsilon^3)$$

for any three points P, $\hat{P}$, $\tilde{P}$ on $\diamond$. Hence, if such deformations exist, then we refer to them as (infinitesimal) isometric deformations of second order.

---

[1] Up to $T(\epsilon)$ at one vertex, corresponding to a trivial translation of F^0.

Differentiation of the edge relation (4.1) and evaluation at $\epsilon = 0$ now produce

$$\epsilon[F_2 - F, F^*] + \frac{\epsilon^2}{2}([F_2 - F, G^*] + [[F_2 - F, F^*], F^*])$$
$$= \epsilon[F_2 - F, F_{\bar{1}}^*] + \frac{\epsilon^2}{2}([F_2 - F, G_{\bar{1}}^*] + [[F_2 - F, F_{\bar{1}}^*], F_{\bar{1}}^*]) + O(\epsilon^3),$$

and an analogous expansion obtains in the case of the second edge relation (4.2). The requirement that the terms are linear in ϵ vanish therefore imposes the conditions

$$[F_2 - F, F^* - F_{\bar{1}}^*] = 0, \quad [F_1 - F, F^* - F_{\bar{2}}^*] = 0. \tag{4.4}$$

These encapsulate nothing but the fact that an isometric deformation of first order exists if and only if there exists a reciprocal-parallel discrete surface F^*. The terms quadratic in ϵ may then be simplified and application of the Jacobi identity results in

$$[F_2 - F, G^* - G_{\bar{1}}^* + [F_{\bar{1}}^*, F^*]] = 0, \tag{4.5a}$$
$$[F_1 - F, G^* - G_{\bar{2}}^* + [F_{\bar{2}}^*, F^*]] = 0. \tag{4.5b}$$

Since the closing condition for the translation T is satisfied for finite isometric deformations and its nature is such that it holds separately for any order in ϵ, the defining relations for the coefficients T^* and S^* which are obtained from the Taylor expansion of (4.3) are compatible. We therefore conclude that a discrete surface F admits an isometric deformation of second order if and only if there exists a reciprocal-parallel surface F^* and another dual surface G^* which obey the pair (4.5).

4.3. Second-order deformations of discrete conjugate nets

In the case of second-order deformations of discrete conjugate nets, progress may be made by exploiting a discrete version of the classical Lelieuvre formulae which encapsulate the Gauss map for surfaces parametrized in terms of asymptotic coordinates [15]. To this end, it is recalled that if a discrete conjugate net F admits an infinitesimal isometric deformation, then the associated reciprocal-parallel net F^* is discrete asymptotic. Thus, if N denotes the unit normals to the planar stars of the discrete asymptotic net as depicted in Figure 6, then there exist lattice functions ρ and σ such that the connection between the discrete asymptotic net and its unit normals is represented by

$$F_1^* - F^* = \rho N_1 \times N, \quad F_2^* - F^* = \sigma N \times N_2.$$

The compatibility ('closing') condition $F_{12}^* = F_{21}^*$ is given by

$$\rho_2 N_{12} \times N_2 - \rho N_1 \times N = \sigma_1 N_1 \times N_{12} - \sigma N \times N_2$$

so that multiplication by N_1 and N_2, respectively, yields

$$\rho \rho_2 = \sigma \sigma_1.$$

The functions ρ and σ may therefore be parametrized according to

$$\rho = \tau \tau_1, \quad \sigma = \tau \tau_2,$$

where τ constitutes an arbitrary lattice function. Introduction of the scaled normal

$$\mathcal{V} = \tau N$$

then reduces the compatibility condition to

$$(\mathcal{V}_{12} + \mathcal{V}) \times (\mathcal{V}_1 + \mathcal{V}_2) = 0.$$

Thus, the following statement may be made:

Theorem 4.1. *If F^* constitutes a discrete asymptotic net, then there exists a scaled normal $\mathcal{V}$ such that the* discrete Lelieuvre formulae [7, 18]

$$F_1^* - F^* = \mathcal{V}_1 \times \mathcal{V}, \quad F_2^* - F^* = \mathcal{V} \times \mathcal{V}_2 \tag{4.6}$$

hold. These imply that $\mathcal{V}$ satisfies the discrete Moutard equation

$$\mathcal{V}_{12} + \mathcal{V} = H(\mathcal{V}_1 + \mathcal{V}_2) \tag{4.7}$$

for some scalar lattice function H. Conversely, for any given solution $(\mathcal{V}, H)$ of the discrete Moutard equation (4.7), the discrete Lelieuvre formulae (4.6) are compatible and uniquely define a discrete asymptotic net F^* with $\mathcal{V}$ being its scaled normal.

Since we are concerned with second-order isometric deformations of discrete conjugate nets, the condition (4.4), which enshrines the existence of an asymptotic reciprocal-parallel net F^*, may be replaced by the discrete Moutard equation (4.7) by virtue of Theorem 2.7. Moreover, the identity

$$AB = \langle A \times B, e \rangle - \langle A, B \rangle \, \mathbb{1}$$

which holds for any $\Lambda, B \in \mathbb{R}^3 \cong su(2)$ shows that the remaining constraints (4.5) are satisfied if and only if there exist functions a and b such that

$$G_1^* - G^* = 2a(F_1^* - F^*) + 2F_1^* \times F^*,$$
$$G_2^* - G^* = 2b(F_2^* - F^*) + 2F_2^* \times F^*.$$

Elimination of G^* leads to the compatibility condition

$$a_2(F_{12}^* - F_2^*) - a(F_1^* - F^*)$$
$$- b_1(F_{12}^* - F_1^*) + b(F_2^* - F^*) = (F_{12}^* - F^*) \times (F_1^* - F_2^*)$$

which may be expressed entirely in terms of $\mathcal{V}$ by virtue of the discrete Lelieuvre formulae. In fact, on use of the discrete Moutard equation, multiplication by $\mathcal{V}_1, \mathcal{V}_2$ and $\mathcal{V}$ results in

$$a_2 - b = \langle \mathcal{V}_{12} + \mathcal{V}, \mathcal{V}_1 \rangle, \quad a - b_1 = \langle \mathcal{V}_{12} + \mathcal{V}, \mathcal{V}_2 \rangle, \quad a_2 - b_1 = \langle \mathcal{V}_1 + \mathcal{V}_2, \mathcal{V} \rangle,$$

respectively. These relations imply that $\Delta_2(a + \langle \mathcal{V}_1, \mathcal{V} \rangle) = 0$ and $\Delta_1(b - \langle \mathcal{V}_2, \mathcal{V} \rangle) = 0$ so that summation yields

$$a = -\langle \mathcal{V}_1, \mathcal{V} \rangle + f(n_1), \quad b = \langle \mathcal{V}_2, \mathcal{V} \rangle - g(n_2)$$

and hence

$$\langle \mathcal{V}_{12} + \mathcal{V}, \mathcal{V}_1 + \mathcal{V}_2 \rangle = f(n_1) + g(n_2).$$

Since the latter may be regarded as a definition of the function H in the discrete Moutard equation, we have established the following theorem:

Theorem 4.2. *A discrete conjugate net F admits an infinitesimal isometric deformation of second order if and only if there exists a (discrete asymptotic) reciprocal-parallel net F^* whose scaled normal $\mathcal{V}$ obeys the vector equation*

$$\mathcal{V}_{12} + \mathcal{V} = \frac{f(n_1) + g(n_2)}{|\mathcal{V}_1 + \mathcal{V}_2|^2}(\mathcal{V}_1 + \mathcal{V}_2)$$

for some functions f and g or, equivalently,

$$\mathcal{V}_{12} + \mathcal{V} = H(\mathcal{V}_1 + \mathcal{V}_2), \quad \Delta_{12} \langle \mathcal{V}_{12} + \mathcal{V}, \mathcal{V}_1 + \mathcal{V}_2 \rangle = 0. \tag{4.8}$$

In particular, if the scaled normal $\mathcal{V}$ of a discrete asymptotic net F^* satisfies the constraint $(4.8)_2$, then the associated reciprocal-parallel discrete conjugate nets F admit an infinitesimal isometric deformation of second order.

4.4. Discussion

In 1892, Bianchi [4] observed that a conjugate net (x, y) on a surface F is preserved by a finite isometric deformation if and only if there exists a correspondence between the conjugate lines on F and the asymptotic lines on another surface F^* such that the Gauss maps N and N^* coincide and the Gaussian curvature of F^* is constrained by

$$\left(\frac{1}{\sqrt{-K^*}}\right)_{xy} = 0.$$

Based on a parameter-dependent linear representation (cf. Section 5), Bianchi [3] derived a Bäcklund transformation for the surfaces F^* which have come to be known as *Bianchi surfaces* [11, 20]. An analytic description of Bianchi surfaces is readily obtained on use of the Lelieuvre formulae [15]

$$F_x^* = \mathcal{V}_x \times \mathcal{V}, \quad F_y^* = \mathcal{V} \times \mathcal{V}_y,$$

where the scaled normal $\mathcal{V}$ obeys the Moutard equation

$$\mathcal{V}_{xy} = h\mathcal{V}. \tag{4.9}$$

The Lelieuvre formulae immediately provide an expression for the Gaussian curvature of F^*, namely $K^* = -1/|\mathcal{V}|^4$. Accordingly, the constraint which defines Bianchi surfaces may be formulated as

$$(|\mathcal{V}|^2)_{xy} = 0. \tag{4.10}$$

It is evident that the pair (4.9), (4.10) which entirely encodes Bianchi surfaces constitutes the natural continuum limit of (4.8). However, the latter has been derived in connection with *second-order* deformations while the classical differential-geometric derivation is based on *finite* deformations. This discrepancy may be partly resolved by referring to the fact that, in the classical setting, second-order isometric deformations of conjugate nets may be shown to be finite. Thus, Bianchi surfaces are, in fact, retrieved by imposing the *a priori* weaker condition of the existence of second-order deformations. Nevertheless, it is emphasized that, in general, this statement *does not* apply in the discrete context, that is, second-order isometric deformations of discrete conjugate nets are not necessarily finite.

As in the smooth setting, the discrete Bianchi system (4.8) has been shown to be integrable [26]. Second-order isometric deformations of discrete conjugate nets may therefore be considered integrable and, remarkably, any discrete conjugate net which admits a *finite* isometric deformation corresponds to a particular solution of the integrable Bianchi system (4.8).

At present, it is not known how to formulate the existence of finite isometric deformations in terms of constraints on the discrete Bianchi system. However, if we assume that the modulus of the scaled normal $\mathcal{V}$ is constant, that is,

$$|\mathcal{V}| = 1$$

without loss of generality, then the discrete Moutard equation (4.8)$_1$ reduces to

$$\mathcal{V}_{12} + \mathcal{V} = \frac{\langle \mathcal{V}, \mathcal{V}_1 + \mathcal{V}_2 \rangle}{1 + \langle \mathcal{V}_1, \mathcal{V}_2 \rangle} (\mathcal{V}_1 + \mathcal{V}_2),$$

and its algebraic consequences

$$\Delta_1 \langle \mathcal{V}_2, \mathcal{V} \rangle = 0, \quad \Delta_2 \langle \mathcal{V}_1, \mathcal{V} \rangle = 0$$

reveal that the constraint (4.8)$_2$ is identically satisfied. The discrete Lelieuvre formulae (4.6) then imply that F^* constitutes a discrete (generalized) Chebyshev net, that is,

$$\Delta_1 |F_2^* - F^*| = 0, \quad \Delta_2 |F_1^* - F^*| = 0,$$

so that discrete K-surfaces together with their reciprocal-parallel isometrically deformable discrete Voss surfaces are retrieved.

5. Integrability of finite deformations

In the preceding, it has been demonstrated that the discrete surfaces which are reciprocal parallel to isometrically deformable conjugate nets are necessarily of discrete Bianchi type. In particular, discrete Voss surfaces correspond to integrable discrete K-surfaces. Here, we investigate in more detail the connection with the theory of integrable systems and demonstrate that the appearance of *integrable* discrete differential geometry in the context of isometric deformations is, in fact, no coincidence. Thus, it has been shown that a discrete quadrilateral surface F is isometrically deformable if and only if there exists an $SU(2)$–valued function $\phi(\epsilon)$ with $\phi(0) = \mathbb{1}$ which obeys the relations (4.1) and (4.2), that is,

$$[F_2 - F, \phi \phi_1^{-1}] = 0, \quad [F_1 - F, \phi \phi_2^{-1}] = 0. \tag{5.1}$$

Now, since $\phi(0) = \mathbb{1}$, the quantity

$$F^* = \phi_\epsilon(0)$$

is $su(2)$-valued and differentiation of (5.1) with respect to the deformation parameter ϵ reproduces the condition for reciprocal-parallelism

$$[F_2 - F, F^* - F_1^*] = 0, \quad [F_1 - F, F^* - F_2^*] = 0.$$

The pair (5.1) then implies that

$$[F_1^* - F^*, \phi_1 \phi^{-1}] = 0, \quad [F_2^* - F^*, \phi_2 \phi^{-1}] = 0,$$

which, in turn, shows that there exist real lattice functions $a(\epsilon), b(\epsilon)$ and $c(\epsilon), d(\epsilon)$ such that

$$\phi_1 = \mathcal{L}(\epsilon)\phi, \qquad\qquad \mathcal{L}(\epsilon) = a(\epsilon)(F_1^* - F^*) + b(\epsilon)\mathbb{1}, \qquad (5.2\text{a})$$
$$\phi_2 = \mathcal{M}(\epsilon)\phi, \qquad\qquad \mathcal{M}(\epsilon) = c(\epsilon)(F_2^* - F^*) + d(\epsilon)\mathbb{1}. \qquad (5.2\text{b})$$

Finally, the compatibility condition $\phi_{12} = \phi_{21}$ produces the *discrete zero-curvature condition*

$$\mathcal{L}_2(\epsilon)\mathcal{M}(\epsilon) = \mathcal{M}_1(\epsilon)\mathcal{L}(\epsilon). \qquad (5.3)$$

The latter encodes a system of lattice equations if the $SU(2)$-valued functions $\mathcal{L}(\epsilon)$ and $\mathcal{M}(\epsilon)$ are assumed to admit the power series expansions

$$\mathcal{L}(\epsilon) = \sum_{k=0}^{\infty} \epsilon^k \mathcal{L}^{(k)}, \quad \mathcal{M}(\epsilon) = \sum_{k=0}^{\infty} \epsilon^k \mathcal{M}^{(k)}.$$

The above analysis shows that any isometrically deformable quadrilateral surface F gives rise to a system of *nonlinear* lattice equations encoded in the discrete zero-curvature condition (5.3) which constitutes the compatibility condition associated with the ϵ-dependent *linear* system (5.2). A parameter-dependent linear representation of the form $(5.2\text{a})_1, (5.2\text{b})_1$ constitutes the key ingredient in the mathematical treatment of a discrete integrable system represented by the corresponding discrete zero-curvature condition (5.3) [1]. In fact, all isometrically deformable quadrilateral surfaces are encapsulated in the following theorem.

Theorem 5.1. *Let $\mathcal{L}(\epsilon)$ and $\mathcal{M}(\epsilon)$ be two parameter-dependent $SU(2)$-valued lattice functions of the form*

$$\mathcal{L}(\epsilon) = a(\epsilon)A + b(\epsilon)\mathbb{1}, \quad \mathcal{M}(\epsilon) = c(\epsilon)B + d(\epsilon)\mathbb{1},$$

which obey the discrete zero-curvature condition

$$\mathcal{L}_2(\epsilon)\mathcal{M}(\epsilon) = \mathcal{M}_1(\epsilon)\mathcal{L}(\epsilon) \qquad (5.4)$$

for some lattice functions $A, B \in su(2)$ and $a(\epsilon), b(\epsilon), c(\epsilon), d(\epsilon) \in \mathbb{R}$ with

$$a(0) = c(0) = 0, \quad b(0) = d(0) = 1 \qquad (5.5)$$

and $a_\epsilon(0) \neq 0, c_\epsilon(0) \neq 0$. Then, there exists a unique $\phi(\epsilon) \in SU(2)$ (up to an irrelevant gauge matrix depending on ϵ) with $\phi(0) = \mathbb{1}$ which satisfies the linear system

$$\phi_1 = \mathcal{L}(\epsilon)\phi, \quad \phi_2 = \mathcal{M}(\epsilon)\phi. \qquad (5.6)$$

If the quadrilateral surface

$$F^* = \phi_\epsilon(0)$$

admits a reciprocal-parallel discrete surface F, then F is isometrically deformable.

Proof. The zero-curvature condition (5.4) guarantees that the linear pair (5.6) is compatible. Its $SU(2)$-valued solution is uniquely determined by the value of $\phi(\epsilon)$ at one vertex which we may choose to be $\mathbb{1}$ without loss of generality. The constraints (5.5) imply that

$\mathcal{L}(0) = \mathcal{M}(0) = \mathbb{1}$ so that $\phi(0) = \mathbb{1}$ everywhere and hence $F^* = \phi_\epsilon(0) \in su(2)$. Differentiation of (5.6) and evaluation at $\epsilon = 0$ produce

$$F_1^* - F^* = a_\epsilon(0)A + b_\epsilon(0), \quad F_2^* - F^* = c_\epsilon(0)B + d_\epsilon(0),$$

wherein $b_\epsilon(0) = d_\epsilon(0) = 0$ since $F^* \in su(2)$. Accordingly, the pair (5.6) gives rise to the commutator relations

$$[F_1^* - F^*, \phi_1\phi^{-1}] = 0, \quad [F_2^* - F^*, \phi_2\phi^{-1}] = 0,$$

which are equivalent to the conditions (5.1) for the existence of an isometric deformation of a reciprocal-parallel surface F. □

5.1. Finite deformations of discrete conjugate nets

In general, there is no guarantee that the quadrilateral surface F^* in Theorem 5.1 admits a reciprocal-parallel counterpart F. This property must therefore be regarded as a constraint on the $SU(2)$-valued lattice functions $\mathcal{L}(\epsilon)$ and $\mathcal{M}(\epsilon)$. In the case of isometric deformations of discrete conjugate nets, this constraint may be implemented explicitly since the quantities A and B may be expressed in terms of the scaled normal $\mathcal{V}$ by virtue of the discrete Lelieuvre formulae (4.6) and the representation (5.2) of $\mathcal{L}(\epsilon)$ and $\mathcal{M}(\epsilon)$. Accordingly, Theorem 5.1 may be simplified and brought into the following form:

Theorem 5.2. *Let $\mathcal{A}(\epsilon)$ and $\mathcal{B}(\epsilon)$ be two parameter-dependent matrix-valued lattice functions of the form*

$$\mathcal{A}(\epsilon) = \alpha(\epsilon)S_1S + \beta(\epsilon)\mathbb{1}, \quad \mathcal{B}(\epsilon) = \gamma(\epsilon)S_2S + \delta(\epsilon)\mathbb{1},$$

which obey the discrete zero-curvature condition

$$\mathcal{A}_2(\epsilon)\mathcal{B}(\epsilon) = \mathcal{B}_1(\epsilon)\mathcal{A}(\epsilon) \tag{5.7}$$

for some lattice functions $S \in su(2)$ and $\alpha(c), \beta(c), \gamma(c), \delta(c) \subset \mathbb{R}$ with

$$\alpha(0) = \gamma(0) = 0$$

and the inequalities

$$\beta(0) > 0, \quad \delta(0) > 0, \quad \alpha_\epsilon(0) \neq 0, \quad \gamma_\epsilon(0) \neq 0, \quad \det\mathcal{A}(\epsilon) > 0, \quad \det\mathcal{B}(\epsilon) > 0.$$

Then, the determinants of $\mathcal{A}(\epsilon)$ and $\mathcal{B}(\epsilon)$ may be parametrized according to

$$\det\mathcal{A}(\epsilon) = \frac{\tau_1^2(\epsilon)}{\tau^2(\epsilon)}, \quad \det\mathcal{B}(\epsilon) = \frac{\tau_2^2(\epsilon)}{\tau^2(\epsilon)}, \quad \tau(\epsilon) > 0,$$

and the linear system

$$\psi_1 = \mathcal{A}(\epsilon)\psi, \quad \psi_2 = \mathcal{B}(\epsilon)\psi \tag{5.8}$$

admits a unique solution (up to an irrelevant gauge matrix depending on ϵ) with

$$\phi(\epsilon) := \frac{\psi(\epsilon)}{\tau(\epsilon)} \in SU(2), \quad \phi(0) = \mathbb{1}.$$

The quadrilateral surface

$$F^* = \phi_\epsilon(0)$$

is discrete asymptotic and its reciprocal-parallel counterparts F constitute isometrically deformable discrete conjugate nets with S being a common normal.

The above theorem states that all isometrically deformable discrete conjugate nets are enshrined in the discrete zero-curvature condition (5.7). Before we analyse this condition in more detail, we present the proof and an illustration of Theorem 5.2.

Proof. It is readily verified that

$$\mathcal{A}^\dagger \mathcal{A} = \mathbb{1} \det \mathcal{A}, \quad \mathcal{B}^\dagger \mathcal{B} = \mathbb{1} \det \mathcal{B},$$

where

$$\det \mathcal{A} = \alpha^2 |S_1|^2 |S|^2 - 2\alpha\beta \langle S_1, S \rangle + \beta^2 \tag{5.9a}$$

$$\det \mathcal{B} = \beta^2 |S_2|^2 |S|^2 - 2\gamma\delta \langle S_2, S \rangle + \delta^2. \tag{5.9b}$$

The discrete zero-curvature condition implies that

$$\det \mathcal{A}_2 \det \mathcal{B} = \det \mathcal{B}_1 \det \mathcal{A},$$

whence there exists a positive lattice function $\tau(\epsilon)$ such that

$$\det \mathcal{A}(\epsilon) = \frac{\tau_1^2(\epsilon)}{\tau^2(\epsilon)}, \quad \det \mathcal{B}(\epsilon) = \frac{\tau_2^2(\epsilon)}{\tau^2(\epsilon)}.$$

Thus, if we introduce the quantities

$$\phi(\epsilon) = \frac{\psi(\epsilon)}{\tau(\epsilon)}, \quad \mathcal{L}(\epsilon) = \frac{\tau(\epsilon)}{\tau_1(\epsilon)} \mathcal{A}(\epsilon), \quad \mathcal{M}(\epsilon) = \frac{\tau(\epsilon)}{\tau_2(\epsilon)} \mathcal{B}(\epsilon),$$

then $\mathcal{L}(\epsilon), \mathcal{M}(\epsilon) \in SU(2)$ and the linear system (5.8) becomes

$$\phi_1 = \mathcal{L}(\epsilon)\phi, \quad \phi_2 = \mathcal{M}(\epsilon)\phi, \tag{5.10}$$

so that Theorem 5.1 applies. Indeed, evaluation of the identities (5.9) at $\epsilon = 0$ reveals that

$$\beta(0) = \frac{\tau_1(0)}{\tau(0)}, \quad \delta(0) = \frac{\tau_2(0)}{\tau(0)},$$

and hence $\mathcal{L}(0) = \mathcal{M}(0) = \mathbb{1}$. It therefore remains to show that the quadrilateral surface $F^* = \phi_\epsilon(0)$ is discrete asymptotic. To this end, it is observed that differentiation of (5.10) yields

$$\phi_{1\epsilon}(0) = \frac{\alpha_\epsilon(0)}{\beta(0)} S_1 S + \left(\beta(\epsilon) \frac{\tau(\epsilon)}{\tau_1(\epsilon)} \right)_\epsilon \bigg|_{\epsilon=0} \mathbb{1} + \phi_\epsilon(0)$$

and an analogous expression for $\phi_{2\epsilon}(0)$. Since $F^* = \phi_\epsilon(0) \in su(2)$, we conclude that

$$F_1^* - F^* = \frac{\alpha_\epsilon(0)}{\beta(0)} S_1 \times S, \quad F_2^* - F^* = \frac{\gamma_\epsilon(0)}{\delta(0)} S_2 \times S,$$

so that S is indeed orthogonal to the stars of the quadrilateral surface F^*. This concludes the proof since, according to Theorem 2.7, any (non-degenerate) discrete asymptotic net F^* admits an infinite number of reciprocal-parallel discrete conjugate nets F.   □

5.2. Finite deformations of discrete Voss surfaces

As an illustration of Theorem 5.2, we make the choice

$$\alpha = \epsilon, \quad \beta = 1, \quad \gamma = -\epsilon, \quad \delta = 1, \quad S^2 = -\mathbb{1} \Leftrightarrow |S| = 1.$$

In this case, the discrete zero-curvature condition associated with the linear system

$$\psi_1 = (\epsilon S_1 S + \mathbb{1})\psi, \quad \psi_2 = (-\epsilon S_2 S + \mathbb{1})\psi$$

reduces to

$$\epsilon[S_{12}(S_1 + S_2) - (S_1 + S_2)S] = 0$$

or, equivalently,

$$(S_{12} + S) \times (S_1 + S_2) = 0, \quad \langle S_{12} - S, S_1 + S_2 \rangle = 0. \tag{5.11}$$

The former relation is equivalent to the discrete Moutard equation

$$S_{12} + S = H(S_1 + S_2)$$

in which the function H is determined by the consistency condition $|S_{12}| = 1$, leading to

$$S_{12} + S = \frac{\langle S, S_1 + S_2 \rangle}{1 + \langle S_1, S_2 \rangle}(S_1 + S_2). \tag{5.12}$$

The algebraic consequences

$$\Delta_1 \langle S_2, S \rangle = 0, \quad \Delta_2 \langle S_1, S \rangle = 0$$

then show that the additional requirement $(5.11)_2$ is identically satisfied. Thus, S represents the Gauss map of discrete K-surfaces as discussed in Section 4.4 and the isometrically deformable reciprocal-parallel discrete surfaces are therefore of discrete Voss type.

The above example highlights the geometric interpretation of the 'eigenfunction' $\phi(\epsilon)$ in the sense of soliton theory as the rotational component of the rigid motion undergone by the quadrilaterals during an isometric deformation. Here, the deformation parameter ϵ plays the role of the 'spectral parameter' [1].

5.3. A discrete nonlinear σ-model

Here, we embark on an analysis of the discrete zero-curvature condition (5.7) given by

$$(\alpha_2 S_{12} S_2 + \beta_2 \mathbb{1})(\gamma S_2 S + \delta \mathbb{1}) = (\gamma_1 S_{12} S_1 + \delta_1 \mathbb{1})(\alpha S_1 S + \beta \mathbb{1}).$$

Therein, we may assume without loss of generality that the normal S obeys the discrete Moutard equation

$$S_{12} + S = H(S_1 + S_2) \tag{5.13}$$

by virtue of Theorems 4.1 and 5.2. Decomposition of the discrete zero-curvature condition into its trace and trace-free parts yields

$$\alpha_2 \gamma |S_2|^2 \langle S_{12}, S \rangle - \alpha_2 \delta \langle S_{12}, S_2 \rangle - \beta_2 \gamma \langle S_2, S \rangle + \beta_2 \delta$$
$$= \gamma_1 \alpha |S_1|^2 \langle S_{12}, S \rangle - \gamma_1 \beta \langle S_{12}, S_1 \rangle - \delta_1 \alpha \langle S_1, S \rangle + \delta_1 \beta$$

and

$$\alpha_2\gamma|S_2|^2 S_{12} \times S - \alpha_2\delta S_{12} \times S_2 - \beta_2\gamma S_2 \times S$$
$$= \gamma_1\alpha|S_1|^2 S_{12} \times S - \gamma_1\beta S_{12} \times S_1 - \delta_1\alpha S_1 \times S.$$

The inner product of the latter with S and $S_1 + S_2$, respectively, gives rise to

$$\alpha_2\delta = -\gamma_1\beta, \quad \beta_2\gamma = -\delta_1\alpha, \tag{5.14}$$

so that the above relations reduce to

$$H(\alpha_2\gamma|S_2|^2 - \gamma_1\alpha|S_1|^2) = \alpha_2\delta + \beta_2\gamma, \tag{5.15a}$$

$$\frac{1}{H}(\alpha_2\delta|S_{12}|^2 + \beta_2\gamma|S^2|) = \beta_2\delta - \delta_1\beta. \tag{5.15b}$$

Now, on the one hand, it is readily verified that the linear system (5.8) is invariant under

$$\psi \to \mu\psi, \quad (\alpha,\beta) \to \frac{\mu_1}{\mu}(\alpha,\beta), \quad (\gamma,\delta) \to \frac{\mu_2}{\mu}(\gamma,\delta).$$

On the other hand, the quadratic relations (5.14) guarantee that there exist 'potentials' μ and ν such that

$$\mu_1 = \alpha\nu, \quad \mu_2 = -\gamma\nu, \quad \nu_1 = \beta\mu, \quad \nu_2 = \delta\mu.$$

Without loss of generality, we may therefore introduce the parametrization

$$\alpha = \frac{\mu}{\nu}, \quad \beta = \frac{\nu_1}{\mu_1}, \quad \gamma = -\frac{\mu}{\nu}, \quad \delta = \frac{\nu_2}{\mu_2}.$$

Finally, if we set

$$\sigma = \frac{\nu}{\mu},$$

then the relations (5.15) adopt the form

$$\sigma_{12} - \sigma = H\left(\frac{|S_2|^2}{\sigma_2} - \frac{|S_1|^2}{\sigma_1}\right), \quad \sigma_2 - \sigma_1 = \frac{1}{H}\left(\frac{|S_{12}|^2}{\sigma_{12}} - \frac{|S|^2}{\sigma}\right). \tag{5.16}$$

Thus, it is required to determine all one-parameter (ϵ) families of solutions of the coupled system (5.13), (5.16) which are such that S is *independent* of ϵ. It is observed that the pair (5.16) may be brought into the form

$$\Delta_2\left(2\langle S_1, S\rangle - \sigma_1\sigma - \frac{|S_1|^2}{\sigma_1}\frac{|S|^2}{\sigma}\right) = 0, \quad \Delta_1\left(2\langle S_2, S\rangle + \sigma_2\sigma + \frac{|S_2|^2}{\sigma_2}\frac{|S|^2}{\sigma}\right) = 0,$$

which provides two first integrals.

In order to proceed, it turns out convenient to define the quantity

$$\tau = \frac{|S|^2}{\sigma}.$$

The system (5.13), (5.16) is then equivalent to the *linear* system

$$S_{12} + S = H(S_1 + S_2),$$
$$\sigma_{12} - \sigma = H(\tau_2 - \tau_1),$$
$$\tau_{12} - \tau = H(\sigma_2 - \sigma_1),$$

subject to the *nonlinear* constraint

$$\sigma\tau = |S|^2.$$

It is noted in passing that the choice

$$\sigma = \frac{1}{f(\epsilon)}, \quad \tau = f(\epsilon)$$

reduces this system to the Gauss map (5.12) of discrete K-surfaces corresponding to isometrically deformable discrete Voss surfaces. In general, the change of variables

$$A = (-1)^{n_1}\frac{\sigma - \tau}{2}, \quad B = (-1)^{n_2}\frac{\sigma + \tau}{2}, \quad \mathsf{V} = \begin{pmatrix} S \\ A \\ B \end{pmatrix}$$

gives rise to the vector equation

$$\mathsf{V}_{12} + \mathsf{V} = H(\mathsf{V}_1 + \mathsf{V}_2), \quad \langle \mathsf{V}, \mathsf{V} \rangle = 0,$$

where the inner product is defined by

$$\langle \mathsf{V}, \tilde{\mathsf{V}} \rangle = \langle S, \tilde{S} \rangle + A\tilde{A} - B\tilde{B}.$$

Since the function H is determined by the requirement that V constitutes a null-vector, discrete conjugate nets which are isometrically deformable are encapsulated in the following theorem:

Theorem 5.3. *Isometrically deformable discrete conjugate nets are encoded in one-parameter (ϵ) families of solutions of the vector equation*

$$\mathsf{V}_{12} + \mathsf{V} = \frac{\langle \mathsf{V}, \mathsf{V}_1 + \mathsf{V}_2 \rangle}{\langle \mathsf{V}_1, \mathsf{V}_2 \rangle}(\mathsf{V}_1 + \mathsf{V}_2), \quad \langle \mathsf{V}, \mathsf{V} \rangle = 0,$$

which are such that the first three components of V are independent of ϵ. Here, the inner product is taken with respect to the metric diag$(1, 1, 1, 1, -1)$,

Remarkably, the above vector equation constitutes the standard integrable discretization of a particular *nonlinear σ-model* (see, e.g., [8, 26]). The complete characterization of its families of solutions which admit the required dependence on a parameter ϵ and are compatible with the technical assumptions made in Theorem 5.2 is the subject of ongoing research.

References

[1] M.J. Ablowitz and H. Segur, *Solitons and the Inverse Scattering Transform*, SIAM, Philadelphia: 1981.

[2] A.D. Alexandrov, *Konvexe Polyeder*, Akad. Verlag, Berlin: 1958.

[3] L. Bianchi, Sopra alcone nuove classi di superficie e di sistemi tripli ortogonali, *Ann. Matem.* **18** (1890), 301–358.

[4] ———, Sulle deformazioni infinitesime delle superficie flessibili ed inestendibili, *Rend. Lincei* **1** (1892), 41–48.

[5] W. Blaschke, Reziproke Kräftepläne zu den Spannungen in einer biegsamen Haut, *International Congress of Mathematicians Cambridge* (1912), 291–294.

[6] A. Bobenko and U. Pinkall, Discrete surfaces with constant negative Gaussian curvature and the Hirota equation, *J. Diff. Geom.* **43** (1996), 527–611.

[7] A.I. Bobenko and W.K. Schief, Affine spheres: discretization via duality relations, *J. Exp. Math.* **8** (1999), 261–280.

[8] A.I. Bobenko and R. Seiler, eds, *Discrete Integrable Geometry and Physics*, Clarendon Press, Oxford: 1999.

[9] R. Bricard, Mémoire sur la théorie de l'octaédre articulé, *J. math. pur. appl., Liouville* **3** (1897), 113–148.

[10] A. Cauchy, Sur les polygones et polyedres, Second Memoire, *J. École Polytechnique* **9** (1813), 87.

[11] B. Cenkl, Geometric deformations of the evolution equations and Bäcklund transformations, *Physica D* **18** (1986), 217–219.

[12] S.E. Cohn-Vossen, Die Verbiegung von Flächen im Grossen, *Fortschr. Math. Wiss.* **1** (1936), 33–76.

[13] M. Dehn, Über die Starrheit konvexer Polyeder, *Math. Ann.* **77** (1916), 466–473.

[14] A. Doliwa, M. Nieszporski, and P.M. Santini, Asymptotic lattices and their integrable reductions. I. The Bianchi-Ernst and the Fubini-Ragazzi lattices, *J. Phys. A: Math. Gen.* **34** (2001), 10423–10439.

[15] L.P. Eisenhart, *A Treatise on the Differential Geometry of Curves and Surfaces*, Dover Publications, New York (1960).

[16] R. Hirota, Nonlinear partial difference equations. III. Discrete sine-Gordon equation, *J. Phys. Japan* **43** (1977), 2079–2086.

[17] A. Kokotsakis, Über bewegliche Polyeder, *Math. Ann.* **107** (1932), 627–647.

[18] B.G. Konopelchenko and U. Pinkall, Projective generalizations of Lelieuvre's formula, *Geometriae Dedicata* **79** (2000), 81–99.

[19] B.G. Konopelchenko and W.K. Schief, Three-dimensional integrable lattices in Euclidean spaces: conjugacy and orthogonality, *Proc. Roy. Soc. London A* **454** (1998), 3075–3104.

[20] D. Levi and A. Sym, Integrable systems describing surfaces of non-constant curvature, *Phys. Lett. A* **149** (1990), 381–387.

[21] H. Liebmann, Über die Verbiegung der geschlossenen Flächen positiver Krümmung, *Math. Ann.* **53** (1900), 81–112; *Habilitationsschrift*, Leipzig: 1899.

[22] V.V. Novozhilov, *Thin Shell Theory*, P. Noordhoff, Groningen: 1964.

[23] R. Sauer, Parallelogrammgitter als Modelle für pseudosphärische Flächen, *Math. Z.* **52** (1950), 611–622.

[24] _____ , *Differenzengeometrie*, Springer Verlag, Berlin-Heidelberg-New York: 1970.

[25] R. Sauer and H. Graf, Über Flächenverbiegungen in Analogie zur Verknickung offener Facettenflache, *Math. Ann.* **105** (1931), 499–535.

[26] W.K. Schief, Isothermic surfaces in spaces of arbitrary dimension: Integrability, discretization and Bäcklund transformations. A discrete Calapso equation, *Stud. Appl. Math.* **106** (2001), 85–137.

[27] A. Voss, Über diejenigen Flächen, auf denen zwei Scharen geodätischer Linien ein conjugirtes System bilden, *Sitzungsber. Bayer. Akad. Wiss., math.-naturw. Klasse* (1888), 95–102.

[28] H. Weyl, Über die Starrheit der Eiflächen und konvexen Polyeder, *S.-B. Preuss. Akad. Wiss.* (1917), 250–266.

[29] W. Wunderlich, Zur Differenzengeometrie der Flächen konstanter negativer Krümmung, *Österreich. Akad. Wiss. Math.-Nat. Kl. S.-B. II* **160** (1951), 39–77.

Wolfgang K. Schief
Institut für Mathematik, MA 8–3
Technische Universität Berlin
Str. des 17. Juni 136
10623 Berlin
Germany
and
Australian Research Council Centre of Excellence for
Mathematics and Statistics of Complex Systems
School of Mathematics
The University of New South Wales
Sydney, NSW 2052
Australia
e-mail: schief@math.tu-berlin.de

Alexander I. Bobenko
Institut für Mathematik, MA 8–3
Technische Universität Berlin
Str. des 17. Juni 136
10623 Berlin
Germany
e-mail: bobenko@math.tu-berlin.de

Tim Hoffmann
Mathematisches Institut
Universität München
Theresienstr. 39
80333 München
Germany
e-mail: hoffmann@mathematik.uni-muenchen.de

Discrete Differential Geometry, A.I. Bobenko, P. Schröder, J.M. Sullivan and G.M. Ziegler, eds.

Oberwolfach Seminars, Vol. 38, 95–115

Discrete Hashimoto Surfaces and a Doubly Discrete Smoke-Ring Flow

Tim Hoffmann

Abstract. In this paper Bäcklund transformations for smooth and discrete Hashimoto surfaces are discussed and a geometric interpretation is given. It is shown that the complex curvature of a discrete space curve evolves with the discrete nonlinear Schrödinger equation (NLSE) of Ablowitz and Ladik, when the curve evolves with the Hashimoto or smoke-ring flow. A doubly discrete Hashimoto flow is derived and it is shown that in this case the complex curvature of the discrete curve obeys Ablovitz and Ladik's doubly discrete NLSE. Elastic curves (curves that evolve by rigid motion under the Hashimoto flow) in the discrete and doubly discrete case are shown to be the same.

Keywords. Doubly discrete smoke-ring flow, Bäcklund transformation, discrete Hashimoto surface, nonlinear Schrödinger equation, discrete elastic curves, isotropic Heisenberg magnet model.

1. Introduction

Many of the surfaces that can be described by integrable equations have been discretized. This paper continues the program by adding Hashimoto surfaces to the list. Hashimoto surfaces are obtained by evolving a regular space curve γ by the *Hashimoto* or *smoke-ring flow*

$$\dot{\gamma} = \gamma' \times \gamma''.$$

As shown by Hashimoto [6] this evolution is directly linked to the famous nonlinear Schrödinger equation (NLSE)

$$i\dot{\Psi} + \Psi'' + \frac{1}{2}|\Psi|^2\Psi = 0.$$

In [1] and [2] Ablowitz and Ladik gave a differential-difference and a difference-difference discretization of the NLSE. In [7] the author shows[1] that they correspond to a Hashimoto flow on discrete curves (i.e., polygons) [3, 4] and a doubly discrete Hashimoto

---

[1] The equivalence for the differential-difference case appeared first in [8].

flow, respectively. This discrete evolution is derived in Section 3.2.2 from a discretization of the Bäcklund transformations for regular space curves and Hashimoto surfaces.

In Section 2 a short review of the smooth Hashimoto flow and its connection to the isotropic Heisenberg magnet model and the nonlinear Schrödinger equation is given. It is shown that the solutions to the auxiliary problems of these integrable equations serve as frames for the Hashimoto surfaces and a Sym formula is derived. In Section 2.2.1 the dressing procedure or Bäcklund transformation is discussed and applied to the vacuum. A geometric interpretation of this transformation as a generalization of the tractrix construction for a curve is given.

In Section 3 the same program is carried out for the Hashimoto flow on discrete curves. Then in Section 4 special double Bäcklund transformations (for discrete curves) are singled out to get a unique evolution which serves as our doubly discrete Hashimoto flow.

Elastic curves (curves that evolve by rigid motion under the Hashimoto flow) are discussed in all these cases. It turns out that discrete elastic curves for both the discrete and the doubly discrete Hashimoto flow coincide.

Throughout this paper we use a quaternionic description. Quaternions are the algebra generated by 1, i, j, and $\mathfrak{k}$ with the relations $i^2 = j^2 = \mathfrak{k}^2 = -1$, $ij = \mathfrak{k}, j\mathfrak{k} = i$, and $\mathfrak{k}i = j$. The real and imaginary parts of a quaternion are defined in an obvious manner: If $q = \alpha + \beta i + \gamma j + \delta \mathfrak{k}$ we set $\Re(q) = \alpha$ and $\Im(q) = \beta i + \gamma j + \delta \mathfrak{k}$. Note that unlike in the complex case the imaginary part is not a real number. We identify the 3-dimensional euclidean space with the imaginary quaternions, i.e., the span of i, j, and $\mathfrak{k}$. Then for two imaginary quaternions q, r the following formula holds:

$$qr = -\langle q, r \rangle + q \times r$$

with $\langle \cdot, \cdot \rangle$ and $\cdot \times \cdot$ denoting the usual scalar and cross products of vectors in 3-space. A rotation of an imaginary quaternion around the axis $r, |r| = 1$, with angle ϕ can be written as conjugation with the unit length quaternion $\left(\cos \frac{\phi}{2} + (\sin \frac{\phi}{2})r \right)$.

Especially when dealing with the Lax representations of the various equations it will be convenient to identify the quaternions with complex 2-by-2 matrices:

$$i = i\sigma_3 = \begin{pmatrix} i & 0 \\ 0 & -i \end{pmatrix}, \quad j = i\sigma_1 = \begin{pmatrix} 0 & i \\ i & 0 \end{pmatrix}, \quad \mathfrak{k} = -i\sigma_2 = \begin{pmatrix} 0 & -1 \\ 1 & 0 \end{pmatrix}.$$

2. The Hashimoto flow, the Heisenberg flow and the nonlinear Schrödinger equation

Let $\gamma : \mathbb{R} \to \mathbb{R}^3 = \Im\mathbb{H}$ be an arc length–parametrized regular curve and $\mathcal{F} : \mathbb{R} \to \mathbb{H}^*$, $\|\mathcal{F}\| = 1$, be a parallel frame for it, i.e.,

$$\mathcal{F}^{-1} i \mathcal{F} = \gamma' = \gamma_x, \tag{2.1}$$

$$(\mathcal{F}^{-1} j \mathcal{F})' \parallel \gamma'. \tag{2.2}$$

The second equation says that $\mathcal{F}^{-1} j \mathcal{F}$ is a parallel section in the normal bundle of γ, which justifies the name. Moreover, let $A = \mathcal{F}' \mathcal{F}^{-1}$ be the logarithmic derivative of $\mathcal{F}$.

Equation (2.2) implies that A must lie in the $j\mathfrak{k}$-plane and thus can be written as

$$A = -\frac{\Psi}{2}\mathfrak{k} \tag{2.3}$$

with $\Psi \in \mathrm{span}(1, i) \cong \mathbb{C}$.

Definition 2.1. We call Ψ the *complex curvature* of γ.

Now let us evolve γ with the following flow:

$$\dot{\gamma} = \gamma' \times \gamma'' = \gamma'\gamma''. \tag{2.4}$$

Here $\dot{\gamma}$ denotes the derivative in time. This is an evolution in binormal direction with velocity equal to the (real) curvature. It is known as the *Hashimoto* or *smoke-ring flow*. Hashimoto was the first to show that under this flow the complex curvature Ψ of γ solves the nonlinear Schrödinger equation (NLSE) [6]

$$i\dot{\Psi} + \Psi'' + \frac{1}{2}|\Psi|^2\Psi = 0, \tag{2.5}$$

or written in terms of A:

$$i\dot{A} + A'' = 2A^3. \tag{2.6}$$

Definition 2.2. The surfaces $\gamma(x, t)$ swept out by the flow given in equation (2.4) are called *Hashimoto surfaces*.

Equation (2.5) arises as the zero-curvature condition $\hat{L}_t - \hat{M}_x + \lfloor\hat{L}, \hat{M}\rfloor = 0$ of the system

$$\begin{aligned}
\hat{\mathcal{F}}_x(\mu) &= \hat{L}(\mu)\hat{\mathcal{F}}(\mu), \\
\hat{\mathcal{F}}_t(\mu) &= \widehat{M}(\mu)\hat{\mathcal{F}}(\mu),
\end{aligned} \tag{2.7}$$

with

$$\begin{aligned}
\hat{L}(\mu) &= \mu i - \frac{\Psi}{2}\mathfrak{k}, \\
\widehat{M}(\mu) &= \frac{|\Psi|^2}{4}i + \frac{\Psi_x}{2}j - 2\mu\hat{L}(\mu).
\end{aligned} \tag{2.8}$$

To make the connection to the description with the parallel frame $\mathcal{F}$ we add torsion to the curve γ by setting

$$A(\mu) = \frac{1}{2}e^{-2\mu x i}\Psi\mathfrak{k}.$$

This gives rise to a family of curves $\gamma(\mu)$, the so-called *associated family* of γ. Now one can gauge the corresponding parallel frame $\mathcal{F}(\mu)$ with $e^{\mu x i}$ and get

$$(e^{\mu x i}\mathcal{F}(\mu))_x = ((e^{\mu x i})_x e^{-\mu x i} + e^{\mu x i}A(\mu)e^{-\mu x i})e^{\mu x i}\mathcal{F}(\mu) = L(\mu)e^{\mu x i}\mathcal{F}(\mu)$$

with $L(\mu) = \hat{L}(\mu)$ as in (2.7). So above $\hat{\mathcal{F}}(\mu) = e^{\mu x i}\mathcal{F}(\mu)$ is for each t_0 a frame for the curve $\gamma(x, t_0)$.

Theorem 2.3 (Sym formula). *Let* $\Psi(x, t)$ *be a solution of the NLSE, equation (2.5). Then up to a euclidean motion the corresponding Hashimoto surface* $\gamma(x, t)$ *can be obtained by*

$$\gamma(x, t) = \widehat{\mathcal{F}}^{-1}\widehat{\mathcal{F}}_\mu|_{\mu=0} \tag{2.9}$$

where $\widehat{\mathcal{F}}$ *is a solution to (2.7).*

Proof. Obviously $\widehat{\mathcal{F}}|_{\lambda=0}(x, t_0)$ is a parallel frame for each $\gamma(x, t_0)$. Therefore, writing $\widehat{\mathcal{F}}(x, t_0)|_{\lambda=0} =: \mathcal{F}(x)$, one easily computes $(\widehat{\mathcal{F}}^{-1}\widehat{\mathcal{F}}_\lambda|_{\lambda=0})_x = \mathcal{F}^{-1}i\mathcal{F} = \gamma_x$ and $(\widehat{\mathcal{F}}^{-1}\widehat{\mathcal{F}}_\lambda|_{\lambda=0})_t = \mathcal{F}^{-1}\Psi\mathfrak{k}\mathcal{F}$. But $\gamma_t = \gamma_x\gamma_{xx} = \mathcal{F}^{-1}\Psi\mathfrak{k}\mathcal{F}$. $\square$

If one differentiates equation (2.4) with respect to x, one gets the so-called isotropic Heisenberg magnet (IHM) model

$$\dot{S} = S \times S'' = S \times S_{xx} \tag{2.10}$$

with $S = \gamma'$. This equation arises as the zero-curvature condition $U_t - V_x + [U, V] = 0$ with matrices

$$\begin{aligned} U(\lambda) &= \lambda S, \\ V(\lambda) &= -2\lambda^2 S - \lambda S'S. \end{aligned} \tag{2.11}$$

In fact, if G is a solution to

$$\begin{aligned} G_x &= U(\lambda)G, \\ G_t &= V(\lambda)G, \end{aligned} \tag{2.12}$$

it can be viewed as a frame for the Hashimoto surface too and one has a similar Sym formula:

$$\gamma(x, t) = G^{-1}G_\lambda|_{\lambda=0}. \tag{2.13}$$

The system (2.12) is known [5] to be gauge-equivalent to (2.7).

2.1. Elastic curves

The stationary solutions of the NLSE (i.e., the curves that evolve by rigid motion under the Hashimoto flow) are known to be the *elastic curves* [3]. They are the critical points of the functional

$$E(\gamma) = \int \kappa^2$$

with $\kappa = |\Psi|$ the curvature of γ. The fact that they evolve by rigid motion under the Hashimoto flow can be used to give a characterization in terms of their complex curvature Ψ only: When the curve evolves by rigid motion Ψ changes by a phase factor only. Thus $\dot{\Psi} = ci\Psi$. Inserted into equation (2.5) this gives

$$\Psi'' = (c - \frac{1}{2}|\Psi|^2)\Psi. \tag{2.14}$$

2.2. Bäcklund transformations for smooth space curves and Hashimoto surfaces

Now we want to describe the dressing procedure or Bäcklund transformation for the IHM model and the Hashimoto surfaces. This is a method to generate new solutions of our equations from a given one in a purely algebraic way. Afterwards we give a geometric interpretation for this transformation.

2.2.1. Algebraic description of the Bäcklund transformation.

Theorem 2.4. *Let G be a solution to equations (2.12) with U and V as in (2.11) (i.e., $U(1)$ solves the IHM model). Choose $\lambda_0, s_0 \in \mathbb{C}$. Then $\tilde{G}(\lambda) := B(\lambda)G(\lambda)$ with $B(\lambda) = (1 + \lambda\rho), \rho \in \mathbb{H}$, defined by the conditions that $\lambda_0, \bar{\lambda}_0$ are the zeroes of $\det(B(\lambda))$ and*

$$\tilde{G}(\lambda_0)\begin{pmatrix} s_0 \\ 1 \end{pmatrix} = 0 \quad and \quad \tilde{G}(\bar{\lambda}_0)\begin{pmatrix} 1 \\ -\bar{s}_0 \end{pmatrix} = 0 \tag{2.15}$$

solves a system of the same type. In particular $\tilde{U}(1) = \tilde{G}_x(1)\tilde{G}^{-1}(1)$ solves again the Heisenberg magnet model (2.10).

Proof. We define $\tilde{U}(\lambda) = \tilde{G}_x\tilde{G}^{-1}$ and $\tilde{V}(\lambda) = \tilde{G}_t\tilde{G}^{-1}$. Equation (2.15) ensures that $\tilde{U}(\lambda)$ and $\tilde{V}(\lambda)$ are smooth at λ_0 and $\bar{\lambda}_0$. Using

$$\tilde{U}(\lambda) = B_x(\lambda)B^{-1}(\lambda) + B(\lambda)U(\lambda)B^{-1}(\lambda)$$

this in turn implies that $\tilde{U}(\lambda)$ has the form $\tilde{U}(\lambda) = \lambda\tilde{S}$ for some $\tilde{S}$.

Since the zeroes of $\det(B(\lambda))$ are fixed we know that $r := \Re(\rho)$ and $l := |\Im(\rho)|$ are constant. Writing $\rho = r + v$, we get $\tilde{S} = S + v_x$ and

$$v_x = \frac{2rl}{r^2 + l^2}\frac{v \times S}{l} + \frac{2l^2}{r^2 + l^2}\frac{\langle v, S \rangle}{l^2}v - \frac{2l^2}{r^2 + l^2}S. \tag{2.16}$$

This can be used to show $|\tilde{S}| = 1$.

Again equation (2.15) ensures that $\tilde{V}(\lambda) = \lambda^2 X + \lambda Y$ for some X and Y. But then the integrability condition $\tilde{U}_t - \tilde{V}_x + [\tilde{U}, \tilde{V}]$ gives up to a factor c and possible constant real parts x and y that X and Y are fixed to be $X = x + c\tilde{S}_x\tilde{S} + d\tilde{S}$ and $Y = y + 2c\tilde{S}$. The additional term $d\tilde{S}$ in X corresponds to the (trivial) tangential flow which always can be added. The form $\tilde{V}(\lambda) = B_t(\lambda)B^{-1}(\lambda) + B(\lambda)V(\lambda)B^{-1}(\lambda)$ gives $c = -1$ and $d = 0$. Thus one ends up with $\tilde{V}(\lambda) = -2\lambda^2\tilde{S} - \lambda\tilde{S}_x\tilde{S}$. □

So we get a four-parameter family (λ_0 and s_0 give two real parameters each) of transformations for our curve γ that are compatible with the Hashimoto flow. They correspond to the four-parameter family of Bäcklund transformations of the NLSE.

Example 2.5. Let us do this procedure in the easiest case: We choose $S \equiv i$ (or $\gamma(x, t) = xi$) which gives

$$G(\lambda) = \exp((\lambda x - 2\lambda^2 t)i) = \begin{pmatrix} e^{i(\lambda x - 2\lambda^2 t)} & 0 \\ 0 & e^{-i(\lambda x - 2\lambda^2 t)} \end{pmatrix}.$$

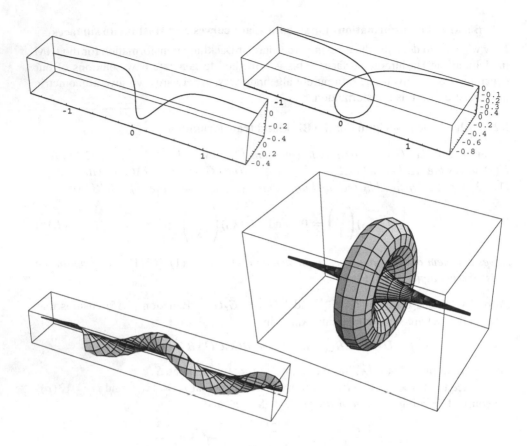

FIGURE 1. Two dressed straight lines and the corresponding Hashi-moto surfaces.

After choosing λ_0 and s_0 and writing $\rho = \begin{pmatrix} a & b \\ -\bar{b} & \bar{a} \end{pmatrix}$ one gets with equation (2.15),

$$-e^{i(\lambda_0 x - 2\lambda_0^2 t)} = \lambda_0(e^{i(\lambda_0 x - 2\lambda_0^2 t)})a + s_0 e^{-i(\lambda_0 x - 2\lambda_0^2 t)}b,$$
$$s_0 e^{-i(\lambda_0 x - 2\lambda_0^2 t)} = \lambda_0(e^{i(\lambda_0 x - 2\lambda_0^2 t)}\bar{b} - s_0 e^{-i(\lambda_0 x - 2\lambda_0^2 t)}\bar{a}). \tag{2.17}$$

These equations can be solved for a and b:

$$a = -\frac{1/\lambda_0 + \psi/\bar{\lambda}_0}{1 + \psi},$$
$$b = \bar{s}_0 e^{2i\bar{\lambda}_0 x - 4i\bar{\lambda}_0^2 t}\frac{1/\bar{\lambda}_0 - 1/\lambda_0}{1 + \psi}, \tag{2.18}$$

where $\psi := s_0 \bar{s}_0 \exp\left(-2i(\lambda_0 - \bar{\lambda}_0)x + 4i(\lambda_0^2 - \bar{\lambda}_0^2)t\right)$. Using the Sym formula (2.13) one can immediately write the formula for the resulting Hashimoto surface $\tilde{\gamma}$:

$$\tilde{\gamma} = \Im(\rho) + \gamma = \begin{pmatrix} \Im(a) + ix & b \\ -\bar{b} & -\Im(a) - ix \end{pmatrix}.$$

The need for taking the imaginary part is due to the fact that we did not normalize $B(\lambda)$ to $\det(B(\lambda)) = 1$.

If one wants to have the result in a plane, $\arg b$ should be constant. This can be achieved by choosing $\lambda \in i\mathbb{R}$. Figure 1 shows the result for $s_0 = 0.5 + i$ and $\lambda_0 = 1 - i$ and $\lambda_0 = -i$, respectively.

Of course one can iterate the dressing procedure to get new curves (or surfaces) and it is a natural question how many one can get. This question immediately leads to the Bianchi permutability theorem:

Theorem 2.6 (Bianchi permutability). *Let $\bar{\gamma}$ and $\hat{\gamma}$ be two Bäcklund transforms of γ. Then there is a unique Hashimoto surface $\hat{\bar{\gamma}}$ that is the Bäcklund transform of $\bar{\gamma}$ and $\hat{\gamma}$.*

Proof. Let $G, \hat{G}$, and $\bar{G}$ be the solutions of (2.12) corresponding to $\gamma, \hat{\gamma}$, and $\bar{\gamma}$. One has $\hat{G} = \hat{B}G$ and $\bar{G} = \bar{B}G$ with $\hat{B} = 1 + \lambda\hat{\rho}$ and $\bar{B} = 1 + \lambda\bar{\rho}$. The ansatz $\hat{\bar{B}}\hat{G} = \hat{\bar{B}}\bar{G}$ leads to the compatability condition $\hat{\bar{B}}\hat{B} = \hat{\bar{B}}\bar{B}$ or

$$(1 + \lambda\hat{\bar{\rho}})(1 + \lambda\hat{\rho}) = (1 + \lambda\hat{\bar{\rho}})(1 + \lambda\bar{\rho}) \tag{2.19}$$

which gives:

$$\hat{\bar{\rho}} = (\hat{\rho} - \bar{\rho})\,\bar{\rho}\,(\hat{\rho} - \bar{\rho})^{-1},$$
$$\hat{\bar{\rho}} = (\hat{\rho} - \bar{\rho})\,\hat{\rho}\,(\hat{\rho} - \bar{\rho})^{-1}. \tag{2.20}$$

Thus $\hat{\bar{B}}$ and $\hat{\bar{B}}$ are completely determined. To show that they give dressed solutions we note that since $\det \hat{\bar{B}} \det \hat{B} = \det \hat{\bar{B}} \det \bar{B}$, the zeroes of $\det \hat{\bar{B}}$ are the same as the ones of $\det \bar{B}$ (and the ones of $\det \hat{\bar{B}}$ coincide with those of $\det \hat{B}$). Therefore they do not depend on x and t. Moreover, at these points the kernel of $\hat{\bar{B}}\hat{G}$ coincides with the one of $\bar{G}$. Thus it does not depend on x or t either. Now Theorem 2.4 gives the desired result. $\square$

2.2.2. Geometry of the Bäcklund transformation. As before let $\gamma : I \to \mathbb{R}^3 = \Im\mathbb{H}$ be an arc length–parametrized regular curve. Moreover, let $v : I \to \mathbb{R}^3 = \Im\mathbb{H}$, $|v| = l$ be a solution to the following system:

$$\hat{\gamma} = \gamma + \frac{1}{2}v,$$
$$\hat{\gamma}' \parallel v.$$

Then $\hat{\gamma}$ is called a *tractrix* of γ. The definition of a twisted tractix below is motivated by the following observation: If we set $\tilde{\gamma} = \gamma + v$, it is again an arc length–parametrized curve and $\hat{\gamma}$ is a tractrix of $\tilde{\gamma}$ too. One can generalize this in the following way:

Lemma 2.7. *Let $v : I \to \Im\mathbb{H}$ be a vector field along γ of constant length l satisfying*

$$v' = 2\sqrt{b - b^2}\,\frac{v \times \gamma'}{l} + 2b\frac{<v, \gamma'>}{l^2}v - 2b\gamma' \qquad (2.21)$$

with $0 \le b \le 1$. Then $\tilde{\gamma} = \gamma + v$ is arc length–parametrized.

Proof. Obviously the above transformation coincides with the dressing described in the last section with $b = \frac{l^2}{r^2 + l^2}$ in formula (2.16). This proves the lemma.   □

So $\Im(\rho)$ from Theorem 2.4 is nothing but the difference vector between the original curve and the Bäcklund transform. Note that in the case $b = 1$ one gets the above tractrix construction, that is, $\widehat{\gamma}' \parallel v$ holds for $\widehat{\gamma} = \gamma + v$.

Definition 2.8. The curve $\widehat{\gamma} = \gamma + \frac{1}{2}v$ with v as in Lemma 2.7 is called a *twisted tractrix* of the curve γ and $\tilde{\gamma} = \gamma + v$ is called a *Bäcklund transform* of γ.

Equation (2.21) gives that v' is perpendicular to v and therefore $|v| \equiv$ const. Since $v = \tilde{\gamma} - \gamma$ we see that the Bäcklund transform is in constant distance to the original curve.

3. The Hashimoto flow, the Heisenberg flow, and the nonlinear Schrödinger equation in the discrete case

In this section we give a short review on the discretization (in space) of the Hashimoto flow, the isotropic Heisenberg magnetic model, and the underlying nonlinear Schrödinger equation. For more details on this topic see [3, 4, 5, 7].

We call a map $\gamma : \mathbb{Z} \to \Im\mathbb{H}$ a discrete regular curve if any two successive points do not coincide. It will be called arc length–parametrized curve, if $|\gamma_{n+1} - \gamma_n| = 1$ for all $n \in \mathbb{Z}$. We will use the notation $S_n := \gamma_{n+1} - \gamma_n$. The binormals of the discrete curve can be defined as $\frac{S_n \times S_{n-1}}{|S_n \times S_{n-1}|}$.

There is a natural discrete analog of a parallel frame:

Definition 3.1. A *discrete parallel frame* is a map $\mathcal{F} : \mathbb{Z} \to \mathbb{H}^*$ with $\|\mathcal{F}_k\| = 1$ satisfying

$$S_n = \mathcal{F}_n^{-1}\mathrm{i}\mathcal{F}_n, \qquad (3.1)$$

$$\Im\left((\mathcal{F}_{n+1}^{-1}\mathrm{j}\mathcal{F}_{n+1})(\mathcal{F}_n^{-1}\mathrm{j}\mathcal{F}_n)\right) \parallel \Im\left(S_{n+1}S_n\right). \qquad (3.2)$$

Again we set $\mathcal{F}_{n+1} = A_n\mathcal{F}_n$, and in complete analogy to the continuous case, (3.2) gives the following form for A:

$$A = \cos\frac{\phi_n}{2} - \sin\frac{\phi_n}{2}\exp\left(\mathrm{i}\sum_{k=0}^{n}\tau_k\right)\mathfrak{k}$$

with $\phi_n = \angle\,(S_n, S_{n+1})$ the folding angles and τ_n the angles between successive binormals. If we drop the condition that $\mathcal{F}$ should be of unit length, we can renormalize A_n to be $1 - \tan\frac{\phi_n}{2}\exp\left(\mathrm{i}\sum_{k=0}^{n}\tau_k\right)\mathfrak{k} =: 1 - \Psi_n\mathfrak{k}$ with $\Psi_n \in \mathrm{span}(1, \mathrm{i}) \cong \mathbb{C}$ and $|\Psi_n| = \kappa_n$ the discrete (real) curvature.

Definition 3.2. We call Ψ the *complex curvature*[2] of the discrete curve γ.

Discretizations of the Hashimoto flow (2.4) (i.e., a Hashimoto flow for a discrete arc length–parametrized curve) and the isotropic Heisenberg model (2.10) are well known [5] (see also [3] for a good discussion of the topic). In particular a discrete version of (2.4) is given by

$$\dot{\gamma}_k = 2 \frac{S_k \times S_{k-1}}{1 + \langle S_k, S_{k-1} \rangle} \tag{3.3}$$

which implies for a discretization of (2.10)

$$\dot{S}_k = 2 \frac{S_{k+1} \times S_k}{1 + \langle S_{k+1}, S_k \rangle} - 2 \frac{S_k \times S_{k-1}}{1 + \langle S_k, S_{k-1} \rangle}. \tag{3.4}$$

Let us state the zero-curvature representation for this equation too: Equation (3.4) is the compatibility condition of $\dot{U}_k = V_{k+1}U_k - U_k V_k$ with

$$U_k = 1 + \lambda S_k,$$

$$V_k = -\frac{1}{1+\lambda^2}\left(2\lambda^2 \frac{S_k + S_{k-1}}{1 + \langle S_k, S_{k-1} \rangle} + 2\lambda \frac{S_k \times S_{k-1}}{1 + \langle S_k, S_{k-1} \rangle}\right). \tag{3.5}$$

The solution to the auxiliary problem

$$G_{k+1} = U_k(\lambda)G_k,$$
$$\dot{G}_k = V_k(\lambda)G_k \tag{3.6}$$

can be viewed as the frame to a discrete Hashimoto surface $\gamma_k(t)$, and one has the same Sym formula as in the continuous case:

Theorem 3.3. *Given a solution G to the system (3.6) the corresponding discrete Hashimoto surface can be obtained up to an euclidean motion by*

$$\gamma_k(t) = (G_k^{-1} \frac{\partial}{\partial \lambda} G_k)|_{\lambda=0}. \tag{3.7}$$

Proof. One has $G_k^{-1} \frac{\partial}{\partial \lambda} G_k|_{\lambda=0} = \sum_{i=0}^{k-1} S_i = \gamma_k$ for fixed time t_0 and

$$(G_k^{-1} \frac{\partial}{\partial \lambda} G_k|_{\lambda=0})_t = (\frac{\partial}{\partial \lambda} V_k(\lambda)|_{\lambda=0}) = 2 \frac{S_k \times S_{k-1}}{1 + \langle S_k, S_{k-1} \rangle}. \qquad \square$$

To complete the analogy to the smooth case we give a discretization of the NLSE that can be found in [1] (see also [5, 11]):

$$-i\dot{\Psi}_k = \Psi_{k+1} - 2\Psi_k + \Psi_{k-1} + |\Psi_k|^2(\Psi_{k+1} + \Psi_{k-1}). \tag{3.8}$$

Theorem 3.4. *Let γ be a discrete arc length–parametrized curve. If γ evolves with the discrete Hashimoto flow (3.3), then its complex curvature Ψ evolves with the discrete nonlinear Schrödinger equation (3.8).*

---

[2]It would be more reasonable to define $A_n = 1 - \frac{\Psi_n}{2} \mathfrak{e}$, which implies $\kappa_n = 2 \tan \frac{\phi_n}{2}$, but notational simplicity makes the given definition more convenient.

A proof of this theorem can be found in [7] and [8]. There is another famous discretization of the NLSE in literature that is related to the dIHM [5, 9]. Again in [7] it is shown that it is in fact gauge equivalent to the above-cited which turns out to be more natural from a geometric point of view.

3.1. Discrete elastic curves

As mentioned in Section 2.1 the stationary solutions of the NLSE (i.e., the curves that evolve by rigid motion under the Hashimoto flow) are known to be the *elastic curves*. They have a natural discretization using this property:

Definition 3.5. A *discrete elastic curve* is a curve γ for which the evolution of γ_n under the Hashimoto flow (3.3) is a rigid motion which means that its tangents evolve under the discrete isotropic Heisenberg model (3.4) by rigid rotation.

In [3] Bobenko and Suris showed the equivalence of this definition to a variational description.

The fact that (3.4) has to be a rigid rotation means that the left-hand side must be $S_n \times p$ with a unit imaginary quaternion p. We will now give a description of elastic curves by their complex curvature function only:

Theorem 3.6. *The complex curvature Ψ_n of a discrete elastic curve γ_n satisfies the following difference equation:*

$$C \frac{\Psi_n}{1 + |\Psi_n|^2} = \Psi_{n+1} + \Psi_{n-1} \tag{3.9}$$

for some real constant C.

Equation (3.9) is a special case of a discrete-time Garnier system (see [10]).

Proof. One can proof the theorem by direct calculations or using the equivalence of the dIHM model and the dNLSE stated in Theorem 3.4. If the curve γ evolves by rigid motion, its complex curvature may vary by a phase factor only: $\Psi(x, t) = e^{i\lambda(t)}\Psi(x, t_0)$ or $\dot{\Psi} = i\dot{\lambda}\Psi$. Plugging this into (3.8) gives

$$-\dot{\lambda}\Psi_k = \Psi_{k+1} - 2\Psi_k + \Psi_{k-1} + |\Psi_k|^2(\Psi_{k+1} + \Psi_{k-1})$$

which is equivalent to (3.9) with $C = 2 - \dot{\lambda}$.   □

As an example, Figure 2 shows two discretizations of the elastic figure eight.

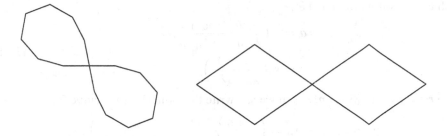

FIGURE 2. Two discretizations of the elastic figure eight.

3.2. Bäcklund transformations for discrete space curves and Hashimoto surfaces

3.2.1. Algebraic description. In complete analogy to Section 2.2.1 we state:

Theorem 3.7. *Let G_k be a solution to (3.6) with U_k and V_k as in (3.5) (i.e., $U(1) - 1$ solves the dIHM model). Choose $\lambda_0, s_0 \in \mathbb{C}$. Then $\tilde{G}_k(\lambda) := B_k(\lambda)G_k(\lambda)$ with $B_k(\lambda) = (1 + \lambda \rho_k), \rho_k \in \mathbb{H}$ defined by the conditions that $\lambda_0, \bar{\lambda}_0$ are the zeroes of $\det(B_k(\lambda))$ and*

$$\tilde{G}_k(\lambda_0)\begin{pmatrix} s_0 \\ 1 \end{pmatrix} = 0 \quad \text{and} \quad \tilde{G}_k(\bar{\lambda}_0)\begin{pmatrix} 1 \\ -\bar{s}_0 \end{pmatrix} = 0 \tag{3.10}$$

solves a system of the same type. In particular

$$\tilde{U}_k(1) - 1 = \tilde{G}_x(1)\tilde{G}^{-1}(1) - 1$$

solves again the discrete Heisenberg magnet model (3.4).

Proof. Analogous to the smooth case. □

Example 3.8. Let us dress the (this time discrete) straight line again: We set $S_n \equiv i$ and get

$$G_n(\lambda) = (1 + \lambda i)^n \exp(-2\frac{\lambda^2}{1+\lambda^2} t i)$$

$$= \begin{pmatrix} (1 + i\lambda)^n e^{-2i\frac{\lambda^2}{1+\lambda^2}t} & 0 \\ 0 & (1 - i\lambda)^n e^{2i\frac{\lambda^2}{1+\lambda^2}t} \end{pmatrix}.$$

After choosing λ_0 and s_0 and writing again $\rho = \begin{pmatrix} a & b \\ -\bar{b} & \bar{a} \end{pmatrix}$ we get with the shorthands

$$p = (1 + i\lambda_0)^n e^{-2i\frac{\lambda_0^2}{1+\lambda_0^2}t} \text{ and } q = (1 - i\lambda_0)^n e^{2i\frac{\lambda_0^2}{1+\lambda_0^2}t} \text{ that}$$

$$p = -\lambda_0(pa + s_0qb),$$
$$q = \lambda_0(p\bar{b} - s_0q\bar{a}),$$

which can be solved for a and b :

$$a = -\left(\frac{1}{\lambda_0}\frac{\bar{p}}{\bar{q}} + \frac{s_0\bar{s}_0}{\bar{\lambda}_0}\frac{q}{p}\right)/r,$$

$$b = \bar{s}_0\left(\frac{1}{\bar{\lambda}_0} - \frac{1}{\lambda_0}\right)/r,$$

(3.11)

where $r := \bar{p}/\bar{q} + s_0\bar{s}_0 q/p$. Again we can write the formula for the curve $\tilde{\gamma}$:

$$\tilde{\gamma}_n = \Im(\rho_n) + \gamma_n = \begin{pmatrix} \Im(a_n) + in & b_n \\ -\bar{b}_n & -\Im(a_n) - in \end{pmatrix}.$$

Figure 3 shows two solutions with $s_0 = 0.5 + i$ and $\lambda_0 = 0.4 - 0.4i$ and $\lambda_0 = -0.4i$,

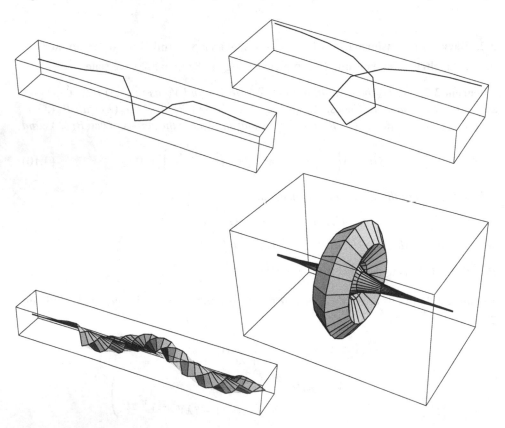

FIGURE 3. Two discrete dressed straight lines and the corresponding Hashimoto surfaces.

respectively. The second one is again planar. Note the strong similarity to the smooth examples in Figure 1.

Of course one has again a permutability theorem:

Theorem 3.9 (Bianchi permutability). *Let $\tilde{\gamma}$ and $\hat{\gamma}$ be two Bäcklund transforms of γ. Then there is a unique discrete Hashimoto surface $\hat{\tilde{\gamma}}$ that is a Bäcklund transform of $\tilde{\gamma}$ and $\hat{\gamma}$.*

Proof. Literally the same as for Theorem 2.6. $\qquad\qquad\qquad\qquad\qquad$ $\square$

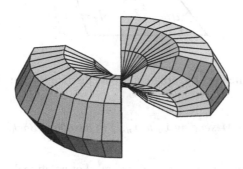

FIGURE 4.   The Hashimoto surface from a discrete elastic eight.

3.2.2. Geometry of the discrete Bäcklund transformation. In this section we want to derive the discrete Bäcklund transformations by mimicking the twisted tractrix construction from Lemma 2.7:

Let $\gamma : \mathbb{Z} \to \Im\mathbb{H}$ be a discrete arc length–parametrized curve. To any initial vector v_n of length l there is a S^1-family of vectors v_{n+1} of length l satisfying $|\gamma_n + v_n - (\gamma_{n+1} + v_{n+1})| = 1$. This is basically folding the parallelogram spanned by v_n and S_n along the diagonal $S_n - v_n$. To single out one of these new vectors let us fix the angle δ_1 between the planes spanned by v_n and S_n and v_{n+1} and S_n (see Figure 5). This furnishes a unique evolution of an initial v_0 along γ. The polygon $\tilde{\gamma}_n = \gamma_n + v_n$ is again a discrete arc length–parametrized curve which we will call a *Bäcklund transform* of γ.

There are two cases in which the elementary quadrilaterals $(\gamma_n, \gamma_{n+1}, \tilde{\gamma}_{n+1}, \tilde{\gamma}_n)$ are planar. One is the parallelogram case. The other can be viewed as a discrete version of the tractrix construction.

Definition 3.10. Let γ be a discrete arc length–parametrized curve. Given δ_1 and v_0, $|v_0| = l$, there is a unique discrete arc length–parametrized curve $\tilde{\gamma}_n = \gamma_n + v_n$ with $|v_n| = l$ and $\angle(\mathrm{span}(v_n, S_n), \mathrm{span}(v_{n+1}, S_n)) = \delta_1$. We call $\tilde{\gamma}$ a *Bäcklund transform* of γ and $\hat{\gamma} = \gamma + \frac{1}{2}v$ is called a *discrete twisted tractrix* for γ (and $\tilde{\gamma}$).

Note that in case of $\delta_1 = \pi$ the cross-ratio is $\mathrm{cr}(\gamma, \tilde{\gamma}, \tilde{\gamma}_+, \gamma_+) = l^2$.

Of course we will show that this notion of Bäcklund transformation coincides with the one from the last section. Let us investigate this Bäcklund transformation in greater detail. For now we do not restrict ourselves to arc length–parametrized curves. We state the following:

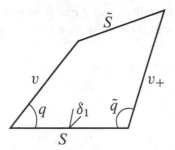

FIGURE 5. An elementary quadrilateral of the discrete Bäcklund transformation.

Lemma 3.11. *The map M sending v_n to v_{n+1} in the above Bäcklund transformation is a Möbius transformation.*

Proof. Let us look at an elementary quadrilateral: For notational simplicity let us write $S = \gamma_{n+1} - \gamma_n$, $\tilde{S} = \tilde{\gamma}_{n+1} - \tilde{\gamma}_n$, $|S| = s$, $v = v_n$, and $v_+ = v_{n+1}$. If we denote the angles $\angle(S, v)$ and $\angle(v_+, S)$ with q and $\tilde{q}$, we get

$$e^{i\tilde{q}} = \frac{ke^{iq} - 1}{e^{iq} - k} \tag{3.12}$$

with $k = \tan \frac{\delta_1}{2} \tan \frac{\delta_2}{2}$ and δ_1 as in Figure 5, δ_2 is the corresponding angle along the edge v. Note that l, s, k, δ_1, and δ_2 are coupled by

$$k = \tan \frac{\delta_1}{2} \tan \frac{\delta_2}{2}, \quad \frac{l}{s} = \frac{\sin \delta_2}{\sin \delta_1}. \tag{3.13}$$

To get an equation for v_+ from this, we need to have all vectors in one plane. So set $\sigma = \cos \frac{\delta_1}{2} + \sin \frac{\delta_1}{2} \frac{S}{s}$. Then conjugation with σ is a rotation around S with angle δ_1. If we replace e^{iq} by $\frac{\sigma v \sigma^{-1}}{l} \left(\frac{S}{s}\right)^{-1}$ and $e^{i\tilde{q}}$ by $-\frac{S}{s}v_+^{-1}l$, equation (3.12) becomes quaternionic but stays valid (one can think of it as a complex equation with different "i"). Equation (3.12) now reads

$$\frac{v_+ S}{ls} = \frac{\frac{s}{l}\sigma v \sigma^{-1}S^{-1} - k}{\frac{ks}{l}\sigma v \sigma^{-1}S^{-1} - 1}.$$

We can write this in homogenous coordinates: $\mathbb{H}^2$ carries a natural right $\mathbb{H}$-module structure, so one can identify a point in $\mathbb{HP}^1$ with a quaternionic line in $\mathbb{H}^2$ by $p \cong (r, s) \iff p = rs^{-1}$. In this picture our equation becomes

$$\begin{pmatrix} \frac{1}{ls}v_+ S \\ 1 \end{pmatrix} \lambda = \begin{pmatrix} \frac{s}{l}\sigma & -kS\sigma \\ \frac{ks}{l}\sigma & -S\sigma \end{pmatrix} \begin{pmatrix} v \\ 1 \end{pmatrix}.$$

Bringing ls and S on the right-hand side gives us finally the matrix

$$\mathcal{M} := \begin{pmatrix} \frac{1}{k}\sigma & -\frac{l}{s}S\sigma \\ \frac{1}{ls}S\sigma & \frac{1}{k}\sigma \end{pmatrix}. \tag{3.14}$$

Since we know that this map sends a sphere of radius l onto itself, we can project this sphere stereographically to get a complex matrix. The matrix

$$P = \begin{pmatrix} 2\mathfrak{i} & -2l \\ \frac{\mathfrak{i}}{l} & \mathfrak{k} \end{pmatrix}$$

projects $l S^2$ onto $\mathbb{C}$. Its inverse is given by

$$P^{-1} = -\frac{1}{4} \begin{pmatrix} \mathfrak{i} & 2l\mathfrak{j} \\ \frac{1}{l} & 2\mathfrak{k} \end{pmatrix}.$$

One easily computes

$$\mathcal{M}_{\mathbb{C}} = P \mathcal{M} P^{-1} = -\frac{1}{4} \begin{pmatrix} \nu + i \Re(S\mathfrak{i}) & 2l \Im(S\mathfrak{i})\mathfrak{j} \\ -\frac{1}{2l} \Im(S\mathfrak{i})\mathfrak{j} & \nu - i \Re(S\mathfrak{i}) \end{pmatrix}$$

with $\nu = i s \left(\tan \frac{\delta_1}{2} i - \frac{1}{k} \right) / \left(\frac{\tan \frac{\delta_1}{2}}{k} i - 1 \right)$. This completes our proof. $\qquad\square$

Using equation (3.13) one can compute

$$\nu = s \tan \frac{\delta_1}{2} \frac{1 - k^2}{\tan^2 \frac{\delta_1}{2} + k^2} + i l = l \tan \frac{\delta_2}{2} \frac{1 - k^2}{\tan^2 \frac{\delta_2}{2} + k^2} + i l. \tag{3.15}$$

So the real part of ν is invariant under the change $s \leftrightarrow l$, $\delta_1 \leftrightarrow \delta_2$. Therefore instead of thinking of $\tilde{S}$ as a transform of S with parameter ν one could view as ν_+ a transform of ν with parameter $\nu + i(s - l)$.

One can gauge $\mathcal{M}_{\mathbb{C}}$ to get rid of the off diagonal $2l$ factors

$$M = \begin{pmatrix} \frac{1}{\sqrt{2l}} & 0 \\ 0 & \sqrt{2l} \end{pmatrix} \mathcal{M}_{\mathbb{C}} \begin{pmatrix} \sqrt{2l} & 0 \\ 0 & \frac{1}{\sqrt{2l}} \end{pmatrix}.$$

Then we can write, in abuse of notation,

$$M = \nu 1 - S.$$

Here $\nu 1$ is no quaternion if ν is complex. The eigenvalues of $\mathcal{M}_{\mathbb{C}}$ and M clearly coincide and M obviously coincides with the Lax matrix U_k of the dIHM model in equation (3.5) up to a factor $\frac{1}{\nu}$ with $\lambda = -\frac{1}{\nu}$.

As promised the next lemma shows that the geometric Bäcklund transformation discussed in this section coincides with the one from the algebraic description.

Lemma 3.12. *Let $S, \nu \in \Im \mathbb{H}$ be nonzero vectors, $|\nu| = l$, $\tilde{S}$ and ν_+ be the evolved vectors in the sense of our Bäcklund transformation with parameter ν $(\Im \nu = l)$. Then*

$$(\lambda 1 + \tilde{S})(\lambda 1 + \Re \nu + \nu) = (\lambda 1 + \Re \nu + \nu_+)(\lambda 1 + S) \tag{3.16}$$

holds for all λ.

Proof. Comparing the orders in λ on both sides of (3.16) gives two equations

$$\tilde{S} + \Re \nu + \nu = \Re \nu + \nu_+ + S, \tag{3.17}$$

$$\tilde{S}(\Re \nu + \nu) = (\Re \nu + \nu_+)S. \tag{3.18}$$

The first holds trivially from construction; the second gives

$$\Re v = (v_+ S - \tilde{S} v)(\tilde{S} - S)^{-1}.$$

This can be checked by elementary calculations using (3.15) for the real part of v. □

As in the continuous case we can deduce that $\Im(\rho_n) = v_n = \tilde{\gamma}_n - \gamma_n$ which gives us the constant distance between the original curve γ_n and its Bäcklund transform $\tilde{\gamma}_n$.

Note that it is not necessary for S to be of constant length. We are able to perform a Bäcklund transformation for arbitrary polygons. This holds for the doubly discrete flow described in the next section as well.

4. The doubly discrete Hashimoto flow

From now on let $\gamma : \mathbb{Z} \to \Im\mathbb{H}$ be periodic or have at least periodic tangents $S_n = \gamma_{n+1} - \gamma_n$ with period N (we will see later that rapidly decreasing boundary conditions are valid also). As before let $\tilde{\gamma}$ be a Bäcklund transform of γ with initial point $\tilde{\gamma}_0 = \gamma_0 + v_0$, $|v_0| = l$. As we have seen the map sending v_n to v_{n+1} is a Möbius transformation and therefore the map sending v_0 to v_N is one too. As such, it has in general two, but at least one fixed point. Thus starting with one of them as initial point the Bäcklund transform $\tilde{\gamma}$ is periodic too (or has periodic tangents S). Clearly this can be iterated to get a discrete evolution of our discrete curve γ.

Lemma 4.1. *Let γ be a discrete curve with periodic tangents S of period N. Then the tangents $\tilde{S}$ of a dressed curve $\tilde{\gamma}$ with the parameters λ_0 and s_0 are again periodic if and only if the vector $(1, s_0)$ is an eigenvector of the monodromy matrix $G_N(\lambda)$ at $\lambda = \lambda_0$.*

Proof. We use the notation from Theorem 3.7. Since $\tilde{\gamma}_n - \gamma_n = v_n = \Im(\rho_n)$ and since $B(\lambda) = 1 + \lambda\rho$ is completely determined by λ_0 and v we have that $B_0(\lambda) = B_N(\lambda)$. On the other hand one can determine $B(\lambda)$ by λ_0 and s_0. Since $G_0(\lambda) = 1$, Condition 3.10 says that $\binom{1}{s_0}$ and $G_n(\lambda_0)\binom{1}{s_0}$ must lie in ker $B_0(\lambda_0)$. □

A Lax representation for this evolution is given by equation (3.18) which is basically the Bianchi permutability of the Bäcklund transformation.

In the following we will show that for the special choice $l = 1$ and $\delta_1 \approx \frac{\pi}{2}$ the resulting evolution can be viewed as a discrete smoke-ring flow. More precisely, one has to apply the transformation twice: once with δ_1 and once with $-\delta_1$. In [7] it is shown that under this evolution the complex curvature of the discrete curve solves the doubly discrete NLSE introduced by Ablowitz and Ladik [2].

Proposition 4.2. *A Möbius transformation that sends a disc into its interior has a fixed point in it.*

Proof. For the Möbius transformation M look at the vector field f given by $f(x) = M(x) - x$. This must have a zero. □

Now we show the following lemma.

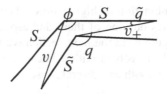

FIGURE 6. An elementary quadrilateral if $l = 1$ and $\delta_1 \approx \frac{\pi}{2}$.

Lemma 4.3. *If* $\angle(-S_-, v) \leq \epsilon$, *with* ϵ *sufficiently small, there exists a* δ_1 *such that* $\angle(-S, v_+) < \epsilon$.

Proof. With notation as in Figure 6 we know $e^{i\tilde{q}} = \frac{ke^{iq}-1}{k-e^{iq}}$ and $q \in [\phi - \epsilon, \phi + \epsilon]$ giving us

$$2i \sin \tilde{q} = 2i \, \Im e^{i\tilde{q}} = 2i \frac{(k^2 - 1) \sin(\phi \pm \epsilon)}{(k^2 + 1) - 2k \cos(\phi \pm \epsilon)}$$

which proves the claim since k goes to 1 if δ_1 tends to $\frac{\pi}{2}$. $\square$

Knowing this one can see that an initial v_0 with $\angle(-S_{N-1}, v_0) \leq \epsilon$ is mapped to a v_N with $\angle(-S_{N-1}, v_N) < \epsilon$. The proposition above gives that there must be a fixed point p_0 with $\angle(-S_{N-1}, p_0) < \epsilon$.

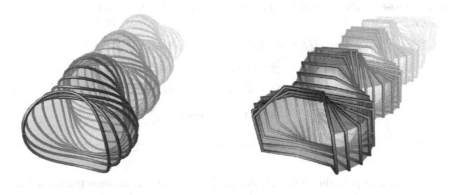

FIGURE 7. An oval curve under the Hashimoto flow and the discrete evolution of its discrete counterpart.

But if $p_n \approx -S_{n-1}$, we get $\tilde{\gamma}_n \approx \gamma_{n-1}$, and $\tilde{\gamma}_n - \gamma_{n-1}$ is almost orthogonal to span(S_{n-2}, S_{n-1}). So it is a discrete version of an evolution in binormal direction—plus a shift. To get rid of this shift, one has to do the transformation twice but with opposite sign for δ_1. Figure 7 shows some stages of the smooth Hashimoto flow for an oval curve and the discrete evolution of its discrete counterpart. In general the Bäcklund transformation can be viewed as a linear combination of higher flows. The special choice of a double Bäcklund transformation singles out one of them and especially cancels the tangential flow part as one can see in Figure 8. In the next section we will see that the curves that move by rigid motion only when evolved with a linear combination of Hashimoto and tangential flow coincide in the smooth and discrete case:

4.1. Discrete elastic curves

As a spin-off of the last section one can easily show that the elastic curves defined in Section 3 as curves that evolve under the Hashimoto flow by rigid motion only do the same for the doubly discrete Hashimoto flow. Again we will use the evolution of the complex curvature of the discrete curve. We mentioned before that in the doubly discrete case the complex curvature evolves with the doubly discrete NLSE given by Ablowitz and Ladik [2, 7].

We start by quoting a special case of their result which can be summarized in the following form (see also [11]):

Theorem 4.4 (Ablowitz and Ladik). *Given*

$$L_n(\mu) = \begin{pmatrix} \mu & q_n \\ -\bar{q}_n & \mu^{-1} \end{pmatrix}$$

and $V_n(\mu)$ with the following μ-dependency:

$$V_n(\mu) = \mu^{-2}V_{-2n} + \mu^{-1}V_{-1n} + V_{0n} + \mu^1 V_{1n} + \mu^2 V_{2n}.$$

Then the zero-curvature condition $V_{n+1}(\mu)L_n(\mu) = \tilde{L}_n(\mu)V_n(\mu)$ gives the following equations:

$$
\begin{aligned}
(\tilde{q}_n - q_n)/i &= \alpha_+ q_{n+1} - \alpha_0 q_n + \bar{\alpha}_0 \tilde{q}_n - \bar{\alpha}_+ \tilde{q}_{n-1} \\
&\quad + (\alpha_+ q_n \mathcal{A}_{n+1} - \bar{\alpha}_+ \tilde{q}_n \bar{\mathcal{A}}_n) \\
&\quad + (-\bar{\alpha}_- \tilde{q}_{n+1} + \alpha_- q_{n-1})(1 + |\tilde{q}_n|^2)\Lambda_n, \qquad (4.1)
\end{aligned}
$$

$$\mathcal{A}_{n+1} - \mathcal{A}_n = \tilde{q}_n \bar{\tilde{q}}_{n-1} - q_{n+1}\bar{q}_n,$$

$$\Lambda_{n+1}(1 + |q_n|^2) = \Lambda_n(1 + |\tilde{q}_n|^2),$$

with constants α_+, α_0 and α_-.

In the case of periodic or rapidly decreasing boundary conditions the natural conditions $\mathcal{A}_n \to 0$, and $\Lambda_n \to 1$ for $n \to \pm\infty$ give formulas for $\mathcal{A}_n$ and Λ_n:

$$\mathcal{A}_n = q_n \bar{q}_{n-1} + \sum_{j=j_0}^{n-1} (q_j \bar{q}_{j-1} - \tilde{q}_j \bar{\tilde{q}}_{j-1}),$$

$$\Lambda_n = \prod_{j=j_0}^{n-1} \frac{1 + |\tilde{q}_j|^2}{1 + |q_j|^2}$$

with $j_0 = 0$ in the periodic case and $j_0 = -\infty$ in case of rapidly decreasing boundary conditions.

Theorem 4.5. *The discrete elastic curves evolve by rigid motion under the doubly discrete Hashimoto flow.*

Proof. Evolving by rigid motion means that the complex curvature of a discrete curve must stay constant up to a possible global phase, i.e., $\tilde{\psi}_n = e^{2i\theta}\Psi_n$. Due to Theorem 4.4 the evolution equation for ψ_n reads

$$\frac{(\tilde{\Psi}_n - \Psi_n)}{i} = \alpha_+\Psi_{n+1} - \alpha_0\Psi_n + \bar{\alpha}_0\tilde{\Psi}_n - \bar{\alpha}_+\tilde{\Psi}_{n-1} + (\alpha_+\Psi_n\mathcal{A}_{n+1}$$
$$- \bar{\alpha}_+\tilde{\Psi}_n\bar{\mathcal{A}}_n) + (-\bar{\alpha}_-\tilde{\Psi}_{n+1} + \alpha_-\Psi_{n-1})(1 + |\tilde{\Psi}_n|^2)\Lambda_n.$$

Using $e^{-i\theta}\tilde{\psi}_n = e^{i\theta}\Psi_n$ gives $\Lambda_n = 1$, $\mathcal{A}_n = e^{2i\theta}\Psi_n\bar{\Psi}_{n-1}$, and finally

$$2\left(\sin\theta + \Re(e^{i\theta}\alpha_0)\right)\frac{\Psi_n}{1 + |\Psi_n|^2}$$
$$= \left(e^{i\theta}\alpha_+ + e^{i\bar{\theta}}\alpha_-\right)\Psi_{n+1} + \left(e^{i\bar{\theta}}\alpha_+ + e^{i\theta}\alpha_-\right)\Psi_{n-1}.$$

So the complex curvature of curves that move by rigid motion solves

$$\mathcal{C}\frac{\Psi_n}{1 + |\Psi_n|^2} = e^{i\mu}\Psi_{n+1} + e^{-i\mu}\Psi_{n-1} \tag{4.2}$$

with some real parameters $\mathcal{C}$ and μ, which clearly holds for discrete elastic curves.   □

The additional parameter μ in equation (4.2) is due to the fact that the Ablowitz-Ladik system is the general double Bäcklund transformation and not only the one with parameters ν and $-\bar{\nu}$. This is compensated by the extra torsion μ, and the resulting curve is in the associated family of an elastic curve. These curves are called *elastic rods* [3]. Another consequence is that when taking the smooth limit of the Ablowitz-Ladik system, one needs to choose the parameters carefully to get the NLSE as limit equation.

4.2. Bäcklund transformations for doubly discrete Hashimoto surfaces

Since the doubly discrete Hashimoto surfaces are build from Bäcklund transformations themselves, the Bianchi permutability theorem (Theorem 3.9) ensures that the Bäcklund transformations for discrete curves give rise to Bäcklund transformations for the doubly discrete Hashimoto surfaces too. Thus everything said in Section 3.2 holds in the doubly discrete case too.

Conclusion

We presented an integrable doubly discrete Hashimoto or Heisenberg flow that arises from the Bäcklund transformation of the (single) discrete flow and showed how the equivalence of the discrete and doubly discrete Heisenberg magnet model with the discrete and doubly discrete nonlinear Schrödinger equation can be understood from the geometric point of view. The fact that the stationary solutions of the dNLSE and the ddNLSE coincide stresses the strong similarity of the both and the power of the concept of integrable discrete geometry. Figure 8 shows one more example of a doubly discrete Hashimoto flow.

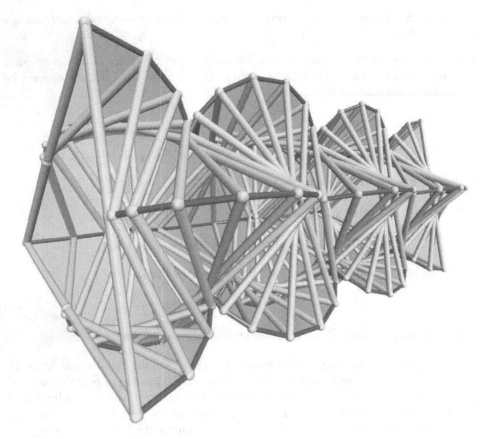

FIGURE 8. The doubly discrete Hashimoto flow on an equilateral triangle with subdivided edges.

References

[1] M.J. Ablowitz and J.F. Ladik, *Nonlinear differential-difference equations and Fourier analysis*, Stud. Appl. Math. **17** (1976), 1011–1018.

[2] _____, *A nonlinear difference scheme and inverse scattering*, Stud. Appl. Math. **55** (1977), 213–229.

[3] A. Bobenko and Y. Suris, *Discrete time Lagrangian mechanics on Lie groups, with an application to the Lagrange top*, Comm. Math. Phys. **204** (1999), 147–188.

[4] A. Doliwa and P.M. Santini, *Geometry of discrete curves and lattices and integrable difference equations*, Discrete integrable geometry and physics (A. Bobenko and R. Seiler, eds.), Oxford University Press, 1999.

[5] L.D. Faddeev and L.A. Takhtajan, *Hamiltonian methods in the theory of solitons*, Springer, 1986.

[6] H. Hashimoto, *A soliton on a vortex filament*, J. Fluid Mech. **51** (1972), 477–485.

[7] T. Hoffmann, *Discrete Amsler surfaces and a discrete Painlevé III equation*, Discrete integrable geometry and physics (A. Bobenko and R. Seiler, eds.), Oxford University Press, 1999, pp. 83–96.

[8] Y. Ishimori, *An integrable classical spin chain*, J. Phys. Soc. Jpn. **51** (1982), no. 11, 3417–3418.

[9] A.G. Izergin and V.E. Korepin, *A lattice model associated with the nonlinear Schrödinger equation*, Dokl. Akad. Nauk SSSR **259** (1981), 76–79, Russian.

[10] Y. Suris, *A discrete-time Garnier system*, Phys. Lett. A **189** (1994), 281–289.

[11] _____, *A note on an integrable discretization of the nonlinear Schrödinger equation*, Inverse Problems **13** (1997), 1121–1136.

Tim Hoffmann
Mathematisches Institut
Universität München
Theresienstr. 39
D-80333 München
Germany
c-mail: hoffmann@mathematik.uni-muenchen.de

Discrete Differential Geometry, A.I. Bobenko, P. Schröder, J.M. Sullivan and G.M. Ziegler, eds.
Oberwolfach Seminars, Vol. 38, 117–133

The Discrete Green's Function

Yuri B. Suris

Abstract. We first discuss discrete holomorphic functions on quad-graphs and their
relation to discrete harmonic functions on planar graphs. Then, the special weights in
the discrete Cauchy–Riemann (and discrete Laplace) equations are considered, coming from quasicrystalline rhombic realizations of quad-graphs. We relate these special
weights to the 3D consistency (integrability) of the discrete Cauchy–Riemann equations, allowing us to extend discrete holomorphic functions to a multidimensional lattice. Discrete exponential functions are introduced and are shown to form a basis in
the space of discrete holomorphic functions growing not faster than exponentially. The
discrete logarithm is constructed and characterized in various ways, including an isomonodromic property. Its real part is nothing but the discrete Green's function.

Keywords. Discrete complex analysis, discrete Cauchy–Riemann equation, discrete
harmonic and holomorphic functions, discrete Laplacian, integrable systems, isomonodromic deformation.

1. Introduction: discrete harmonic and holomorphic functions

Our goal is to give a short presentation of the main ideas and results of the paper [1]. More
specifically, only the linear part of the story will be touched upon here, while there will
be no space for the nonlinear part, dealing with circle patterns, or for the relation between
them.

There is currently much interest in finding discrete counterparts of various structures
of classical (continuous, smooth) mathematics. Here, we are dealing with the discretization of classical complex analysis, more precisely, with discrete analogs of the notions of
harmonic and holomorphic functions.

Recall that a harmonic function $u : \mathbb{R}^2 \simeq \mathbb{C} \to \mathbb{R}$ is characterized by the relation:

$$\Delta u = \frac{\partial^2 u}{\partial x^2} + \frac{\partial^2 u}{\partial y^2} = 0. \tag{1.1}$$

The conjugate harmonic function $v : \mathbb{R}^2 \simeq \mathbb{C} \to \mathbb{R}$ is defined by the Cauchy–Riemann equations:

$$\frac{\partial v}{\partial y} = \frac{\partial u}{\partial x}, \qquad \frac{\partial v}{\partial x} = -\frac{\partial u}{\partial y}. \tag{1.2}$$

Equivalently, $f = u + iv : \mathbb{R}^2 \simeq \mathbb{C} \to \mathbb{C}$ is holomorphic, and satisfies the Cauchy–Riemann equation:

$$\frac{\partial f}{\partial y} = i \frac{\partial f}{\partial x}. \tag{1.3}$$

Thus, the real and the imaginary parts of a holomorphic function are harmonic, and any real-valued harmonic function can be considered as the real part of a holomorphic function.

A standard classical way to discretize these notions, going back to Ferrand [5] and Duffin [3], is the following. The discrete Laplace operator acts on functions $u : \mathbb{Z}^2 \to \mathbb{R}$ according to the formula

$$(\Delta u)_{m,n} = u_{m+1,n} + u_{m-1,n} + u_{m,n+1} + u_{m,n-1} - 4u_{m,n}. \tag{1.4}$$

A function u is called *discrete harmonic* if $\Delta u = 0$. The natural domain of definition of the conjugate discrete harmonic function $v : (\mathbb{Z}^2)^* \to \mathbb{R}$ is the *dual lattice*, as in Figure 1.

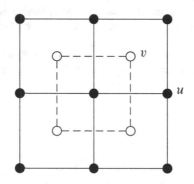

FIGURE 1. The regular square lattice and its dual.

The *discrete Cauchy–Riemann equations* are defined in Figure 2. The corresponding *discrete holomorphic* function $f : \mathbb{Z}^2 \cup (\mathbb{Z}^2)^* \to \mathbb{C}$ is defined on the superposition of the original square lattice $\mathbb{Z}^2$ and the dual one $(\mathbb{Z}^2)^*$, by the formula $f = \begin{cases} u, & \bullet, \\ iv, & \circ, \end{cases}$ which comes to replace the smooth one $f = u + iv$. Remarkably, the discrete Cauchy–Riemann equation for f is one and the same for both pictures above, as shown in Figure 3.

This discretization of the Cauchy–Riemann equations apparently preserves the most number of important structural features, and the corresponding theory has been developed in [5, 3].

A pioneering step in the direction of a further generalization of the notions of discrete harmonic and discrete holomorphic functions was undertaken by Duffin [4], where

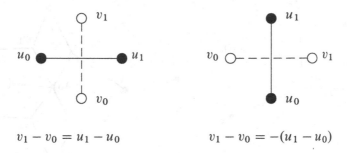

$$v_1 - v_0 = u_1 - u_0 \qquad v_1 - v_0 = -(u_1 - u_0)$$

FIGURE 2. The discrete Cauchy–Riemann equations in terms of u, v.

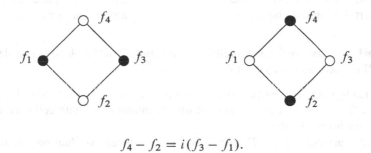

$$f_4 - f_2 = i(f_3 - f_1).$$

FIGURE 3. The discrete Cauchy–Riemann equations in terms of f.

the combinatorics of $\mathbb{Z}^2$ was given up in favor of arbitrary planar graphs with rhombic faces. A far reaching generalization of these ideas was given by Mercat [9], who extended the theory to discrete Riemann surfaces. The corresponding definitions are as follows.

Discrete harmonic functions can be defined for an arbitrary graph. Denote by $V(\mathcal{G})$, $E(\mathcal{G})$ and $\vec{E}(\mathcal{G})$ the sets of vertices, undirected and directed edges of a graph $\mathcal{G}$, respectively. Let there be given a complex-valued function $\nu : E(\mathcal{G}) \to \mathbb{C}$ on the edges (the most interesting case is that of real positive weights $\nu : E(\mathcal{G}) \to \mathbb{R}_+$). Then the *Laplacian* Δ corresponding to the weight function ν is the operator acting on functions $f : V(\mathcal{G}) \to \mathbb{C}$ by

$$(\Delta f)(x_0) = \sum_{x \sim x_0} \nu(x_0, x)(f(x) - f(x_0)). \tag{1.5}$$

Here the summation is extended over the set of all vertices x connected to x_0 by an edge, as in Figure 4.

Definition 1.1. A function $f : V(\mathcal{G}) \to \mathbb{C}$ is called *discrete harmonic* (with respect to the weights ν) if $\Delta f = 0$.

Discrete holomorphic functions live on quad-graphs.

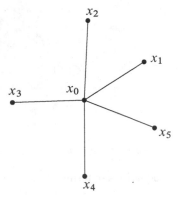

FIGURE 4. The star of the
vertex x_0 in the graph $\mathcal{G}$.

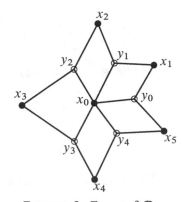

FIGURE 5. Faces of $\mathcal{D}$
around the vertex x_0.

Definition 1.2. A strongly regular cell decomposition $\mathcal{D}$ of the plane $\mathbb{C}$ is called a *quad-graph* if all its faces are quadrilaterals.

So, quad-graphs are not just graphs, but are additionally assumed to be embedded in the plane $\mathbb{C}$. A more general version of this definition deals with cell decompositions of an arbitrary oriented surface.

An arbitrary quad-graph $\mathcal{D}$ embedded in $\mathbb{C}$ produces two dual polytopal (not necessarily quadrilateral) cell decompositions $\mathcal{G}$ and $\mathcal{G}^*$ of $\mathbb{C}$. Indeed, $\mathcal{D}$ is automatically bipartite: its vertices $V(\mathcal{D})$ are decomposed into two complementary halves, $V(\mathcal{D}) = V(\mathcal{G}) \sqcup V(\mathcal{G}^*)$ ("black" and "white" vertices), such that the ends of each edge from $E(\mathcal{D})$ are of different colours. Edges of $\mathcal{G}$ (resp. of $\mathcal{G}^*$) connect "black" (resp. "white") vertices along the diagonals of each face of $\mathcal{D}$. A flower of adjacent quadrilaterals from $F(\mathcal{D})$ around the common "black" vertex $x_0 \in V(\mathcal{G})$ (say) produces a star of this vertex in $\mathcal{G}$; see Figure 5.

Let there be given a function $\nu : E(\mathcal{G}) \sqcup E(\mathcal{G}^*) \to \mathbb{C}$ on diagonals of the quadrilaterals from $F(\mathcal{D})$, satisfying the relation

$$\nu(e^*) = 1/\nu(e) \tag{1.6}$$

on the dual diagonals of each quadrilateral. Again, the most interesting case corresponds to real positive weights $\nu : E(\mathcal{G}) \sqcup E(\mathcal{G}^*) \to \mathbb{R}_+$.

Definition 1.3. A function $f : V(\mathcal{D}) \to \mathbb{C}$ is called *discrete holomorphic* (with respect to the weights ν) if for any positively oriented quadrilateral $(x_0, y_0, x_1, y_1) \in F(\mathcal{D})$ there holds:

$$\frac{f(y_1) - f(y_0)}{f(x_1) - f(x_0)} = i\nu(x_0, x_1) = -\frac{1}{i\nu(y_0, y_1)}. \tag{1.7}$$

These equations are called the *discrete Cauchy–Riemann equations*.

To establish a relation between discrete harmonic functions on a graph $\mathcal{G}$ and discrete holomorphic functions, one should start with a *planar* graph $\mathcal{G}$ with an additional structure of a cell decomposition of $\mathbb{C}$ (not necessarily quadrilateral). In this case, the above construction can be reversed. To any such $\mathcal{G}$ there corresponds canonically a *dual cell decomposition* $\mathcal{G}^*$ (it is only defined up to isotopy, but can be fixed uniquely with the help of the Voronoi/Delaunay construction). If one assigns a direction to an edge $\mathfrak{e} \in E(\mathcal{G})$, then it will be assumed that the dual edge $\mathfrak{e}^* \in E(\mathcal{G}^*)$ is also .directed, in a way consistent with the orientation of the underlying surface, namely so that the pair $(\mathfrak{e}, \mathfrak{e}^*)$ is oriented directly at its crossing point. This orientation convention implies that $\mathfrak{e}^{**} = -\mathfrak{e}$. The *double* $\mathcal{D}$ is a quad-graph, constructed from $\mathcal{G}$ and its dual $\mathcal{G}^*$ as follows. The set of vertices of the double $\mathcal{D}$ is $V(\mathcal{D}) = V(\mathcal{G}) \sqcup V(\mathcal{G}^*)$. Each pair of dual edges, say, $\mathfrak{e} = (x_0, x_1) \in E(\mathcal{G})$ and $\mathfrak{e}^* = (y_0, y_1) \in E(\mathcal{G}^*)$, defines a quadrilateral (x_0, y_0, x_1, y_1). These quadrilaterals constitute the faces of the cell decomposition (quad-graph) $\mathcal{D}$. The edges of $\mathcal{D}$ belong neither to $E(\mathcal{G})$ nor to $E(\mathcal{G}^*)$.

Theorem 1.4. *If a function $f : V(\mathcal{D}) \to \mathbb{C}$ is discrete holomorphic, then its restrictions to $V(\mathcal{G})$ and to $V(\mathcal{G}^*)$ are discrete harmonic.*

Conversely, any discrete harmonic function $f : V(\mathcal{G}) \to \mathbb{C}$ admits a family of discrete holomorphic extensions to $V(\mathcal{D})$, differing by an additive constant on $V(\mathcal{G}^*)$. Such an extension is uniquely defined by its value at one arbitrary vertex $y \in V(\mathcal{G}^*)$.

One special class of quad-graphs and one special class of weight functions ν play a prominent role in the theory of discrete holomorphic functions, namely the quad-graphs that are embedded so that all faces are *rhombi*, and the weights that can be interpreted as *tangents of half-angles* of these rhombi. The major part of the theory in [4, 9] was developed exactly for this case. In [7], an explicit formula (an integral representation) was given for the Green's function of a Laplacian operators with such weights. It is our purpose here to give an explanation of the special role played by rhombic quad-graphs and the weights originating from them.

2. Rhombically embedded quad-graphs

The paper [8] studies *rhombic embeddings* of a quad-graph $\mathcal{D}$ in $\mathbb{C}$, which map each face of $\mathcal{D}$ to a rhombus. A combinatorial criterion for the existence of a rhombic embedding of a given quad-graph $\mathcal{D}$ found in [8] is as follows.

Definition 2.1. A *strip* S in $\mathcal{D}$ is a sequence of quadrilateral faces $q_j \in F(\mathcal{D})$ such that any pair q_{j-1}, q_j is adjacent along the edge $\mathfrak{a}_j = q_{j-1} \cap q_j$, and $\mathfrak{a}_j, \mathfrak{a}_{j+1}$ are opposite edges of q_j. The edges $\mathfrak{a}_j$ are called the *traverse edges* of the strip S.

Theorem 2.2 ([8]). *A planar quad-graph $\mathcal{D}$ admits a rhombic embedding in $\mathbb{C}$ if and only if the following two conditions are satisfied:*

- *No strip crosses itself or is periodic.*
- *Two distinct strips cross each other at most once.*

Given a rhombic embedding $p : V(\mathcal{D}) \to \mathbb{C}$, its directed edges are given by

$$\alpha(x, y) = p(y) - p(x), \quad \forall (x, y) \in \vec{E}(\mathcal{D}). \tag{2.1}$$

Thus, the function $\alpha : \vec{E}(\mathcal{D}) \to \mathbb{S}^1$ is such that $\alpha(-\mathfrak{a}) = -\alpha(\mathfrak{a})$ for any $\mathfrak{a} \in \vec{E}(\mathcal{D})$, and the values of α on two opposite and equally directed edges of any quadrilateral from $F(\mathcal{D})$ are equal to one another. Any function with these properties will be called a *labeling of directed edges* of $\mathcal{D}$.

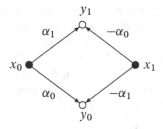

FIGURE 6. Labeling of directed edges.

Given any labeling $\alpha : \vec{E}(\mathcal{D}) \to \mathbb{S}^1$, formula (2.1) correctly defines a function $p : V(\mathcal{D}) \to \mathbb{C}$ and assures that the p-image of any quadrilateral face of $\mathcal{D}$ is a rhombus. The rhombus angles are naturally assigned to the edges of $\mathcal{G}$ and $\mathcal{G}^*$; see Figure 7. In the

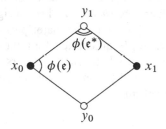

FIGURE 7. A rhombic embedding of a quadrilateral $(x_0, y_0, x_1, z_1) \in F(\mathcal{D})$, with $\mathfrak{e} = (x_0, x_1)$, $\mathfrak{e}^* = (y_0, y_1)$.

notation of Figures 6 and 7, the rhombus angles for a given labeling are given by

$$\tan \frac{\phi(\mathfrak{e})}{2} = i \frac{\alpha_0 - \alpha_1}{\alpha_0 + \alpha_1}, \qquad \tan \frac{\phi(\mathfrak{e}^*)}{2} = i \frac{\alpha_0 + \alpha_1}{\alpha_1 - \alpha_0}. \tag{2.2}$$

According to [8], in order that a function $\phi : E(\mathcal{G}) \sqcup E(\mathcal{G}^*) \to (0, \pi)$ can be interpreted as angles of a rhombus embedding, it is necessary and sufficient that the following conditions be satisfied: first,

$$\phi(\mathfrak{e}^*) = \pi - \phi(\mathfrak{e}), \qquad \forall \mathfrak{e} \in E(\mathcal{G}). \tag{2.3}$$

and second, for all $x_0 \in V(\mathcal{G})$ and for all $y_0 \in V(\mathcal{G}^*)$,

$$\sum_{\mathfrak{e} \in \mathrm{star}(x_0; \mathcal{G})} \phi(\mathfrak{e}) = 2\pi, \qquad \sum_{\mathfrak{e}^* \in \mathrm{star}(y_0; \mathcal{G}^*)} \phi(\mathfrak{e}^*) = 2\pi. \tag{2.4}$$

3. 3D consistent Cauchy–Riemann equations

We now consider the special properties of the discrete Cauchy–Riemann equations (1.7) with the special choice of weights

$$\nu(e) = \tan\frac{\phi(e)}{2},$$

defined by a rhombic embedding $p : V(\mathcal{D}) \to \mathbb{C}$ in accordance to formulas (2.2) and (2.1). It turns out that the root of these special properties lies in the *3D consistency* of the equations.

Recall the notion of the 3D consistency [2] for general equations on quad-graphs

$$\Phi(f(x_0), f(y_0), f(x_1), f(y_1)) = 0, \tag{3.1}$$

relating fields f sitting on the four vertices of an arbitrary (oriented) face (x_0, y_0, x_1, y_1) of a quad-graph $\mathcal{D}$. Here the function Φ may depend on some parameters, and it is supposed that equation (3.1) is uniquely solvable for any of the fields in terms of the other three. The discrete Cauchy–Riemann equations are just a specific linear instance of (3.1), the role of parameters being played by the weights ν. We extend the planar quad-graph $\mathcal{D}$ into the third dimension. Formally speaking, we consider a second copy $\widehat{\mathcal{D}}$ of $\mathcal{D}$ and add edges connecting each vertex $x \in V(\mathcal{D})$ with its copy $\widehat{x} \in V(\widehat{\mathcal{D}})$. On this way we obtain a "three-dimensional quad-graph" $\mathbf{D}$, whose elementary building blocks are cubes $(x_0, y_0, x_1, y_1, \widehat{x}_0, \widehat{y}_0, \widehat{x}_1, \widehat{y}_1)$, as shown in Figure 8.

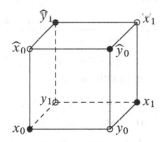

FIGURE 8. Elementary cube of $\mathbf{D}$.

Definition 3.1. Equation (3.1) is called *3D consistent* if it can be imposed on all faces of any elementary cube of $\mathbf{D}$, in such a manner that opposite faces carry one and the same equation (i.e., the same parameters).

This should be understood as follows. Consider an elementary cube of $\mathbf{D}$, as in Figure 8. Suppose that the values of the function f are given at the vertex x_0 and at its three neighbors y_0, y_1, and $\widehat{x}_0$. Then equation (3.1) uniquely determines the values of f at x_1, $\widehat{y}_0$, and $\widehat{y}_1$. After that, (3.1) delivers three *a priori* different values for the value of the field f at the vertex $\widehat{x}_1$, coming from the faces $(y_0, x_1, \widehat{x}_1, \widehat{y}_0)$, $(x_1, y_1, \widehat{y}_1, \widehat{x}_1)$, and $(\widehat{x}_0, \widehat{y}_0, \widehat{x}_1, \widehat{y}_1)$, respectively. The 3D consistency means that these three values for $f(\widehat{x}_1)$ actually coincide, independent of the choice of initial conditions.

As discussed in detail in [2], the 3D consistency of the given system (3.1) allows one to construct Bäcklund transformations and to find in an algorithmic way a zero-curvature representation for it, which are traditionally considered as main attributes of integrability.

In applying the notion of the 3D consistency to the discrete Cauchy–Riemann equation (1.7), we assume that the opposite diagonals of the quad-graph $\mathcal{D}$ (on the ground floor) and of its copy $\widehat{\mathcal{D}}$ (on the first floor) carry the given weights ν, satisfying (1.6), and ask for weights on the diagonals of the vertical faces of $\mathbf{D}$ (equal on the opposite diagonals of any elementary cube and satisfying (1.6)), such that the discrete Cauchy–Riemann equations are fulfilled on all faces.

Theorem 3.2. *The function* $\nu : E(\mathcal{G}) \sqcup E(\mathcal{G}^*) \to \mathbb{C}$ *can be extended to the diagonals of the vertical faces of* $\mathbf{D}$, *giving a 3D consistent system of discrete Cauchy–Riemann equations, if and only if the following condition is satisfied for all* $x_0 \in V(\mathcal{G})$ *and for all* $y_0 \in V(\mathcal{G}^*)$:

$$\prod_{e \in \text{star}(x_0;\mathcal{G})} \frac{1 + i\nu(e)}{1 - i\nu(e)} = 1, \qquad \prod_{e^* \in \text{star}(y_0;\mathcal{G}^*)} \frac{1 + i\nu(e^*)}{1 - i\nu(e^*)} = 1. \tag{3.2}$$

It is not difficult to see that the integrability condition (3.2) for the function $\nu : E(\mathcal{G}) \sqcup E(\mathcal{G}^*) \to \mathbb{C}$ is equivalent to the existence of a labeling $\alpha : \vec{E}(\mathcal{D}) \to \mathbb{C}$ of directed edges of $\mathcal{D}$ such that, in notation of Figure 6,

$$\nu(y_0, y_1) = \frac{1}{\nu(x_0, x_1)} = i \frac{\alpha_1 + \alpha_0}{\alpha_1 - \alpha_0}. \tag{3.3}$$

Let $p : V(\mathcal{D}) \to \mathbb{C}$ be a parallelogram realization of $\mathcal{D}$ defined by the labeling α. Then the 3D consistent Cauchy–Riemann equations read:

$$\frac{f(y_1) - f(y_0)}{f(x_1) - f(x_0)} = \frac{\alpha_1 - \alpha_0}{\alpha_1 + \alpha_0} = \frac{p(y_1) - p(y_0)}{p(x_1) - p(x_0)}. \tag{3.4}$$

In other words, if f is a discrete holomorphic function, then the quotient of diagonals of the f-image of any quadrilateral $(x_0, y_0, x_1, y_1) \in F(\mathcal{D})$ is equal to the quotient of diagonals of the corresponding parallelogram.

Under this condition, the values of the weights ν on the diagonals of the vertical faces, leading to a 3D consistent collection of equations on $\mathbf{D}$, read:

$$\nu(y, \widehat{x}) = \frac{1}{\nu(x, \widehat{y})} = i \frac{\lambda + \alpha}{\lambda - \alpha}, \tag{3.5}$$

where $\alpha = \alpha(x, y)$, and $\lambda \in \mathbb{C}$ is an arbitrary number having the interpretation of the label carried by all vertical edges of $\mathbf{D}$: $\lambda = \alpha(x, \widehat{x}) = \alpha(y, \widehat{y})$.

The most interesting case is when ν takes values in $\mathbb{R}_+$. In this case we will use the notation

$$\nu(e) = \tan \frac{\phi(e)}{2}, \quad \phi(e) \in (0, \pi). \tag{3.6}$$

The condition $\nu(e^*) = 1/\nu(e)$ is translated in this case into (2.3). The integrability condition (3.2) takes in this case the form

$$\prod_{e \in \text{star}(x_0; \mathcal{G})} \exp(i\phi(e)) = 1, \qquad \prod_{e^* \in \text{star}(y_0; \mathcal{G}^*)} \exp(i\phi(e^*)) = 1. \qquad (3.7)$$

The latter condition is a generalization of (2.4), and is equivalent to saying that the system of angles $\phi : E(\mathcal{G}) \sqcup E(\mathcal{G}^*) \to (0, \pi)$ comes from a realization of the quad-graph $\mathcal{D}$ by a *rhombic ramified embedding* in $\mathbb{C}$. The flowers of such an embedding can wind around a vertex more than once. The labels α take values in $\mathbb{S}^1$, and have a geometric interpretation as the edges of a rhombic realization of $\mathcal{D}$.

4. Extending discrete holomorphic functions to a multidimensional lattice

To best exploit the analytic possibilities provided by the 3D consistency of the discrete Cauchy–Riemann equations, we now restrict our considerations to quasicrystalline quad-graphs $\mathcal{D}$.

Definition 4.1. A rhombic embedding $p : V(\mathcal{D}) \to \mathbb{C}$ of a quad-graph $\mathcal{D}$ is called *quasicrystalline* if the set of values of the corresponding labeling $\alpha : \vec{E}(\mathcal{D}) \to \mathbb{S}^1$, defined by (2.1), is finite, say, $A = \{\pm\alpha_1, \ldots, \pm\alpha_d\}$.

It will be of a central importance for us that any quasicrystalline rhombic embedding p of a quad-graph $\mathcal{D}$ can be realized as a certain two-dimensional subcomplex (combinatorial surface) $\Omega_{\mathcal{D}}$ of a multidimensional regular square lattice $\mathbb{Z}^d$. The vertices of $\Omega_{\mathcal{D}}$ are given by the map $P : V(\mathcal{D}) \to \mathbb{Z}^d$ constructed as follows. Fix some $x_0 \in V(\mathcal{D})$, and set $P(x_0) = \mathbf{0}$. For all other vertices of $\mathcal{D}$, their images in $\mathbb{Z}^d$ are defined recurrently by the property:

- For any two neighbors $x, y \in V(\mathcal{D})$, if $p(y) - p(x) = \pm\alpha_i \in A$, then $P(y) - P(x) = \pm\mathbf{e}_i$,

where $\mathbf{e}_i$ is the ith coordinate vector of $\mathbb{Z}^d$. The edges and faces of $\Omega_{\mathcal{D}}$ correspond to edges and faces of $\mathcal{D}$, so that the combinatorics of $\Omega_{\mathcal{D}}$ is that of $\mathcal{D}$.

Extend the labeling $\alpha : \vec{E}(\mathcal{D}) \to \mathbb{C}$ to all edges of $\mathbb{Z}^d$, assuming that all edges parallel to (and directed as) $\mathbf{e}_k$ carry the label α_k. Now, the 3D consistency of the discrete Cauchy–Riemann equations allows us to impose them not only on $\Omega_{\mathcal{D}}$, but on the whole of $\mathbb{Z}^d$.

Definition 4.2. A function $f : \mathbb{Z}^d \to \mathbb{C}$ is called *discrete holomorphic* if it satisfies, on each elementary square of $\mathbb{Z}^d$, the equation

$$\frac{f(\mathbf{n} + \mathbf{e}_j + \mathbf{e}_k) - f(\mathbf{n})}{f(\mathbf{n} + \mathbf{e}_j) - f(\mathbf{n} + \mathbf{e}_k)} = \frac{\alpha_j + \alpha_k}{\alpha_j - \alpha_k}. \qquad (4.1)$$

Obviously, for any discrete holomorphic function $f : \mathbb{Z}^d \to \mathbb{C}$, its restriction to $V(\Omega_{\mathcal{D}}) \sim V(\mathcal{D})$ is a discrete holomorphic function on $\mathcal{D}$.

We want to *reverse* the procedure, i.e., to extend an arbitrary discrete holomorphic function on $\mathcal{D}$ to $\mathbb{Z}^d$, keeping the property of being discrete holomorphic. An elementary step of such an extension consists of finding f at the eighth vertex of an elementary 3D cube from the known values at seven vertices; see Figure 9. This can be alternatively viewed as a flip (elementary transformation) on the set of rhombically embedded quad-graphs $\mathcal{D}$, or on the set of the corresponding surfaces $\Omega_{\mathcal{D}}$ in $\mathbb{Z}^d$. The 3D consistency assures that any quad-graph $\mathcal{D}$ (or any corresponding surface $\Omega_{\mathcal{D}}$) obtainable from the original one by such flips, carries a unique solution of (4.1), which is an extension of the original one.

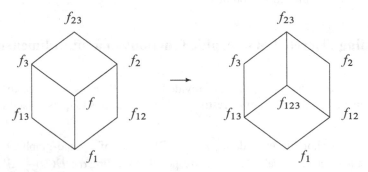

FIGURE 9. Elementary flip.

Definition 4.3. For a given set $V \subset \mathbb{Z}^d$, its *hull* $\mathcal{H}(V)$ is the minimal set $\mathcal{H} \subset \mathbb{Z}^d$ containing V and satisfying the condition: if three vertices of an elementary square belong to $\mathcal{H}$, then so does the fourth vertex.

It is not difficult to show by induction that for an arbitrary connected subcomplex of $\mathbb{Z}^d$ with the set of vertices V, its hull is a *brick*

$$\Pi_{\mathbf{a},\mathbf{b}} = \left\{ \mathbf{n} = (n_1, \ldots, n_d) \in \mathbb{Z}^d : a_k \leq n_k \leq b_k, \ k = 1, \ldots, d \right\}, \qquad (4.2)$$

where

$$a_k = a_k(V) = \min_{\mathbf{n} \in V} n_k, \qquad b_k = b_k(V) = \max_{\mathbf{n} \in V} n_k, \qquad k = 1, \ldots, d, \qquad (4.3)$$

and in case that the n_k are unbounded from below or from above on V, we set $a_k(V) = -\infty$, resp. $b_k(V) = \infty$.

Combinatorially, all points of the hull $\mathcal{H}(V(\Omega_{\mathcal{D}}))$ can be reached by the flips of Figure 9. However, there might be obstructions for extending solutions of the discrete Cauchy–Riemann equations from a combinatorial surface (two-dimensional subcomplex of $\mathbb{Z}^d$) to its hull. Indeed, the surface Ω shown in Figure 10 supports discrete holomorphic functions which cannot be extended to the whole of $\mathcal{H}(V(\Omega))$: the recursive extension process will lead to contradictions. The reason for this is the non-monotonicity of Ω: it contains pairs of points which cannot be connected by a path in Ω with all edges lying in one octant. However, such surfaces do not come from rhombic embeddings, and in the case of $\Omega_{\mathcal{D}}$ there will be no contradictions.

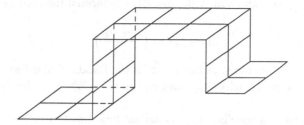

FIGURE 10. A non-monotone surface in $\mathbb{Z}^3$.

Theorem 4.4. *Let the combinatorial surface $\Omega_{\mathcal{D}}$ in $\mathbb{Z}^d$ come from a rhombic embedding of a quad-graph $\mathcal{D}$, and let its hull be $\mathcal{H}(V(\Omega_{\mathcal{D}})) = \Pi_{\mathbf{a},\mathbf{b}}$. An arbitrary discrete holomorphic function on $\Omega_{\mathcal{D}}$ can be uniquely extended to a discrete holomorphic function on $\Pi_{\mathbf{a},\mathbf{b}}$.*

5. Discrete exponential functions

A discrete exponential function on quad-graphs $\mathcal{D}$ was defined and studied in [9, 7]. Its definition, for a given rhombic embedding $p : V(\mathcal{D}) \to \mathbb{C}$, is as follows: fix a point $x_0 \in V(\mathcal{D})$. For any other point $x \in V(\mathcal{D})$, choose some path $\{\mathfrak{a}_j\}_{j=1}^n \subset \vec{E}(\mathcal{D})$ connecting x_0 to x, so that $\mathfrak{a}_j = (x_{j-1}, x_j)$ and $x_n = x$. Let the slope of the jth edge be $\alpha_j = p(x_j) - p(x_{j-1}) \in \mathbb{S}^1$. Then

$$e(x; z) = \prod_{j=1}^{n} \frac{z + \alpha_j}{z - \alpha_j}.$$

Clearly, this definition depends on the choice of the point $x_0 \in V(\mathcal{D})$, but not on the path connecting x_0 to x. A question posed by Kenyon [7] is whether discrete exponential functions are dense in the space of discrete holomorphic functions on $\mathcal{D}$.

An extension of the discrete exponential function from $\Omega_{\mathcal{D}}$ to the whole of $\mathbb{Z}^d$ is the *discrete exponential function*, given by the following simple formula:

$$e(\mathbf{n}; z) = \prod_{k=1}^{d} \left(\frac{z + \alpha_k}{z - \alpha_k} \right)^{n_k}. \tag{5.1}$$

For $d = 2$, this function was considered in [5, 3]. The discrete Cauchy–Riemann equations for the discrete exponential function are easily checked: they are equivalent to a simple identity:

$$\left(\frac{z + \alpha_j}{z - \alpha_j} \cdot \frac{z + \alpha_k}{z - \alpha_k} - 1 \right) \Big/ \left(\frac{z + \alpha_j}{z - \alpha_j} - \frac{z + \alpha_k}{z - \alpha_k} \right) = \frac{\alpha_j + \alpha_k}{\alpha_j - \alpha_k}.$$

At a given $\mathbf{n} \in \mathbb{Z}^d$, the discrete exponential function is rational with respect to the parameter z, with poles at the points $\epsilon_1 \alpha_1, \ldots, \epsilon_d \alpha_d$, where $\epsilon_k = \operatorname{sign} n_k$.

Equivalently, one can identify the discrete exponential function by its initial values on the axes:

$$e(n\mathbf{e}_k; z) = \left(\frac{z + \alpha_k}{z - \alpha_k}\right)^n. \tag{5.2}$$

Still another characterization says that $e(\cdot; z)$ is the Bäcklund transformation of the zero solution of the discrete Cauchy–Riemann equations on $\mathbb{Z}^d$, with the "vertical" parameter z.

We now show that the discrete exponential functions form a basis in some natural class of functions (growing not faster than exponentially), thus answering in affirmative the above-mentioned question by Kenyon.

Theorem 5.1. *Let f be a discrete holomorphic function on $V(\mathcal{D}) \sim V(\Omega_\mathcal{D})$, satisfying*

$$|f(\mathbf{n})| \le \exp(C(|n_1| + \cdots + |n_d|)), \qquad \forall \mathbf{n} \in V(\Omega_\mathcal{D}), \tag{5.3}$$

for some $C \in \mathbb{R}$. Extend it to a discrete holomorphic function on $\mathcal{H}(V(\Omega_\mathcal{D}))$. Then inequality (5.3) holds for all $\mathbf{n} \in \mathcal{H}(V(\Omega_\mathcal{D}))$, possibly with some larger constant C. There exists a function g defined on the disjoint union of small neighborhoods around the points $\pm\alpha_k \in \mathbb{C}$ and holomorphic on each one of these neighborhoods, such that

$$f(\mathbf{n}) - f(\mathbf{0}) = \frac{1}{2\pi i} \int_\Gamma g(\lambda)e(\mathbf{n}; \lambda)d\lambda, \qquad \forall \mathbf{n} \in \mathcal{H}(V(\Omega_\mathcal{D})), \tag{5.4}$$

where Γ is a collection of $2d$ small loops, each one running counter-clockwise around one of the points $\pm\alpha_k$.

Proof. The proof is constructive and consists of three steps.

(i) Extend f from $V(\Omega_\mathcal{D})$ to $\mathcal{H}(V(\Omega_\mathcal{D}))$; inequality (5.3) propagates in the extension process, if the constant C is chosen large enough.

(ii) Introduce the restrictions $f_n^{(k)}$ of $f : \mathcal{H}(V(\Omega_\mathcal{D})) \to \mathbb{C}$ to the coordinate axes:

$$f_n^{(k)} = f(n\mathbf{e}_k), \qquad a_k(\Omega_\mathcal{D}) \le n \le b_k(\Omega_\mathcal{D}).$$

(iii) Set $g(\lambda) = \sum_{k=1}^d (g_k(\lambda) + g_{-k}(\lambda))$, where the functions $g_{\pm k}(\lambda)$ vanish everywhere except in small neighborhoods of the points $\pm\alpha_k$, respectively, and are given there by convergent series

$$g_k(\lambda) = \frac{1}{2\lambda}\left(f_1^{(k)} - f(0) + \sum_{n=1}^\infty \left(\frac{\lambda - \alpha_k}{\lambda + \alpha_k}\right)^n \left(f_{n+1}^{(k)} - f_{n-1}^{(k)}\right)\right), \tag{5.5}$$

and a similar formula for $g_{-k}(\lambda)$. Formula (5.4) is then easily verified by computing the residues $\lambda = \pm\alpha_k$. □

It is important to observe that the data $f_n^{(k)}$, necessary for the construction of $g(\lambda)$, are *not among* the values of f on $V(\mathcal{D}) \sim V(\Omega_\mathcal{D})$ known initially, but are encoded in the extension process.

6. The discrete logarithm

We first give a construction of the discrete logarithm on $\mathcal{D}$ which is equivalent to Kenyon's one [7]. This function is defined, after fixing some point $x_0 \in V(\mathcal{D})$, by the formula

$$f(x) = \frac{1}{2\pi i} \int_\Gamma \frac{\log(\lambda)}{2\lambda} e(x; \lambda) d\lambda, \qquad \forall x \in V(\mathcal{D}). \tag{6.1}$$

Here the integration path Γ is the same as in Theorem 5.1, and fixing x_0 is necessary for the definition of the discrete exponential function on $\mathcal{D}$. To make (6.1) a valid definition, one has to specify a branch of $\log(\lambda)$ in a neighborhood of each point $\pm\alpha_k$. This choice depends on x, and is done as follows.

Assume, without loss of generality, that the circular order of the points $\pm\alpha_k$ on the positively oriented unit circle $\mathbb{S}^1$ is the following: $\alpha_1, \dots, \alpha_d, -\alpha_1, \dots, -\alpha_d$. We set $\alpha_{k+d} = -\alpha_k$ for $k = 1, \dots, d$, and then define α_r for all $r \in \mathbb{Z}$ by $2d$-periodicity. For each $r \in \mathbb{Z}$, assign to $\alpha_r = \exp(i\theta_r) \in \mathbb{S}^1$ a certain value of the argument $\theta_r \in \mathbb{R}$: choose a value θ_1 of the argument of α_1 arbitrarily, and then extend it according to the rule

$$\theta_{r+1} - \theta_r \in (0, \pi), \qquad \forall m \in \mathbb{Z}.$$

Clearly, there holds $\theta_{r+d} = \theta_r + \pi$, and therefore also $\theta_{r+2d} = \theta_r + 2\pi$. It will be convenient to consider the points α_r, supplied with the arguments θ_r, as belonging to the Riemann surface Λ of the logarithm (a branched covering of the complex λ-plane).

For each $m \in \mathbb{Z}$, define the "sector" U_m on the embedding plane $\mathbb{C}$ of the quad-graph $\mathcal{D}$ as the set of all points of $V(\mathcal{D})$ which can be reached from x_0 along paths with all edges from $\{\alpha_m, \dots, \alpha_{m+d-1}\}$. Two sectors U_{m_1} and U_{m_2} have a non-empty intersection if and only if $|m_1 - m_2| < d$. The union $U = \bigcup_{m=-\infty}^{\infty} U_m$ is a branched covering of the quad-graph $\mathcal{D}$, and serves as the definition domain of the discrete logarithm.

The definition (6.1) of the latter should be read as follows: for $x \in U_m$, the poles of $e(x; \lambda)$ are exactly the points $\alpha_m, \dots, \alpha_{m+d-1} \in \Lambda$. The integration path Γ consists of d small loops on Λ around these points, and $i \arg(\lambda) = \Im \log(\lambda)$ takes values in a small open neighborhood of the interval

$$[i\theta_m, i\theta_{m+d-1}] \tag{6.2}$$

of length less than π. If m increases by $2d$, the interval (6.2) is shifted by $2\pi i$. As a consequence, the function f is discrete holomorphic, and its restriction to the "black" points $V(\mathcal{G})$ is discrete harmonic everywhere on U except at the point x_0:

$$\Delta f(x) = \delta_{x_0 x}. \tag{6.3}$$

Thus, the functions g_k in the integral representation (5.4) of an arbitrary discrete holomorphic function, defined originally in disjoint neighborhoods of the points α_r, are, in the case of the discrete logarithm, actually restrictions of a single analytic function $\log(\lambda)/(2\lambda)$ to these neighborhoods. This allows one to deform the integration path Γ into a connected contour lying on a single leaf of the Riemann surface of the logarithm, and then to use standard methods of complex analysis in order to obtain asymptotic expressions for the discrete logarithmic function. In particular, one can show [7] that at the "black" points $V(\mathcal{G})$ there holds:

$$f(x) \sim \log|x - x_0|, \qquad x \to \infty. \tag{6.4}$$

Properties (6.3) and (6.4) can be formulated as follows:

Theorem 6.1. *The discrete logarithm on $\mathcal{D}$, restricted to the set of vertices $V(\mathcal{G})$ of the "black" graph $\mathcal{G}$, coincides with the* discrete Green's function *on $\mathcal{G}$.*

Now we extend the discrete logarithm to $\mathbb{Z}^d$. Introduce, in addition to the unit vectors $\mathbf{e}_k \in \mathbb{Z}^d$ (corresponding to $\alpha_k \in \mathbb{S}^1$), their opposites $\mathbf{e}_{k+d} = -\mathbf{e}_k$, $k \in [1, d]$ (corresponding to $\alpha_{k+d} = -\alpha_k$), and then define $\mathbf{e}_r$ for all $r \in \mathbb{Z}$ by $2d$-periodicity. Then

$$S_m = \bigoplus_{r=m}^{m+d-1} \mathbb{Z}\mathbf{e}_r \subset \mathbb{Z}^d \tag{6.5}$$

is a d-dimensional octant containing exactly the part of $\Omega_\mathcal{D}$ which is the P-image of the sector $U_m \subset \mathcal{D}$. Clearly, only $2d$ different octants appear among S_m (out of 2^d possible d-dimensional octants). Define $\widetilde{S}_m$ as the octant S_m equipped with the interval (6.2) of values for $\Im \log(\alpha_r)$. By definition, $\widetilde{S}_{m_1}$ and $\widetilde{S}_{m_2}$ intersect if the underlying octants S_{m_1} and S_{m_2} have a non-empty intersection spanned by the common coordinate semi-axes $\mathbb{Z}\mathbf{e}_r$, and $\Im \log(\alpha_r)$ for these common semi-axes match. It is easy to see that $\widetilde{S}_{m_1}$ and $\widetilde{S}_{m_2}$ intersect if and only if $|m_1 - m_2| < d$. The union $\widetilde{S} = \bigcup_{m=-\infty}^{\infty} \widetilde{S}_m$ is a branched covering of the set $\bigcup_{m=1}^{2d} S_m \subset \mathbb{Z}^d$.

Definition 6.2. *The* discrete logarithm *on $\widetilde{S}$ is given by the formula*

$$f(\mathbf{n}) = \frac{1}{2\pi i} \int_\Gamma \frac{\log(\lambda)}{2\lambda} e(\mathbf{n}; \lambda) d\lambda, \qquad \forall \mathbf{n} \in \widetilde{S}, \tag{6.6}$$

where for $\mathbf{n} \in \widetilde{S}_m$ the integration path Γ consists of d loops around $\alpha_m, \ldots, \alpha_{m+d-1}$ on Λ, and $\Im \log(\lambda)$ on Γ is chosen in a small open neighborhood of the interval (6.2).

The discrete logarithm on $\mathcal{D}$ can be described as the restriction of the discrete logarithm on $\widetilde{S}$ to a branched covering of $\Omega_\mathcal{D} \sim \mathcal{D}$. This holds for an *arbitrary* quasicrystalline quad-graph whose set of edge slopes coincides with $A = \{\pm\alpha_1, \ldots, \pm\alpha_d\}$.

Now we are in a position to give an alternative definition of the discrete logarithm. Clearly, it is completely characterized by its values $f(n\mathbf{e}_r)$, $r \in [m, m + d - 1]$, on the coordinate semi-axes of an arbitrary octant $\widetilde{S}_m$. Let us stress once more that the points $n\mathbf{e}_r$ do not lie, in general, on the original quad-surface $\Omega_\mathcal{D}$.

Theorem 6.3. *The values $f_n^{(r)} = f(n\mathbf{e}_r)$, $r \in [m, m + d - 1]$, of the discrete logarithm on $\widetilde{S}_m \subset \widetilde{S}$ are given by:*

$$f_n^{(r)} = \begin{cases} 2\left(1 + \frac{1}{3} + \cdots + \frac{1}{n-1}\right), & n \text{ even}, \\ \log(\alpha_r), & n \text{ odd}. \end{cases} \tag{6.7}$$

Here the values $\log(\alpha_r)$ are chosen in the interval (6.2).

Observe that values (6.7) at even (resp. odd) points imitate the behavior of the real (resp. imaginary) part of the function $\log(\lambda)$ along the semi-lines $\arg(\lambda) = \arg(\alpha_r)$. This can be easily extended to the whole of $\widetilde{S}$. Restricted to black points $\mathbf{n} \in \widetilde{S}$ (those with $n_1 + \cdots + n_d$ even), the discrete logarithm models the real part of the logarithm. In particular, it is real-valued and does not branch: its values on $\widetilde{S}_m$ depend on m (mod $2d$) only. In other words, it is a well-defined function on S_m. On the contrary, the discrete logarithm restricted to white points $\mathbf{n} \in \widetilde{S}$ (those with $n_1 + \cdots + n_d$ odd) takes purely imaginary values, and increases by $2\pi i$, as m increases by $2d$. Hence, this restricted function models the imaginary part of the logarithm.

7. Isomonodromic property of the discrete logarithm

Thus, the discrete logarithm can be characterized by the values (6.7) on the coordinate axes. This can be interpreted as the recurrent relation for $f_n^{(r)}$:

Theorem 7.1. *For the discrete logarithm, the values $f_n^{(r)} = f(n\mathbf{e}_r)$, $r \in [m, m+d-1]$, on the coordinate semi-axes of $\widetilde{S}_m \subset \widetilde{S}$ solve the difference equation*

$$n(f_{n+1} - f_{n-1}) = 1 - (-1)^n, \tag{7.1}$$

with the initial conditions

$$f_0^{(r)} = f(\mathbf{0}) = 0, \qquad f_1^{(r)} = f(\mathbf{e}_r) = \log(\alpha_r). \tag{7.2}$$

At any point $\mathbf{n} \in \overset{\circ}{S}$, the following constraint holds:

$$\sum_{l=1}^{d} n_l \left(f(\mathbf{n} + \mathbf{e}_l) - f(\mathbf{n} - \mathbf{e}_l) \right) = 1 - (-1)^{n_1 + \cdots + n_d}. \tag{7.3}$$

It turns out that the recurrence relations (7.1) and the constraint (7.3) following from them are characteristic for an important class of solutions of the discrete Cauchy–Riemann equations, namely for the isomonodromic ones. Recall the definition of this class, which involves an additional structural attribute of integrable systems—the zero-curvature representation.

The discrete Cauchy–Riemann equations (3.4) admit a zero-curvature representation in $GL_2(\mathbb{C})[\lambda]$, with transition matrices along $(x, y) \in \vec{E}(\mathcal{D})$ given by

$$L(y, x, \alpha; \lambda) = \begin{pmatrix} \lambda + \alpha & -2\alpha(f(x) + f(y)) \\ 0 & \lambda - \alpha \end{pmatrix}, \quad \text{where} \quad \alpha = p(y) - p(x). \tag{7.4}$$

Similarly, the discrete Cauchy–Riemann equations (4.1) admit a zero-curvature representation with the transition matrices along $(\mathbf{n}, \mathbf{n} + \mathbf{e}_k) \in \vec{E}(\mathbb{Z}^d)$ given by

$$L_k(\mathbf{n}; \lambda) = \begin{pmatrix} \lambda + \alpha_k & -2\alpha_k(f(\mathbf{n} + \mathbf{e}_k) + f(\mathbf{n})) \\ 0 & \lambda - \alpha_k \end{pmatrix}. \tag{7.5}$$

The *moving frame* $\Psi(\cdot, \lambda) : \mathbb{Z}^d \rightarrow GL_2(\mathbb{C})[\lambda]$ is defined by prescribing some $\Psi(\mathbf{0}; \lambda)$, and by extending it recurrently according to the formula

$$\Psi(\mathbf{n} + \mathbf{e}_k; \lambda) = L_k(\mathbf{n}; \lambda)\Psi(\mathbf{n}; \lambda). \tag{7.6}$$

Finally, define the matrices $A(\cdot; \lambda) : \mathbb{Z}^d \rightarrow GL_2(\mathbb{C})[\lambda]$ by

$$A(\mathbf{n}; \lambda) = \frac{d\Psi(\mathbf{n}; \lambda)}{d\lambda}\Psi^{-1}(\mathbf{n}; \lambda). \tag{7.7}$$

These matrices are defined uniquely after fixing some $A(\mathbf{0}; \lambda)$.

Definition 7.2. A discrete holomorphic function $f : \mathbb{Z}^d \rightarrow \mathbb{C}$ is called *isomonodromic*[1] if, for some choice of $A(\mathbf{0}; \lambda)$, the matrices $A(\mathbf{n}; \lambda)$ are meromorphic in λ, with poles whose positions and orders do not depend on $\mathbf{n} \in \mathbb{Z}^d$.

It is clear how to extend this definition to functions on the covering $\widetilde{S}$. In the following statement, we restrict ourselves to the octant $S_1 = (\mathbb{Z}_+)^d$ for notational simplicity.

Theorem 7.3. *The discrete logarithm is isomonodromic: for a proper choice of $A(\mathbf{0}; \lambda)$, the matrices $A(\mathbf{n}; \lambda)$ at any point $\mathbf{n} \in (\mathbb{Z}_+)^d$ have simple poles only:*

$$A(\mathbf{n}; \lambda) = \frac{A^{(0)}(\mathbf{n})}{\lambda} + \sum_{l=1}^{d} \left(\frac{B^{(l)}(\mathbf{n})}{\lambda + \alpha_l} + \frac{C^{(l)}(\mathbf{n})}{\lambda - \alpha_l} \right), \tag{7.8}$$

with

$$A^{(0)}(\mathbf{n}) = \begin{pmatrix} 0 & (-1)^{n_1 + \cdots + n_d} \\ 0 & 0 \end{pmatrix}, \tag{7.9}$$

$$B^{(l)}(\mathbf{n}) = n_l \begin{pmatrix} 1 & -(f(\mathbf{n}) + f(\mathbf{n} - \mathbf{e}_l)) \\ 0 & 0 \end{pmatrix}, \tag{7.10}$$

$$C^{(l)}(\mathbf{n}) = n_l \begin{pmatrix} 0 & f(\mathbf{n} + \mathbf{e}_l) + f(\mathbf{n}) \\ 0 & 1 \end{pmatrix}. \tag{7.11}$$

This result is a basis for applying the well-developed techniques of the theory of isomonodromic solutions [6] to the asymptotic study of the discrete logarithm.

8. Conclusions

The discrete Green's function on a quasicrystalline quad-graph is the real part (i.e., the restriction to $V(\mathcal{G})$) of the discrete logarithm. The latter can be extended to a function on a branched covering of certain octants $S_m \subset \mathbb{Z}^d$, $m = 1, \ldots, 2d$. On each octant, the discrete logarithm is discrete holomorphic, with the distinctive property of being isomonodromic. This function is uniquely defined either by the integral representation (6.6), or

---

[1] This term originates in the theory of integrable nonlinear differential equations, where it is used for solutions with a similar analytic characterization [6].

by the values on the coordinate semi-axes (6.7), or else by the initial values (7.2) and the isomonodromic constraint (7.3).

References

[1] A. Bobenko, Ch. Mercat, and Yu. Suris, *Linear and nonlinear theories of discrete analytic functions. Integrable structure and isomonodromic Green's function*, J. Reine Angew. Math. **583** (2005), 117–161.

[2] A. Bobenko and Yu. Suris, *Integrable equations on quad-graphs*, Internat. Math. Res. Notices **11** (2002), 573–611.

[3] R.J. Duffin, *Basic properties of discrete analytic functions*, Duke Math. J. **23** (1956), 335–363.

[4] ———, *Potential theory on a rhombic lattice*, J. Combinatorial Theory **5** (1968), 258–272.

[5] J. Ferrand, *Fonctions preharmoniques et functions preholomorphes*, Bull. Sci. Math., 2nd ser. **68** (1944), 152–180.

[6] A. Its and V. Novokshenov, *The isomonodromic deformation method in the theory of Painlevé equations*, Lecture Notes Math., vol. 1191, Berlin: Springer-Verlag, 1986.

[7] R. Kenyon, *The Laplacian and Dirac operators on critical planar graphs*, Invent. Math. **150** (2002), 409–439.

[8] R. Kenyon and J.-M. Schlenker, *Rhombic embeddings of planar quad-graphs*, Trans. AMS **357** (2004), 3443–3458.

[9] Ch. Mercat, *Discrete Riemann surfaces and the Ising model*, Commun. Math. Phys. **218** (2001), 177–216.

Yuri B. Suris
Zentrum Mathematik
Technische Universität München
Botzmannstr. 3
85747 Garching bei München
Germany
e-mail: suris@ma.tum.de

Part II

Curvatures of Discrete Curves and Surfaces

Discrete Differential Geometry, A.I. Bobenko, P. Schröder, J.M. Sullivan and G.M. Ziegler, eds.
Oberwolfach Seminars, Vol. 38, 137–161
© 2008 Birkhäuser Verlag Basel/Switzerland

Curves of Finite Total Curvature

John M. Sullivan

Abstract. We consider the class of curves of finite total curvature, as introduced by Milnor. This is a natural class for variational problems and geometric knot theory, and since it includes both smooth and polygonal curves, its study shows us connections between discrete and differential geometry. To explore these ideas, we consider theorems of Fáry/Milnor, Schur, Chakerian and Wienholtz.

Keywords. Curves, finite total curvature, Fáry/Milnor theorem, Schur's comparison theorem, distortion.

Here we introduce the ideas of discrete differential geometry in the simplest possible setting: the geometry and curvature of curves, and the way these notions relate for polygonal and smooth curves. The viewpoint has been partly inspired by work in geometric knot theory, which studies geometric properties of space curves in relation to their knot type, and looks for optimal shapes for given knots.

After reviewing Jordan's definition of the length of a curve, we consider Milnor's analogous definition [Mil50] of total curvature. In this unified treatment, polygonal and smooth curves are both contained in the larger class of FTC (finite total curvature) curves. We explore the connection between FTC curves and BV functions. Then we examine the theorems of Fáry/Milnor, Schur and Chakerian in terms of FTC curves. We consider relations between total curvature and Gromov's distortion, and then we sketch a proof of a result by Wienholtz in integral geometry. We end by looking at ways to define curvature density for polygonal curves.

A companion article [DS08] examines more carefully the topology of FTC curves, showing that any two sufficiently nearby FTC graphs are isotopic. The article [Sul08], also in this volume, looks at curvatures of smooth and discrete surfaces; the discretizations are chosen to preserve various integral curvature relations.

Our whole approach in this survey should be compared to that of Alexandrov and Reshetnyak [AR89], who develop much of their theory for curves having one-sided tangents everywhere, a class somewhat more general than FTC.

1. Length and total variation

We want to consider the geometry of curves. Of course curves—unlike higher-dimensional manifolds—have no local *intrinsic* geometry. So we mean the *extrinsic* geometry of how the curve sits in some ambient space M. Usually M will be in euclidean d-space $\mathbb{E}^d$, but the study of space curves naturally leads also to the study of curves on spheres. Thus we also allow M to be a smooth Riemannian manifold; for convenience we embed M isometrically into some $\mathbb{E}^d$. (Some of our initial results would still hold with M being any path-metric space; compare [AR89]. Here, however, our curves will be quite arbitrary but not our ambient space.)

A curve is a one-dimensional object, so we start by recalling the topological classification of one-manifolds: A compact one-manifold (allowing boundary) is a finite disjoint union of components, each homeomorphic to an interval $\mathbb{I} := [0, L]$ or to a circle $\mathbb{S}^1 := \mathbb{R}/L\mathbb{Z}$. Then a *parametrized curve* in M is a continuous map from a compact one-manifold to M. That is, each of its components is a (parametrized) *arc* $\gamma : \mathbb{I} \to M$ or *loop* $\gamma : \mathbb{S}^1 \to M$. A loop can be viewed as an arc whose end points are equal and thus identified. A *curve* in M is an equivalence class of parametrized curves, where the equivalence relation is given by orientation-preserving reparametrization of the domain.

(An unoriented curve would allow arbitrary reparametrization. Although we will usually not care about the orientation of our curves, keeping it around in the background is convenient, fixing, for instance, a direction for the unit tangent vector of a rectifiable curve.)

Sometimes we want to allow reparametrizations by arbitrary monotone functions that are not necessarily homeomorphisms. Intuitively, we can collapse any time interval on which the curve is constant, or, conversely, stop for some time at any point along the curve. Since there might be infinitely many such intervals, the easiest formalization of these ideas is in terms of Fréchet distance [Fré05].

The *Fréchet distance* between two curves is the infimum, over all strictly monotonic reparametrizations, of the maximum pointwise distance. (This has also picturesquely been termed, perhaps originally in [AG95], the "dog-leash distance": the minimum length of leash required for a dog who walks forwards along one curve while the owner follows the other curve.) Two curves whose Fréchet distance is zero are equivalent in the sense we intend: homeomorphic reparametrizations that approach the infimal value zero will limit to the more general reparametrization that might eliminate or introduce intervals of constancy. See also [Gra46, § X.7].

Given a connected parametrized curve γ, a choice of

$$0 \leq t_1 < t_2 < \cdots < t_n \leq L$$

gives us the vertices $v_i := \gamma(t_i)$ of an *inscribed polygon* P, whose edges are the minimizing geodesics $e_i := v_i v_{i+1}$ in M between consecutive vertices. (If γ is a loop, then indices i are to be taken modulo n, that is, we consider an inscribed polygonal loop.) We will write $P < \gamma$ to denote that P is a polygon inscribed in γ.

The edges are uniquely determined by the vertices when $M = \mathbb{E}^d$ or, more generally, whenever M is simply connected with non-positive sectional curvature (and thus a CAT(0) space). When minimizing geodesics are not unique, however, as in the case when M is a sphere and two consecutive vertices are antipodal, some edges may need to be separately specified. The *mesh* of P (relative to the given parametrization of γ) is

$$\text{Mesh}(P) := \max_i (t_{i+1} - t_i).$$

(Here, of course, for a loop the nth value in this maximum is $(t_1 + L) - t_n$.)

The *length* of a polygon is simply the sum of the edge lengths:

$$\text{Len}(P) = \text{Len}_M(P) := \sum_i d_M(v_i, v_{i+1}).$$

This depends only on the vertices and not on which minimizing geodesics have been picked as the edges, since by definition all minimizing geodesics have the same length. If $M \subset N \subset \mathbb{E}^d$, then a given set of vertices defines (in general) different polygonal curves in M and N, with perhaps greater length in M.

We are now ready to define the *length* of an arbitrary curve:

$$\text{Len}(\gamma) := \sup_{P < \gamma} \text{Len}(P).$$

When γ itself is a polygonal curve, it is easy to check that this definition does agree with the earlier one for polygons. This fact is essentially the definition of what it means for M to be a path metric space: the distance $d(v, w)$ between any two points is the minimum length of paths connecting them. By this definition, the length of a curve $\gamma \subset M \subset \mathbb{E}^d$ is independent of M; length can be measured in $\mathbb{E}^d$ since even though the inscribed polygons may be different in M, their supremal length is the same.

This definition of length for curves originates with Jordan [Jor93] and independently Scheeffer [Sch85], and is thus often called "Jordan length". (See also [Ces56, §2].) For C^1-smooth curves it can easily be seen to agree with the standard integral formula.

Lemma 1.1. *Given a polygon P, if P' is obtained from P by deleting one vertex v_k, then* $\text{Len}(P') \leq \text{Len}(P)$. *We have equality here if and only if v_k lies on a minimizing geodesic from v_{k-1} to v_{k+1}.*

Proof. This is simply the triangle inequality applied to the triple v_{k-1}, v_k, v_{k+1}. □

A curve is called *rectifiable* if its length is finite. (From the beginning, we have considered only compact curves. Thus we do not need to distinguish rectifiable and locally rectifiable curves.)

Proposition 1.2. *A curve is rectifiable if and only if it admits a Lipschitz parametrization.*

Proof. If γ is K-Lipschitz on $[0, L]$, then its length is at most KL, since the Lipschitz bound gives this directly for any inscribed polygon. Conversely, a rectifiable curve can be reparametrized by its arc length

$$s(t) := \text{Len}\big(\gamma|_{[0,t]}\big)$$

and this arc length parametrization is 1-Lipschitz. □

If the original curve was constant on some time interval, the reparametrization here will not be one-to-one. A non-rectifiable curve has no Lipschitz parametrization, but might have a Hölder-continuous one. (For a nice choice of parametrization for an arbitrary curve, see [Mor36].)

Given this proposition, the standard theory of Lipschitz functions shows that a rectifiable curve γ has almost everywhere a well-defined unit tangent vector $T = \gamma'$, its derivative with respect to its arc length parameter s. Given a rectifiable curve, we will most often use this arc length parametrization. The domain is then $s \in [0, L]$ or $s \in \mathbb{R}/L\mathbb{Z}$, where L is the length.

Consider now an arbitrary function f from $\mathbb{I}$ (or $\mathbb{S}^1$) to M, *not* required to be continuous. We can apply the same definition of inscribed polygon P, with vertices $v_i = f(t_i)$, and thus the same definition of length $\mathrm{Len}(f) = \sup \mathrm{Len}(P)$. This length of f is usually called the *total variation* of f, and f is said to be BV (of *bounded variation*) when this is finite.

For a fixed ambient space $M \subset \mathbb{E}^d$, the total variation of a discontinuous f as a function to M may be greater than its total variation in $\mathbb{E}^d$. The supremal ratio here is

$$\sup_{p,q \in M} \frac{d_M(p,q)}{d_{\mathbb{E}^d}(p,q)},$$

which Gromov called the *distortion* of the embedding $M \subset \mathbb{E}^d$. (See [Gro81, pp. 6–9], [Gro83, p. 114] and [Gro78], as well as [KS97, DS04].) When M is compact and smoothly embedded (like $\mathbb{S}^{d-1}$), this distortion is finite; thus f is BV in M if and only if it is BV in $\mathbb{E}^d$.

The class of BV functions (here, from $\mathbb{I}$ to M) is often useful for variational problems. Basic facts about BV functions can be found in the original book [Car18] by Carathéodory or in many analysis texts like [GP83, Sect. 2.19], [Boa96, Chap. 3] or [Ber98]. For more details and higher dimensions, see for instance [Zie89, AFP00].

Here, we recall one nice characterization: f is BV if and only if it has a weak (distributional) derivative. Here, a weak derivative means an $\mathbb{E}^d$-valued Radon measure μ which plays the role of $f' \, dt$ in integration by parts, meaning that

$$\int_0^L f\varphi' \, dt = -\int_0^L \varphi \, \mu$$

for every smooth test function φ vanishing at the end points. (This characterization of BV functions is one form of the Riesz representation theorem.)

Proposition 1.3. *If f is BV, then f has well-defined right and left limits*

$$f_{\pm}(t) := \lim_{\tau \to t^{\pm}} f(\tau)$$

everywhere. Except at countably many jump points of f, we have $f_-(t) = f(t) = f_+(t)$.

Sketch of proof. We consider separately each of the d real-valued coordinate functions f^i. We decompose the total variation of f^i into positive and negative parts, each of which is bounded. This lets us write f^i as the difference of two monotonically increasing functions. (This is its so-called *Jordan decomposition*.) An increasing function can only have countably many (jump) discontinuities. (Alternatively, one can start by noting that a real-valued function without, say, a left-limit at t has infinite total variation even locally.) □

In functional analysis, BV functions are often viewed as equivalence classes of functions differing only on sets of measure zero. Then we replace total variation with *essential total variation*, the infimal total variation over the equivalence class. A minimizing representative will be necessarily continuous wherever $f_- = f_+$. A unique representative can be obtained by additionally requiring left (or right, or upper, or lower) semicontinuity at the remaining jump points.

Our definition of curve length is not very practical, being given in terms of a supremum over all possible inscribed polygons. But it is easy to find a sequence of polygons guaranteed to capture the supremal length:

Proposition 1.4. *Suppose P_k is a sequence of polygons inscribed in a curve γ such that* $\text{Mesh}(P_k) \to 0$. *Then* $\text{Len}(\gamma) = \lim \text{Len}(P_k)$.

Proof. By definition, $\text{Len}(\gamma) \geq \text{Len}(P_k)$, so $\text{Len}(\gamma) \geq \overline{\lim}(P_k)$. Suppose that $\text{Len}(\gamma) > \underline{\lim}(P_k)$. Passing to a subsequence, for some $\varepsilon > 0$ we have $\text{Len}(\gamma) \geq \text{Len}(P_k) + 2\varepsilon$. Then by the definition of length, there is an inscribed P_0 (with, say, n vertices) such that $\text{Len}(P_0) \geq \text{Len}(P_k) + \varepsilon$ for all k.

The common refinement of P_0 and P_k is of course at least as long as P_0. But this refinement is P_k with a fixed number (n) of vertices inserted; for each k, these n insertions together add length at least ε to P_k. For large enough k, these n insertions are at disjoint places along P_k, so their effect on the length is independent of the order in which they are performed. Passing again to a subsequence, there is thus some vertex $v_0 = \gamma(t_0)$ of P_0 such that if P_k^0 is P_k with v_0 inserted, we have $\text{Len}(P_k^0) > \text{Len}(P_k) + \varepsilon/n$.

But γ is continuous, in particular at t_0. So there exists some $\delta > 0$ such that for $t \in [t_0 - \delta, t_0 + \delta]$ we have $d_M(\gamma(t), \gamma(t_0)) < \varepsilon/2n$. Choosing k large enough that $\text{Mesh}(P_k) < \delta$, the vertices of P_k immediately before and after v_0 will be within this range, so $\text{Len}(P_k^0) < \text{Len}(P_k) + \varepsilon/n$, a contradiction. □

Although we have stated this proposition only for continuous curves γ, the same holds for BV functions f, as long as f is semicontinuous at each of its jump points.

An analogous statement does not hold for polyhedral approximations to surfaces. First, an inscribed polyhedron (whose vertices lie "in order" on the surface) can have greater area than the original surface, even if the mesh size (the diameter of the largest triangle) is small. Second, not even the limiting value is guaranteed to be correct. Although Serret had proposed [Ser68, p. 293] defining surface area as a limit of polyhedral areas, claiming this limit existed for smooth surfaces, Schwarz soon found a counterexample, now known as the "Schwarz lantern" [Sch90]: seemingly nice triangular meshes inscribed in a cylinder, with mesh size decreasing to zero, can have area approaching infinity.

Lebesgue [Leb02] thus defined surface area as the *lim inf* of such converging polyhedral areas. (See [AT72, Ces89] for an extensive discussion of related notions.) One can also rescue the situation with the additional requirement that the shapes of the triangles stay bounded (so that their normals approach that of the smooth surface), but we will not explore this here. (See also [Ton21]. In this volume, the companion article [Sul08] treats curvatures of smooth and discrete surfaces, and [War08] considers convergence issues.)

Historically, such difficulties led to new approaches to defining length and area, such as Hausdorff measure. These measure-theoretic approaches work well in all dimensions, and lead to generalizations of submanifolds like the currents and varifolds of geometric measure theory (see [Mor88]). We have chosen here to present the more "old-fashioned" notion of Jordan length for curves because it nicely parallels Milnor's definition of total curvature, which we consider next.

2. Total curvature

Milnor [Mil50] defined a notion of total curvature for arbitrary curves in euclidean space. Suppose P is a polygon in M with no two consecutive vertices equal. Its *turning angle* at an interior vertex v_n is the angle $\varphi \in [0, \pi]$ between the oriented tangent vectors at v_n to the two edges $v_{n-1}v_n$ and v_nv_{n+1}. (Here, by saying interior vertices, we mean to exclude the end points of a polygonal arc, where there is no turning angle; every vertex of a polygonal loop is interior. The supplement of the turning angle, sometimes called an interior angle of P, will not be of interest to us.)

If M is an oriented surface, for instance if $M = \mathbb{E}^2$ or $\mathbb{S}^2$, then we can also define a *signed turning angle* $\varphi \in [-\pi, \pi]$ at v_n, except that the sign is ambiguous when $\varphi = \pm\pi$.

To find the *total curvature* $\mathrm{TC}(P)$ of a polygon P, we first collapse any sequence of consecutive equal vertices to a single vertex. Then $\mathrm{TC}(P)$ is simply the sum of the turning angles at all interior vertices.

Here, we mainly care about the case when P is in $M = \mathbb{E}^d$. Then the unit tangent vectors along the edges, in the directions $v_{n+1} - v_n$, are the vertices of a polygon in $\mathbb{S}^{d-1}$ called the *tantrix* of P. (The word is a shortening of "tangent indicatrix".) The total curvature of P is the length of its tantrix in $\mathbb{S}^{d-1}$.

Lemma 2.1. (See [Mil50, Lemma 1.1] and [Bor47].) *Suppose P is a polygon in $\mathbb{E}^d$. If P' is obtained from P by deleting one vertex v_n then $\mathrm{TC}(P') \leq \mathrm{TC}(P)$. We have equality here if $v_{n-1}v_nv_{n+1}$ are collinear in that order, or if $v_{n-2}v_{n-1}v_nv_{n+1}v_{n+2}$ lie convexly in some two-plane, but never otherwise.*

Proof. Deleting v_n has the following effect on the tantrix: two consecutive vertices (the tangents to the edges $v_{n-1}v_n$ and v_nv_{n+1}) are coalesced into a single one (the tangent to the edge $v_{n-1}v_{n+1}$). It lies on a great circle arc connecting the original two, as in Figure 1. Using the triangle inequality twice, the length of the tantrix decreases (strictly, unless the tantrix vertices $v_{n-1}v_n$ and v_nv_{n+1} coincide, or the relevant part lies along a single great circle in $\mathbb{S}^{d-1}$). □

Corollary 2.2. *If P is a polygon in $\mathbb{E}^d$ and $P' < P$, then $\mathrm{TC}(P') \leq \mathrm{TC}(P)$.*

Proof. Starting with P, first insert the vertices of P'; since each of these lies along an edge of P, these insertions have no effect on the total curvature. Next delete the vertices not in P'; this can only decrease the total curvature. □

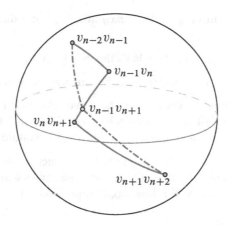

FIGURE 1. Four consecutive edges of a polygon P in space give four vertices and three connecting edges (shown here as solid lines) of its tantrix on the sphere. When the middle vertex v_n of P is deleted, two vertices of the tantrix get collapsed to a single new one (labeled $v_{n-1}v_{n+1}$); it lies somewhere along edge connecting the two original vertices. The two new edges of the tantrix are shown as dashed lines. Applying the triangle inequality twice, we see that the length of the new tantrix (the total curvature of the new polygon) is no greater.

Definition. For any curve $\gamma \subset \mathbb{E}^d$ we follow Milnor [Mil50] to define

$$\mathrm{TC}(\gamma) := \sup_{P < \gamma} \mathrm{TC}(P).$$

We say that γ has *finite total curvature* (or that γ is *FTC*) if $\mathrm{TC}(\gamma) < \infty$.

When γ is itself a polygon, this definition agrees with the first one by Corollary 2.2. Our curves are compact, and thus lie in bounded subsets of $\mathbb{E}^d$. It is intuitively clear then that a compact curve of infinite length must have infinite total curvature; that is, that all FTC curves are rectifiable. This follows rigorously by applying the quantitative estimate of Proposition 6.1 below to finely inscribed polygons, using Propositions 1.4 and 3.1.

Various properties follow very easily from this definition. For instance, if the total curvature of an arc is less than π, then the arc cannot stray too far from its end points. In particular, define the *spindle* of an angle θ with end points p and q to be the body of revolution bounded by a circular arc of total curvature θ from p to q that has been revolved about $\overleftrightarrow{pq}$. (The spindle is convex, looking like an American football, for $\theta \le \pi$, and is a round ball for $\theta = \pi$.)

Lemma 2.3. *Suppose γ is an arc from p to q of total curvature $\varphi < \pi$. Then γ is contained in the spindle of angle 2φ from p to q.*

Proof. Suppose $x \in \gamma$ is outside the spindle. Consider the planar polygonal arc pxq inscribed in γ. In that plane, since x is outside the circular arc of total curvature 2φ, by elementary geometry, the turning angle of pxq at x is greater than φ, contradicting the definition of $\text{TC}(\gamma)$. $\square$

We also immediately recover Fenchel's theorem [Fen29]:

Theorem 2.4 (Fenchel). *Any closed curve in $\mathbb{E}^d$ has total curvature at least 2π.*

Proof. Pick any two distinct points p, q on the curve. (We didn't intend the theorem to apply to the constant curve!) The inscribed polygonal loop from p to q and back has total curvature 2π, so the original curve has at least this much curvature. $\square$

For this approach to Fenchel's theorem to be satisfactory, we do need to verify (as Milnor did [Mil50]) that our definition of total curvature agrees with the usual one, $\int \kappa \, ds$, for smooth curves. For us, this will follow from Proposition 3.1.

3. First variation of length

We can characterize FTC curves as those with BV tangent vectors. This relates to the variational characterization of curvature in terms of first variation of length. (The discussion in this section is based on [CF+04, Sect. 4].)

We have noted that an FTC curve γ is rectifiable, hence has a tangent vector T defined almost everywhere. We now claim that the total curvature of γ is exactly the length (or, more precisely, the essential total variation) of this tantrix T as a curve in $\mathbb{S}^{d-1}$. We have already noted this for polygons, so the general case seems almost obvious from the definitions. However, while the tantrix of a polygon inscribed in γ is a spherical polygon, it is not inscribed in T; instead its vertices are averages of small pieces of T. Luckily, this is close enough to allow the argument of Proposition 1.4 to go through again: Just as for length, in order to compute total curvature it suffices to take any limit of finer and finer inscribed polygons.

Proposition 3.1. *Suppose γ is a curve in $\mathbb{E}^d$. If P_k is a sequence of polygons inscribed in γ with $\text{Mesh}(P_k) \to 0$, then $\text{TC}(\gamma) = \lim \text{TC}(P_k)$. This equals the essential total variation of its tantrix $T \subset \mathbb{S}^{d-1}$.*

We leave the proof of this proposition as an exercise. The first statement essentially follows as in the proof of Proposition 1.4: if it failed, there would be one vertex v_0 along γ whose insertion would cause a uniform increase in total curvature for all polygons in a convergent subsequence, contradicting the fact that sufficiently small arcs before and after v have arbitrarily small total curvature. The second statement follows by measuring both $\text{TC}(\gamma)$ and the total variation through limits of (different but nearby) fine polygons.

To summarize, a rectifiable curve γ has finite total curvature if and only if its unit tangent vector $T = \gamma'(s)$ is a function of bounded variation. (Thus the space of FTC curves could be called $W^{1,BV}$ or BV^1.) If γ is FTC, it follows that at every point of γ there are well-defined left and right tangent vectors $T_\pm$; these are equal and opposite except at countably many points, the *corners* of γ.

Now, to investigate curvature from a variational point of view, suppose we consider a *continuous deformation* γ_t of a curve $\gamma \subset \mathbb{E}^d$: fixing any parametrization of γ, this means a continuous family γ_t of parametrized curves with $\gamma_0 = \gamma$. (If we reparametrize γ, we must apply the same reparametrization to each γ_t.)

We assume that position of each point is (at least C^1) smooth in time; the initial velocity of γ_t will then be given by some (continuous, $\mathbb{E}^d$-valued) *vector field* ξ along γ.

Let γ be a rectifiable curve parametrized by arc length s, with unit tangent vector $T = \gamma'(s)$ (defined almost everywhere). Suppose γ_t is a variation of $\gamma = \gamma_0$ whose initial velocity $\xi(s)$ is a Lipschitz function of arc length. Then the arc length derivative $\xi' = \partial\xi/\partial s$ is defined almost everywhere along γ, and a standard first-variation calculation shows that

$$\delta_\xi \mathrm{Len}(\gamma) := \frac{d}{dt}\Big|_0 \mathrm{Len}(\gamma_t) = \int_\gamma \langle T, \xi' \rangle \, ds.$$

If γ is smooth enough, we can integrate this by parts to get

$$\delta_\xi \mathrm{Len}(\gamma) = -\int_\gamma \langle T', \xi \rangle \, ds - \sum_{x \in \partial\gamma} \langle \pm T, \xi \rangle,$$

where, in the boundary term, the sign is chosen to make $\pm T$ point inward at x. In fact, not much smoothness is required: as long as γ is FTC, we know that its unit tangent vector T is BV, so we can interpret T' as a measure, and this first-variation formula holds in the following sense: the weak (distributional) derivative $\mathcal{K} := T'$ is an $\mathbb{E}^d$-valued Radon measure along γ which we call the *curvature force*.

On a C^2 arc of γ, the curvature force is $\mathcal{K} = dT = \kappa N \, ds$ and is absolutely continuous with respect to the arc length or Hausdorff measure $ds = \mathcal{H}^1$. The curvature force has an atom (a point mass or Dirac delta) at each corner $x \in \gamma$, with $\mathcal{K}\{x\} = T_+(x) + T_-(x)$. Note that at such a corner, the mass of $\mathcal{K}$ is not the turning angle θ at x. Instead,

$$|\mathcal{K}|\{x\} = |\mathcal{K}\{x\}| = 2\sin(\theta/2).$$

Therefore, the total mass (or total variation) $|\mathcal{K}|(\gamma)$ of the curvature force $\mathcal{K}$ is somewhat less than the total curvature of γ: at each corner it counts $2\sin(\theta/2)$ instead of θ. Whereas $\mathrm{TC}(\gamma)$ was the length (or total variation) $\mathrm{Len}_{\mathbb{S}^{d-1}}(T)$ of the tantrix T viewed as a (discontinuous) curve on the sphere $\mathbb{S}^{d-1}$, we recognize this total mass as its length $\mathrm{Len}_{\mathbb{E}^d}(T)$ in euclidean space. Thus we call it the euclidean total curvature of γ:

$$\mathrm{TC}^*(\gamma) := \mathrm{Len}_{\mathbb{E}^d}(T) = |\mathcal{K}|(\gamma).$$

Returning to the first variation of length, we say that a vector field ξ along γ is *smooth* if $\xi(s)$ is a smooth function of arc length. The first variation $\delta\mathrm{Len}(\gamma)$ can be viewed as a linear functional on the space of smooth vector fields ξ along γ. As such a *distribution*, it has degree zero, by definition, if $\delta_\xi \mathrm{Len}(\gamma) = \int_\gamma \langle T, \xi' \rangle \, ds$ is bounded by $C \sup_\gamma |\xi|$ for some constant C. This happens exactly when we can perform the integration by parts above.

We collect the results of this section as:

Proposition 3.2. *Given any rectifiable curve γ, the following conditions are equivalent:*

(a) *γ is FTC.*

(b) *There exists a curvature force $\mathcal{K} = dT$ along γ such that*

$$\delta_\xi \text{Len}(\gamma) = -\int_\gamma \langle \xi, \mathcal{K} \rangle - \sum_{\partial\gamma} \langle \xi, \pm T \rangle.$$

(c) *The first variation $\delta \text{Len}(\gamma)$ has distributional degree zero.*   □

Of course, just as not all continuous functions are BV, not all C^1 curves are FTC. However, given an FTC curve, it is piecewise C^1 exactly when it has finitely many corners, and is C^1 when it has no corners, that is, when $\mathcal{K}$ has no atoms. The FTC curve is furthermore $C^{1,1}$ when T is Lipschitz, or, equivalently, when $\mathcal{K}$ is absolutely continuous (with respect to arc length s) and has bounded Radon/Nikodym derivative $d\mathcal{K}/ds = \kappa N$.

4. Total curvature and projection

The Fáry/Milnor theorem says that a knotted curve in $\mathbb{E}^3$ has total curvature at least 4π, twice that of an unknotted round circle. The different proofs given by Fáry [Fár49] and Milnor [Mil50] can both be interpreted in terms of a proposition about the average total curvature of different projections of a curve.

The Grassmannian $G_k \mathbb{E}^d$ of k-planes in d-space is compact, with a unique rotation-invariant probability measure $d\mu$. For $p \in G_k \mathbb{E}^d$, we denote by π_p the orthogonal projection to p. When we speak about averaging over all projections, we mean using $d\mu$. This proposition is essentially due to Fáry [Fár49], though he only stated the case $d = 3$, $k = 2$.

Proposition 4.1. *Given a curve K in $\mathbb{E}^d$, and some fixed $k < d$, the total curvature of K equals the average total curvature of its projections to k-planes. That is,*

$$\text{TC}(K) = \int_{G_k \mathbb{E}^d} \text{TC}(\pi_p(K)) \, d\mu.$$

Proof. By definition of total curvature and Proposition 3.1, we may reduce to the case where K is a polygon. (To interchange the limit of ever finer inscribed polygons with the average over the Grassmannian, we use the Lebesgue monotone convergence theorem.) Since the total curvature of a polygon is the sum of its turning angles, it suffices to consider a single angle. So let P_θ be a three-vertex polygonal arc with a single turning angle of $\theta \in [0, \pi]$. Defining

$$f_k^d(\theta) := \int_{p \in G_k \mathbb{E}^d} \text{TC}(\pi_p(P_\theta)) \, d\mu,$$

the rotation-invariance of μ shows this is independent of the position of P_θ, and our goal is to show $f_k^d(\theta) = \theta$.

First note that f_k^d is continuous. It is also additive:

$$f_k^d(\alpha + \beta) = f_k^d(\alpha) + f_k^d(\beta)$$

when $0 \le \alpha, \beta \le \alpha + \beta \le \pi$. This follows by cutting the single corner of P_θ into two corners of turning angles α and β. Any projection of the resulting convex planar arc is still convex and planar, so additivity holds in each projection, and thus also holds after averaging.

A continuous, additive function is linear: $f_k^d(\theta) = c_k^d \theta$ for some constant c_k^d. Thus we just need to show $c_k^d = 1$. But we can easily evaluate f at $\theta = \pi$, where P_π is a "cusp" in which the incoming and outgoing edges overlap. Clearly every projection of P_π is again such a cusp with turning angle π (except for a set of measure zero where the projection is a single point). Thus $f_k^d(\pi) = \pi$, so $c_k^d = 1$ and we are done. $\qquad\square$

A curve in $\mathbb{E}^1$ has only cusps for corners, so its total curvature will be a non-negative multiple of π. A loop in $\mathbb{E}^1$ has total curvature a positive multiple of 2π. (In particular, the loop in $\mathbb{E}^1$ is a real-valued function on $\mathbb{S}^1$, and its total curvature is 2π times the number of local maxima.)

This proposition could be used to immediately reduce the d-dimensional case of Fenchel's theorem (Theorem 2.4) to the k-dimensional case. Historically, this could have been useful. For instance, the theorem is trivial in $\mathbb{E}^1$ by the previous paragraph. In $\mathbb{E}^2$, the idea that a simple closed curve has total signed curvature $\pm 2\pi$ essentially dates back to Riemann (compare [Che89, §1]). Fenchel's proof [Fen29] (which was his 1928 doctoral dissertation in Berlin) was for $d = 3$, and the first proof for general dimensions seems to be that of Borsuk [Bor47].

For alternative proofs of Fenchel's theorem—as well as comparisons among them—see [Hor71] and [Che89, §4], and the references therein, especially [Lie29] and [Fen51]. Voss [Vos55] related the total curvature of a space curve to the total Gauss curvature of a tube around it, and thereby gave new a new proof of the Fáry/Milnor theorem as well as Fenchel's theorem.

Milnor's proof [Mil50] of the Fáry/Milnor theorem can be rephrased as a combination of the case $k = 1$ of Proposition 4.1 with the following:

Lemma 4.2. *If $K \subset \mathbb{E}^3$ is nontrivially knotted, then any projection of K to $\mathbb{E}^1$ has total curvature at least 4π.*

Sketch of proof. If there were some projection direction in which the total curvature were only 2π, then the corresponding (linear) height function on $\mathbb{E}^3$ would have only one local minimum and one local maximum along K. Then at each intermediate height, there are exactly two points of K. Connecting each such pair with a straight segment, we form a spanning disk showing K is unknotted. $\qquad\square$

(This proof isn't quite complete as stated: at some intermediate heights, one or even both strands of K might have a whole subarc at that constant height. One could still patch in a disk, but easier is to follow Milnor and at the very beginning replace K by an isotopic inscribed polygon. Compare [DS08].)

Fáry [Fár49], on the other hand, used $k = 2$ in Proposition 4.1, having proved:

Lemma 4.3. *Any nontrivial knot diagram (a projection to $\mathbb{E}^2$ of a knotted curve $K \subset \mathbb{E}^3$) has total curvature at least 4π.*

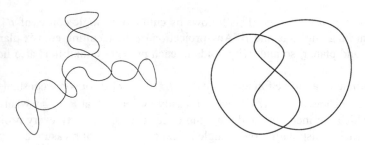

FIGURE 2. A diagram in which every region is adjacent to the outside is in fact unknotted (left), so a knot diagram has a doubly enclosed region (right).

Sketch of proof. Any knot diagram divides the plane into regions: one unbounded and several bounded. If every bounded region is adjacent to the unbounded region, the only possibility is a tree-like diagram as in Figure 2 (left); this is clearly unknotted no matter how we choose over- and under-crossings.

Thus every nontrivial knot diagram D has some region R which is "doubly enclosed" by the curve, not necessarily in the sense of oriented winding numbers, but in the sense that any ray outwards from a point $p \in R$ must cut the curve twice. Then R is part of the *second hull* of the curve [CK$^+$03], and the result follows by Lemmas 5 and 6 there. To summarize the arguments there (which parallel those of [Fár49]), note first that the cone over D from p has cone angle at least 4π at p; by Gauss/Bonnet, this cone angle equals the total signed geodesic curvature of D in the cone, which is at most its total (unsigned) curvature.   □

Either of these lemmas, combined with the appropriate case of Proposition 4.1, immediately yields the Fáry/Milnor theorem.

Corollary 4.4. *The total curvature of a nontrival knot $K \subset \mathbb{E}^3$ is at least 4π.*   □

While Fáry stated the appropriate case of Proposition 4.1 pretty much as such, we note that Milnor didn't speak about total curvature of projections to lines, but instead only about extrema of height functions. (A reinterpretation more like ours can already be found, for instance, in [Tan98].)

Denne [Den04] has given a beautiful new proof of Fáry/Milnor analogous to the easy proof we gave for Fenchel's Theorem 2.4. Indeed, the hope that there could be such a proof had led us to conjecture in [CK$^+$03] that every knot K has an *alternating quadrisecant*. A quadrisecant, by definition, is a line in space intersecting the knot in four points p_i. The p_i form an inscribed polygonal loop, whose total curvature (since it lies in a line) is either 2π or 4π, depending on the relative ordering of the p_i along the line and along K. The quadrisecant is called alternating exactly when the curvature is 4π. Denne proved this conjecture: every nontrivially knotted curve in space has an alternating quadrisecant. This gives as an immediate corollary not only the Fáry/Milnor theorem, but also a new proof that a knot has a *second hull* [CK$^+$03].

We should also note that a proof of Fáry/Milnor for knots in hyperbolic as well as euclidean space was given by Brickell and Hsiung [BH74]; essentially they construct a point on the knot that also lies in its second hull. The theorem is now also known for knots in an arbitrary Hadamard manifold, that is, a simply connected manifold of non-positive curvature: Alexander and Bishop [AB98] found a finite sequence of inscriptions—first a polygon P_1 inscribed in K, then inductively P_{i+1} inscribed in P_i—ending with a $pqpq$ quadrilateral, while Schmitz [Sch98] comes close to constructing a quadrisecant.

The results of this section would not be valid for TC* in place of TC: there is no analog to Proposition 4.1, and the Fáry/Milnor theorem would fail.

We conclude this section by recalling that another standard proof of Fenchel's theorem uses the following integral-geometric lemma due to Crofton (see [Che89, §4] and also [San89, San04]) to conclude that a spherical curve of length less than 2π is contained in an open hemisphere:

Lemma 4.5. *The length of a curve* $\gamma \subset \mathbb{S}^{d-1}$ *equals* π *times the average number of intersections of* γ *with great hyperspheres* $\mathbb{S}^{d-2}$.

Proof. It suffices to prove this for polygons (appealing again to the monotone convergence thereom). Hence it suffices to consider a single great-circle arc. But clearly for such arcs, the average number of intersections is proportional to the length. When the length is π, (almost) every great circle is intersected exactly once. $\quad\square$

This lemma, applied to the tantrix, is equivalent to the case $k = 1$ of our Proposition 4.1. Indeed, when projecting γ to the line in direction v, the total curvature we see counts the number of times the tantrix intersects the great sphere normal to v. We note also that knowing this case $k = 1$ (for all d) immediately implies all other cases of Proposition 4.1, since a projection from $\mathbb{E}^d$ to $\mathbb{E}^1$ can be factored as projections $\mathbb{E}^d \to \mathbb{E}^k \to \mathbb{E}^1$.

Finally, we recall an analogous statement of the famous Cauchy/Crofton formula. Its basic idea dates back to Buffon's 1777 analysis [Buf77] of his needle problem (compare [KR97, Chap. 1]). Cauchy obtained the formula by 1841 [Cau41] and generalized it to find the surface area of a convex body. Crofton's 1868 paper [Cro68] on geometric probability includes this among many integral-geometric formulas for plane curves. (See [AD97] for a treatment like ours for rectifiable curves, and also [dC76, §1.7C] and [San89].)

Lemma 4.6 (Cauchy/Crofton). *The length of a plane curve equals* $\pi/2$ *times the average length of its projections to lines.*

Proof. Again, we can reduce first to polygons, then to a single segment. So the result certainly holds for some constant; to check the constant is $\pi/2$, it is easiest to compute it for the unit circle, where every projection has length 4. $\quad\square$

In all three of our integral-geometric arguments (4.1, 4.5, 4.6) we proved a certain function was linear and then found the constant of proportionality by computing one (perhaps sometimes surprising) example. In the literature one also finds proofs where the integrals (over the circle or, more generally, the Grassmannian) are simply computed explicitly. Although the trigonometric integrals are not too difficult, that approach seems to obscure the geometric essence of the argument.

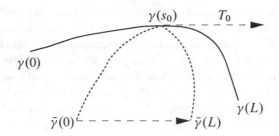

FIGURE 3. Schur's Theorem compares the space curve γ (solid) with a planar curve $\bar{\gamma}$ (dotted) that has pointwise no less curvature. The proof places them with a common tangent vector T_0 (at arc length s_0) pointing in the direction connecting the end points of $\bar{\gamma}$.

5. Schur's comparison theorem

Schur's comparison theorem [Sch21] is a well-known result saying that straightening an arc will increase the distance between its end points.

Chern, in §5 of his beautiful essay [Che89] in his MAA book, gives a proof for C^2 curves and remarks (without proof) that it also applies to piecewise smooth curves. In [CKS02] we noted that Chern's proof actually applies to $C^{1,1}$ curves, that is, to curves with a Lipschitz tangent vector, or with bounded curvature density. In fact, the natural class of curves to which the proof applies is FTC curves.

Theorem 5.1 (Schur's Comparison Theorem). *Let $\bar{\gamma} \subset \mathbb{E}^2$ be a planar arc such that joining the end points of $\bar{\gamma}$ results in a convex (simple, closed) curve, and let $\gamma \subset \mathbb{E}^d$ be an arc of the same length L. Suppose that $\bar{\gamma}$ has nowhere less curvature than γ with respect to the common arc length parameter $s \in [0, L]$, that is, that for any subinterval $I \subset [0, L]$ we have*

$$\mathrm{TC}(\bar{\gamma}|_I) \geq \mathrm{TC}(\gamma|_I).$$

(Equivalently, $|\mathcal{K}| - |\bar{\mathcal{K}}|$ is a non-negative measure.) Then the distance between the end points is greater for γ:

$$|\gamma(L) - \gamma(0)| \geq |\bar{\gamma}(L) - \bar{\gamma}(0)|.$$

Proof. By convexity, we can find an s_0 such that the (or, in the case of a corner, some supporting) tangent direction T_0 to $\bar{\gamma}$ at $\bar{\gamma}(s_0)$ is parallel to $\bar{\gamma}(L) - \bar{\gamma}(0)$. Note that the convexity assumption implies that the total curvature of either half of $\bar{\gamma}$ (before or after s_0) is at most π. Now move γ by a rigid motion so that it shares this same tangent vector at s_0, as in Figure 3. (If γ has a corner at s_0, so does $\bar{\gamma}$. In this case we want to arrange that the angle from T_0 to each one-sided tangent vector $T_\pm$ is at least as big for $\bar{\gamma}$ as for γ.)

By choice of T_0 we have

$$|\bar{\gamma}(L) - \bar{\gamma}(0)| = \langle \bar{\gamma}(L) - \bar{\gamma}(0), T_0 \rangle = \int_0^L \langle \bar{T}(s), T_0 \rangle \, ds,$$

while for γ we have

$$|\gamma(L) - \gamma(0)| \geq \langle \gamma(L) - \gamma(0), T_0 \rangle = \int_0^L \langle T(s), T_0 \rangle \, ds.$$

Thus it suffices to prove (for almost every s) that

$$\langle T(s), T_0 \rangle \geq \langle \bar{T}(s), T_0 \rangle.$$

But starting from s_0 and moving outwards in either direction, $\bar{T}$ moves straight along a great circle arc, at speed given by the pointwise curvature; in total it moves less than distance π. At the same time, T moves at the same or lower speed, and perhaps not straight but on a curved path. Clearly then $T(s)$ is always closer to T_0 than $\bar{T}(s)$ is, as desired. More precisely,

$$\langle \bar{T}(s), T_0 \rangle = \cos \mathrm{TC}(\bar{\gamma}|_{[s_0,s]}) \leq \cos \mathrm{TC}(\gamma|_{[s_0,s]}) \leq \langle T(s), T_0 \rangle. \qquad \square$$

The special case of Schur's theorem when γ and $\bar{\gamma}$ are polygons is usually called Cauchy's arm lemma. It was used in Cauchy's proof [Cau13] of the rigidity of convex polyhedra, although Cauchy's own proof of the arm lemma was not quite correct, as discovered 120 years later by Steinitz. The standard modern proof of the arm lemma (due to Schoenberg; see [AZ98] or [Cro97, p. 235]) is quite different from the proof we have given here. For more discussion of the relation between Schur's theorem and Cauchy's lemma, see [Con82, O'R00].

The history of this result is somewhat complicated. Schur [Sch21] considered only the case where γ and $\bar{\gamma}$ have pointwise equal curvature: twisting a convex plane curve out of the plane by adding torsion will increase its chord lengths. He considers both polygonal and smooth curves. He attributes the orginal idea (only for the case where $\bar{\gamma}$ is a circular arc) to unpublished work of H.A. Schwarz in 1884. The full result, allowing the space curve to have less curvature, is evidently due to Schmidt [Sch25]. See also the surveys by Blaschke in [Bla21] and [Bla24, §28–30].

In Schur's theorem, it is irrelevant whether we use the spherical or euclidean version of total curvature. If we replace TC by TC* throughout, the statement and proof remain unchanged, since the curvature comparison is pointwise.

6. Chakerian's packing theorem

A less familiar result due to Chakerian (and cited for instance as [BS99, Lemma 1.1]) captures the intuition that a long rope packed into a small ball must have large curvature.

Proposition 6.1. *A connected FTC curve γ contained in the unit ball in $\mathbb{E}^d$ has length no more than $2 + \mathrm{TC}^*(\gamma)$. (If γ is closed, the 2 can be omitted.)*

Proof. Use the arc length parametrization $\gamma(s)$. Then

$$\mathrm{Len}(\gamma) = \int 1 \, ds = \int \langle T, T \rangle \, ds$$

$$= \langle T, \gamma \rangle \big|_{\mathrm{endpts}} - \int \langle \gamma, d\mathcal{K} \rangle$$

$$\leq 2 + \int d|\mathcal{K}| = 2 + \mathrm{TC}^*(\gamma). \qquad \square$$

Chakerian [Cha64] gave exactly this argument for C^2 curves and then used a limit argument (rounding the corners of inscribed polygons) to get a version for all curves. Note, however, that this limiting procedure gives the bound with TC^* replaced by TC; this is of course equivalent for C^1 curves but weaker for curves with corners. For closed curves, Chakerian noted that equality holds in $\mathrm{Len} \leq \mathrm{TC}$ only for a great circle (perhaps traced multiple times). In our sharper bound $\mathrm{Len} \leq \mathrm{TC}^*$, we have equality also for a regular n-gon inscribed in a great circle.

(We recall that we appealed to this theorem in Section 2 to deduce that FTC curves are rectifiable. This is not circular reasoning: we first apply the proof above to polygons, then deduce that FTC curves are rectifiable and indeed have BV tangents, and fianlly apply the proof above in general.)

Chakerian had earlier [Cha62] given a quite different proof (following Fáry) that $\mathrm{Len} \leq \mathrm{TC}$. We close by interpreting that first argument in our framework. Start by observing that in $\mathbb{E}^1$, where curvature is quantized, it is obvious that for a closed curve in the unit ball (which is just a segment of length 2)

$$\mathrm{Len} \leq \mathrm{TC}^* = \tfrac{2}{\pi} \mathrm{TC}.$$

Combining this with Cauchy/Crofton (Lemma 4.6) and our Proposition 4.1 gives immediately $\mathrm{Len} \leq \mathrm{TC}$ for curves in the unit disk in $\mathbb{E}^2$. With a little care, the same is true for curves that fail to close by some angular holonomy. (The two end points are at equal radius, and we do include in the total curvature the angle they make when they are rotated to meet.) Rephrased, the length of a curve γ in a unit neighborhood of the cone point on a cone surface of arbitrary cone angle is at most the total (unsigned) geodesic curvature of γ in the cone. Finally, given any curve γ in the unit ball in $\mathbb{E}^d$, Chakerian considers the cone over γ from the origin. The length is at most the total curvature in the cone, which is at most the total curvature in space. Rather than trying to consider cones over arbitrary FTC curves, we can prove the theorem for polygons and then take a limit.

7. Distortion

We have already mentioned Gromov's distortion for an embedded submanifold. In the case of a curve $\gamma \subset \mathbb{E}^d$, this *distortion* is

$$\delta(\gamma) := \sup_{p \neq q \in \gamma} \delta(p, q), \qquad \delta(p, q) := \frac{\mathrm{Len}(p, q)}{|p - q|},$$

where $\text{Len}(p, q)$ is the (shorter) arc length distance along γ. Here, we discuss some relations between distortion and total curvature; many of these appeared in the first version of [DS04], but later improvements to the main argument there made the discussion of FTC curves unnecessary.

Examples like a steep logarithmic spiral show that arcs of infinite total curvature can have finite distortion, even distortion arbitrarily close to 1. However, there is an easy bound the other way:

Proposition 7.1. *Any arc of total curvature $\alpha < \pi$ has distortion at most $\sec(\alpha/2)$.*

Proof. First, note that it suffices to prove this for the end points of the arc. (If the distortion were realized by some other pair (p, q), we would just replace the original arc by the subarc from p to q.)

Second, note by that Schur's Theorem 5.1 we may assume the arc is convex and planar: we replace any given arc by the locally convex planar arc with the same pointwise curvature. Because the total curvature is less than π, the planar arc is globally convex in the sense of Theorem 5.1, and the theorem shows that the end-point separation has only decreased.

Now fix points p and q in the plane; for any given tangent direction at p, there is a unique triangle pxq with exterior angle α at x. Any convex arc of total curvature α from p (with the given tangent) to q lies within this triangle. By the Cauchy/Crofton formula of Lemma 4.6, its length is then at most that of the polygonal arc pxq. Varying now the tangent at p, the locus of points x is a circle, and it is easy to see that the length is maximized in the symmetric situation, with $\delta = \sec(\alpha/2)$. □

This result might be compared with the bound [KS97, Lemma 5.1] on distortion for a $C^{1,1}$ arc with bounded curvature density $\kappa \leq 1$. By Schur's Theorem 5.1 such an arc of length $2a \leq 2\pi$ can be compared to a circle, and thus has distortion at most $a/\sin a$.

For any curve γ, the distortion is realized either by a pair of distinct points or in a limit as the points approach, simply because $\gamma \times \gamma$ is compact. In general, the latter case might be quite complicated. On an FTC curve, however, we now show that the distortion between nearby pairs behaves very nicely. Define $\alpha(r) \leq \pi$ to be the turning angle at the point $r \in \gamma$, with $\alpha = 0$ when r is not a corner.

Lemma 7.2. *On an FTC curve γ, we have*

$$\overline{\lim_{p,q \to r}} \, \delta(p, q) = \sec \frac{\alpha(r)}{2},$$

with this limit realized by symmetric pairs (p, q) approaching r from opposite sides.

Proof. The existence of one-sided tangent vectors $T_{\pm}$ at r is exactly enough to make this work, since the quotient in the definition of $\delta(p, q)$ is similar to the difference quotients defining $T_{\pm}$. Indeed, near r the curve looks very much like a pair of rays with turning angle α. Thus the *lim sup* is the same as the distortion of these rays, which is $\sec(\alpha/2)$, realized by any pair of points symmetrically spaced about the vertex. □

This leads us to define $\delta(r, r) := \sec\left(\alpha(r)/2\right)$, giving a function $\delta : \gamma \times \gamma \to [1, \infty]$ that is upper semicontinuous. The compactness of γ then immediately gives:

Corollary 7.3. *On an FTC curve γ, there is a pair (p, q) of (not necessarily distinct) points on γ which realize the distortion $\delta(\gamma) = \delta(p, q)$.* $\qquad\square$

Although distortion is not a continuous functional on the space of rectifiable curves, it is lower semicontinuous. A version of the next lemma appeared as [KS97, Lem. 2.2]:

Lemma 7.4. *Suppose curves γ_j approach a limit γ in the sense of Fréchet distance. Then $\delta(\gamma) \leq \underline{\lim} \, \delta(\gamma_j)$.*

Proof. The distortion for any fixed pair of points is lower semicontinuous because the arc length between them is. (And length will indeed jump down in a limit unless the tangent vectors also converge in a certain sense. See [Ton21, Chap. 2, §29].) The supremum of a family of lower semicontinuous functions is again lower semicontinuous. $\qquad\square$

8. A projection theorem of Wienholtz

In [KS97], we made the following conjecture:

Any closed curve γ in $\mathbb{E}^d$ of length L has some orthogonal projection to $\mathbb{E}^{d-1}$ of diameter at most L/π.

This yields an easy new proof of Gromov's result (see [KS97, DS04]) that a closed curve has distortion at least $\pi/2$, that of a circle. Indeed, consider the height function along γ in the direction on some projection of small diameter. For any point $p \in \gamma$, consider the *antipodal* point p^* halfway around γ, at arc length $L/2$. Since the height difference between p and p^* is continuous and changes sign, it equals zero for some (p, p^*). The distance between the projected images of these points is at most the diameter, at most L/π. But since the heights were equal, this distance is the same as their distance $|p - p^*|$ in $\mathbb{E}^d$.

For $d = 2$, we noted that our conjecture follows immediately from Cauchy/Crofton: a closed plane curve of length L has average width L/π and thus has width at most L/π in some direction. But for higher d, the analogs of Cauchy/Crofton give a weaker result. (A curve of length L in $\mathbb{E}^3$, for instance, has projections to $\mathbb{E}^2$ of average length $\pi L/4$, and thus has some planar projection of diameter at most $\pi L/8$.)

In a series of Bonn preprints from 1999, Daniel Wienholtz proved our conjecture and in fact somewhat more: a closed curve in $\mathbb{E}^d$ of length L has some orthogonal projection to $\mathbb{E}^{d-1}$ which lies in a ball of diameter L/π. Because Wienholtz's work has unfortunately remained unpublished, we outline his arguments here.

Proposition 8.1. *Given any closed curve γ in $\mathbb{E}^d$ for $d \geq 3$, there is some slab containing γ, bounded by parallel hyperplanes h_1 and h_2, with points $a_i, b_i \in \gamma \cap h_i$ occuring along γ in the order $a_1 a_2 b_1 b_2$. (We call the h_i a pair of parallel interleaved bitangent support planes for γ.)*

Proof. Suppose not. Then for any unit vector $v \in \mathbb{S}^{d-1}$, we can divide the circle parameterizing γ into two complementary arcs $\alpha(v)$, $\beta(v)$, such that the (global) maximum of the height function in direction v is achieved only (strictly) within α, and the minimum is achieved only (strictly) within β. In fact, these arcs can be chosen to depend continuously on v. Now let $a(v) \in \mathbb{S}^1$ be the midpoint of $\alpha(v)$. Consider the continuous map $a: \mathbb{S}^{d-1} \to \mathbb{S}^1 \subset \mathbb{E}^{d-1}$. By one version of the Borsuk/Ulam theorem (see [Mat03]), there must be some v such that $a(v) = a(-v)$. But the height functions for v and $-v$ are negatives of each other, so maxima in direction v live in $\alpha(v)$ and in $\beta(-v)$, while minima live in $\beta(v)$ and $\alpha(-v)$. This is impossible if $a(v)$ is the midpoint of both $\alpha(v)$ and $\alpha(-v)$. (In fact, a sensible choice for α makes a antipodally equivariant: $a(-v) = -a(v)$, allowing direct application of another version of Borsuk/Ulam.)   □

Lemma 8.2. *If γ is a curve in $\mathbb{E}^{m+n}$ of length L, and its projections to $\mathbb{E}^m$ and $\mathbb{E}^n$ have lengths a and b, then $a^2 + b^2 \le L^2$.*

Proof. By Proposition 1.4, it suffices to prove this for polygons. Let $a_i, b_i \ge 0$ be the lengths of the two projections of the ith edge, and consider the polygonal arc in $\mathbb{E}^2$ with successive edge vectors (a_i, b_i). Its total length is $\sum \sqrt{a_i^2 + b_i^2} = L$, but the distance between its end points is $\sqrt{a^2 + b^2}$.   □

Theorem 8.3. *Any closed curve γ in $\mathbb{E}^d$ of length L lies in a cylinder of diameter L/π.*

Proof. As we have noted, the case $d = 2$ follows directly from the Cauchy/Crofton formula (Lemma 4.6), since γ has width at most L/π in some direction. We prove the general case by induction. So given γ in $\mathbb{E}^d$, find two parallel interleaved bitangent support planes with normal v, as in Proposition 8.1. Let τ be the distance between these planes, the thickness of the slab in which γ lies. Project γ to a curve $\bar{\gamma}$ in the plane orthogonal to v, and call its length $\bar{L}$. By induction, $\bar{\gamma}$ lies in a cylinder of radius $\bar{L}/2\pi$. Clearly, γ lies in a parallel cylinder of radius r, centered in the middle of the slab, where $r^2 = (\bar{L}/2\pi)^2 + (\tau/2)^2$. So we need to show that $(\bar{L}/2\pi)^2 + (\tau/2)^2 \le (L/2\pi)^2$, i.e., $\bar{L}^2 + \pi^2 \tau^2 \le L^2$. In fact, since the length of the projection of γ to the one-dimensional space in direction v is at least 4τ, Lemma 8.2 gives us $\bar{L}^2 + 4^2\tau^2 \le L^2$, which is better than we needed.   □

If we are willing to settle for a slightly worse bound in the original conjecture, Wienholtz also shows that we can project in a known direction:

Proposition 8.4. *Suppose a closed curve $\gamma \subset \mathbb{E}^d$ has length L, and $p_1, p_2 \in \gamma$ are points realizing its diameter. Then its projection $\bar{\gamma}$ to the plane orthogonal to $p_1 - p_2$ has diameter at most $L/2\sqrt{2}$. This estimate is sharp for a square.*

Proof. Let a_1, a_2 be the preimages of a pair of points realizing the diameter D of the projected curve $\bar{\gamma}$. We may assume γ is a quadrilateral with vertices a_1, p_1, a_2, p_2, since any other curve would be longer. (This reduces the problem to some affine $\mathbb{E}^3$ containing these four points.)

Along γ, first suppose the a_i are interleaved with the p_i, so that the quadrilateral is $a_1 p_1 a_2 p_2$, as in Figure 4 (left). We can now reduce to $\mathbb{E}^2$: rotating a_1, a_2 independently about the line $\overleftrightarrow{p_1 p_2}$ fixes the length, but maximizes the diameter D when the points are coplanar, with a_i on opposite sides of the line. Now let R be the reflection across this line $\overleftrightarrow{p_1 p_2}$, and consider the vector $R(p_1 - a_1) + (a_1 - p_2) + (p_1 - a_2) + R(a_2 - p_2)$.

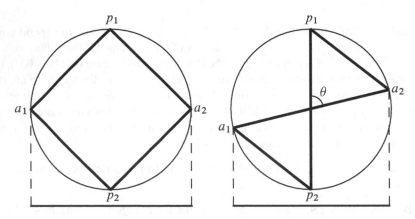

FIGURE 4. These quadrilaterals show the sharp bounds for the two cases in Proposition 8.4. The original diameter is $p_1 p_2$ in both cases, and the projected diameter is D (shown at the bottom). The square $a_1 p_1 a_2 p_2$ (left) has length $2\sqrt{2}D$. The bowtie $a_1 a_2 p_1 p_2$ (right) has length $2D/\sin\theta + D/\cos\frac{\theta}{2}$, minimized at about $3.33D$ for $\theta \approx 76°$ as shown, but in any case clearly greater than $3D$.

Its length is at most L, but its component in the direction $p_1 - p_2$ is at least $2D$, as is its component in the perpendicular direction. Therefore $L \geq 2\sqrt{2}D$.

Otherwise, write the quadrilateral as $a_1 a_2 p_1 p_2$ as in Figure 4 (right), and let θ be the angle between the vectors $a_2 - a_1$ and $p_1 - p_2$. Suppose (after rescaling) that $1 = |a_2 - a_1| \leq |p_1 - p_2|$. Then the projected diameter is $D = \sin\theta$. By the triangle inequality, the two remaining sides have lengths summing to at least

$$|p_1 - a_2 + a_1 - p_2| \geq 2\sin(\theta/2).$$

(Equality here holds, for instance, when the quadrilateral is a symmetric bowtie.) Thus

$$L/D \geq \frac{2 + 2\sin(\theta/2)}{\sin\theta} = \frac{2}{\sin\theta} + \frac{1}{\cos(\theta/2)} \geq 2 + 1 > 2\sqrt{2}. \qquad \square$$

9. Curvature density

We have found a setting which treats polygons and smooth curves in a unified way as two special cases of the more general class of finite total curvature curves. Many standard results on curvature, like Schur's comparison theorem, work nicely in this class.

However, there is some ambiguity in how to measure curvature at a corner, reflected in our quantities TC and TC*. A corner of turning angle θ is counted either as θ or as $2\sin(\theta/2)$, respectively.

At first, TC seems more natural: if we round off a corner into any convex planar arc, its curvature is θ. And the nice behavior of TC under projection (Proposition 4.1) explains why it is the right quantity for results like the Fáry/Milnor theorem.

But from a variational point of view, TC*, which measures the mass of the curvature force $\mathcal{K}$ at a corner, is sometimes most natural. Proposition 6.1 is an example of a result whose sharp form involves TC*. An arbitrary rounding of a corner, whether or not it is convex or planar, will have the same value of

$$\mathcal{K} = T_+ - T_- = 2\sin(\theta/2)\, N.$$

When we do choose a smooth, convex, planar rounding, we note that

$$\int |\kappa N|\, ds = \int \kappa\, ds = \theta, \qquad \text{while} \qquad \left| \int \kappa N\, ds \right| = 2\sin(\theta/2).$$

For us, the curvature of an FTC curve has been given by the measure $\mathcal{K}$. For a polygon, of course, this vanishes along the edges and has an atom at each vertex. But sometimes we wish to view a polygon as an approximation to a smooth curve and thus spread this curvature out into a smooth density. For instance the elastic energy, measured as $\int \kappa^2\, ds$ for smooth curves, blows up if measured directly on a polygon; instead of squaring $\mathcal{K}$, we should find a smoothed *curvature density* κ and square that.

For simplicity, we will consider here only the case of equilateral polygons in $\mathbb{E}^d$, where each edge has length 1. To each vertex v, we allocate the neighborhood N_v consisting of the nearer halves of the two edges incident to v, with total length 1. Depending on whether we are thinking of TC or TC*, we see total curvature either θ or $2\sin(\theta/2)$ at v, and so it would be natural to use either θ or $2\sin(\theta/2)$ as the curvature density κ along N_v. The latter has also a geometric interpretation: if uvw are consecutive vertices of the equilateral polygon, then the circle through those three points has curvature density $\kappa = 2\sin(\theta/2)$.

However, from a number of points of view, there is another even better measure of curvature density. Essentially, what we have ignored above is that when we round off the corners of a polygon to make a smooth curve, we tend to make the curve shorter. Thus, while θ or $2\sin(\theta/2)$ might be the correct total curvature for a neighborhood of v, perhaps it should get averaged over a length less than 1.

Let us consider a particularly simple smoothing, which gives a $C^{1,1}$ and piecewise circular curve. Given a polygon P, replace the neighborhood N_v of each vertex v by an "inscribed" circular arc, tangent at each end point. This arc turns a total angle θ, but since it is shorter than N_v, its curvature density is $\kappa = 2\tan(\theta/2)$.

As a simple example, suppose P is a regular n-gon in the plane with edges of length 1 and turning angles $2\pi/n$. Its inscribed circle has curvature density $2\tan(\theta/2)$, while its circumscribed circle has (smaller) curvature density $2\sin(\theta/2)$. (Of course, the value θ lies between these two.)

Using the formula $2\tan(\theta/2)$ for the curvature density along a polygon P has certain advantages. For instance if $\kappa(P) \leq C$, then we know there is a nearby $C^{1,1}$ curve (the smoothing by inscribed circular arcs we used above) with this same curvature bound. The fact that $2\tan(\theta/2)$ blows up for $\theta = \pi$ reflects the fact that a polygonal corner of turning angle π is really a cusp. For instance, when the turning angle of a polygon in the plane passes through π, the total signed curvature (or *turning number*) jumps. For a smooth curve, such a jump cannot happen, unless the bending energy blows up because of a cusp.

A bending energy for polygons based on $\kappa = 2\tan(\theta/2)$ will similarly blow up if we try to change the turning number.

Of course, the bending energy for curves is one conserved quantity for the integrable system related to the Hasimoto or smoke-ring flow. In the theory of discrete integrable systems, it seems clear due to work of Hoffmann and others that $2\tan(\theta/2)$ is the right notion of curvature density for equilateral polygons. See [HK04, Hof08].

Acknowledgments

My thoughts on curves of finite total curvature have developed over the course of writing various collaborative papers [KS97, CK$^+$03, CF$^+$04, DS04, DDS06] in geometric knot theory. Thus I owe a great debt to my coauthors Jason Cantarella, Elizabeth Denne, Yuanan Diao, Joe Fu, Greg Kuperberg, Rob Kusner and Nancy Wrinkle. I also gratefully acknowledge helpful conversations with many other colleagues, including Stephanie Alexander, Dick Bishop, Mohammad Ghomi and Günter M. Ziegler.

References

[AB98]   Stephanie B. Alexander and Richard L. Bishop, *The Fáry–Milnor theorem in Hadamard manifolds*, Proc. Amer. Math. Soc. **126**:11 (1998), 3427–3436.

[AD97]   Sahbi Ayari and Serge Dubuc, *La formule de Cauchy sur la longueur d'une courbe*, Canad. Math. Bull. **40**:1 (1997), 3–9.

[AFP00]   Luigi Ambrosio, Nicola Fusco, and Diego Pallara, *Functions of bounded variation and free discontinuity problems*, Clarendon/Oxford, 2000.

[AG95]   Helmut Alt and Michael Godau, *Computing the Fréchet distance between two polygonal curves*, Internat. J. Comput. Geom. Appl. **5**:1-2 (1995), 75–91, Proc. 8th ACM Symp. Comp. Geom. (Berlin, 1992).

[AR89]   Aleksandr D. Alexandrov and Yuri G. Reshetnyak, *General theory of irregular curves*, Math. Appl. (Soviet) 29, Kluwer, Dordrecht, 1989.

[AT72]   Louis I. Alpert and Leopoldo V. Toralballa, *An elementary definition of surface area in* $\mathbb{E}^{n+1}$ *for smooth surfaces*, Pacific J. Math. **40** (1972), 261–268.

[AZ98]   Martin Aigner and Günter M. Ziegler, *Proofs from The Book*, Springer, 1998.

[Ber98]   Sterling K. Berberian, *Fundamentals of real analysis*, Springer, 1998.

[BH74]   Frederick Brickell and Chuan-Chih Hsiung, *The total absolute curvature of closed curves in Riemannian manifolds*, J. Differential Geometry **9** (1974), 177–193.

[Bla21]   Wilhelm Blaschke, *Ungleichheiten von H. A. Schwarz und A. Schur für Raumkurven mit vorgeschriebener Krümmung.*, Hamb. Abh. **1** (1921), 49–53.

[Bla24]   ———, *Vorlesungen über Differentialgeometrie*, Springer, Berlin, 1924.

[Boa96]   Ralph P. Boas, Jr., *A primer of real functions*, fourth ed., Carus Math. Manus., no. 13, Math. Assoc. Amer., 1996.

[Bor47]   Karol Borsuk, *Sur la courbure totale des courbes fermées*, Ann. Soc. Polon. Math. **20** (1947), 251–265 (1948).

[BS99]   Gregory R. Buck and Jonathon K. Simon, *Thickness and crossing number of knots*, Topol. Appl. **91**:3 (1999), 245–257.

[Buf77]   Georges-Louis Leclerc, Comte de Buffon, *Essai d'arithmétique morale*, Histoire na-
          turelle, générale er particulière, Suppl. 4, 1777, pp. 46–123; www.buffon.cnrs.fr/
          ice/ice_book_detail-fr-text-buffon-buffon_hn-33-7.html.

[Car18]   Constantin Carathéodory, *Vorlesungen über reele Funktionen*, Teubner, 1918, reprinted
          2004 by AMS/Chelsea.

[Cau13]   Augustin-Louis Cauchy, *Deuxième mémoire sur les polygones et les polyèdres*, J. École
          Polytechnique **16** (1813), 87–98.

[Cau41]   _____, *Notes sur divers théorèmes relatifs à la rectification des courbes, et à la quad-
          rature des surfaces*, C. R. Acad. Sci. Paris **13** (1841), 1060–1063, reprinted in *Oeuvres
          complètes 6*, Gauthier-Villars, 1888, pp. 369–375.

[Ces56]   Lamberto Cesari, *Surface area*, Ann. Math. Stud., no. 35, Princeton, 1956.

[Ces89]   _____, *Surface area*, Global Differential Geometry (S.S. Chern, ed.), Math. Assoc.
          Amer., 1989, pp. 270–302.

[CF+04]   Jason Cantarella, Joseph H.G. Fu, Robert B. Kusner, John M. Sullivan, and Nancy C.
          Wrinkle, *Criticality for the Gehring link problem*, Geometry and Topology **10** (2006),
          2055–2115; arXiv.org/math.DG/0402212.

[Cha62]   Gulbank D. Chakerian, *An inequality for closed space curves*, Pacific J. Math. **12** (1962),
          53–57.

[Cha64]   _____, *On some geometric inequalities*, Proc. Amer. Math. Soc. **15** (1964), 886–888.

[Che89]   Shiing Shen Chern, *Curves and surfaces in euclidean space*, Global Differential Geom-
          etry (S.S. Chern, ed.), Math. Assoc. Amer., 1989, pp. 99–139.

[CK+03]   Jason Cantarella, Greg Kuperberg, Robert B. Kusner, and John M. Sullivan, *The second
          hull of a knotted curve*, Amer. J. Math **125** (2003), 1335–1348; arXiv.org/math.
          GT/0204106.

[CKS02]   Jason Cantarella, Robert B. Kusner, and John M. Sullivan, *On the minimum rope-
          length of knots and links*, Invent. Math. **150** (2002), 257–286; arXiv.org/math.
          GT/0103224.

[Con82]   Robert Connelly, *Rigidity and energy*, Invent. Math. **66** (1982), 11–33.

[Cro68]   Morgan W. Crofton, *On the theory of local probability*, Phil. Trans. R. Soc. London **158**
          (1868), 181–199.

[Cro97]   Peter Cromwell, *Polyhedra*, Cambridge, 1997.

[dC76]   Manfredo P. do Carmo, *Differential geometry of curves and surfaces*, Prentice-Hall,
          1976.

[DDS06]   Elizabeth Denne, Yuanan Diao, and John M. Sullivan, *Quadrisecants give new lower
          bounds for the ropelength of a knot*, Geometry and Topology **10** (2006), 1–26; arXiv.
          org/math.DG/0408026.

[Den04]   Elizabeth Denne, *Alternating quadrisecants of knots*, Ph.D. thesis, Univ. Illinois, Urbana,
          2004, arXiv.org/math.GT/0510561.

[DS04]   Elizabeth Denne and John M. Sullivan, *The distortion of a knotted curve*, preprint, 2004,
          arXiv.org/math.GT/0409438v1.

[DS08]   _____, *Convergence and isotopy type for graphs of finite total curvature*, Discrete Dif-
          ferential Geometry (A.I. Bobenko, P. Schröder, J.M. Sullivan, G.M. Ziegler, eds.), Ober-
          wolfach Seminars, vol. 38, Birkhäuser, 2008, this volume, pp. 163–174; arXiv.org/
          math.GT/0606008.

[Fár49]   István Fáry, *Sur la courbure totale d'une courbe gauche faisant un nœud*, Bull. Soc. Math. France **77** (1949), 128–138.

[Fen29]   Werner Fenchel, *Über Krümmung und Windung geschlossener Raumkurven*, Math. Ann. **101** (1929), 238–252; `www-gdz.sub.uni-goettingen.de/cgi-bin/ digbib.cgi?PPN235181684_0101`.

[Fen51]   Werner Fenchel, *On the differential geometry of closed space curves*, Bull. Amer. Math. Soc. **57** (1951), 44–54.

[Fré05]   Maurice Fréchet, *Sur l'écart de deux courbes et sur les courbes limites*, Trans. Amer. Math. Soc. **6** (1905), 435–449.

[GP83]   Casper Goffman and George Pedrick, *First course in functional analysis*, second ed., Chelsea, 1983.

[Gra46]   Lawrence M. Graves, *Theory of functions of real variables*, McGraw Hill, 1946.

[Gro78]   Mikhael Gromov, *Homotopical effects of dilatation*, J. Diff. Geom. **13** (1978), 303–310.

[Gro81]   _____, *Structures métriques pour les variétés riemanniennes*, Cedic, Paris, 1981, edited by J. Lafontaine and P. Pansu.

[Gro83]   _____, *Filling Riemannian manifolds*, J. Diff. Geom. **18** (1983), 1–147.

[HK04]   Tim Hoffmann and Nadja Kutz, *Discrete curves in $\mathbb{C}P^1$ and the Toda lattice*, Stud. Appl. Math. **113**:1 (2004), 31–55; `arXiv.org/math.DG/0208190`.

[Hof08]   Tim Hoffmann, *Discrete Hashimoto surfaces and a doubly discrete smoke-ring flow*, Discrete Differential Geometry (A.I. Bobenko, P. Schröder, J.M. Sullivan, G.M. Ziegler, eds.), Oberwolfach Seminars, vol. 38, Birkhäuser, 2008, this volume, pp. 95–115; `arXiv.org/math.DG/0007150`.

[Hor71]   Roger A. Horn, *On Fenchel's theorem*, Amer. Math. Monthly **78** (1971), 380–381.

[Jor93]   Camille Jordan, *Cours d'analyse de l'école polytechnique*, Gauthier-Villars, 1893.

[KR97]   Daniel A. Klain and Gian-Carlo Rota, *Introduction to geometric probability*, Cambridge, 1997.

[KS97]   Robert B. Kusner and John M. Sullivan, *On distortion and thickness of knots*, Topology and Geometry in Polymer Science (Whittington, Sumners, and Lodge, eds.), IMA Vol. 103, Springer, 1997, pp. 67–78; `arXiv.org/dg-ga/9702001`.

[Leb02]   Henri Lebesgue, *Intégrale, longueur, aire*, Annali di Mat. pura appl. **7** (1902), 231–359.

[Lie29]   Heinrich Liebmann, *Elementarer Beweis des Fenchelschen Satzes über die Krümmung geschlossener Raumkurven*, Sitz.ber. Akad. Berlin (1929), 392–393.

[Mat03]   Jiří Matoušek, *Using the Borsuk–Ulam theorem*, Springer, Berlin, 2003.

[Mil50]   John W. Milnor, *On the total curvature of knots*, Ann. of Math. **52** (1950), 248–257.

[Mor36]   Marston Morse, *A special parametrization of curves*, Bull. Amer. Math. Soc. **42** (1936), 915–922.

[Mor88]   Frank Morgan, *Geometric measure theory: A beginner's guide*, Academic Press, 1988.

[O'R00]   Joseph O'Rourke, *On the development of the intersection of a plane with a polytope*, Comput. Geom. Theory Appl. **24**:1 (2003), 3–10; `arXiv.org/cs.CG/0006035v3`.

[San89]   Luis A. Santaló, *Integral geometry*, Global Differential Geometry, Math. Assoc. Amer., 1989, pp. 303–350.

[San04]   _____, *Integral geometry and geometric probability*, second ed., Cambridge, 2004.

[Sch85] Ludwig Scheeffer, *Allgemeine Untersuchungen über Rectification der Curven*, Acta Math. **5** (1884–85), 49–82.

[Sch90] Hermann Amandus Schwarz, *Sur une définition erronée de l'aire d'une surface courbe*, Ges. math. Abhandl., vol. 2, Springer, 1890, pp. 309–311 and 369–370.

[Sch21] Axel Schur, *Über die Schwarzsche Extremaleigenschaft des Kreises unter den Kurven konstanter Krümmung*, Math. Annalen **83** (1921), 143–148; www-gdz.sub. uni-goettingen.de/cgi-bin/digbib.cgi?PPN235181684_0083.

[Sch25] Erhard Schmidt, *Über das Extremum der Bogenlänge einer Raumkurve bei vorgeschriebenen Einschränkungen ihrer Krümmung*, Sitz.ber. Akad. Berlin (1925), 485–490.

[Sch98] Carsten Schmitz, *The theorem of Fáry and Milnor for Hadamard manifolds*, Geom. Dedicata **71**:1 (1998), 83–90.

[Ser68] Joseph-Alfred Serret, *Cours de calcul différentiel et intégral*, vol. 2, Gauthier-Villars, Paris, 1868; historical.library.cornell.edu/cgi-bin/cul.math/ docviewer?did=05270001&seq=311.

[Sul08] John M. Sullivan, *Curvatures of smooth and discrete surfaces*, Discrete Differential Geometry (A.I. Bobenko, P. Schröder, J.M. Sullivan, G.M. Ziegler, eds.), Oberwolfach Seminars, vol. 38, Birkhäuser, 2008, this volume, pp. 175–188; arXiv.org/0710.4497.

[Tan98] Kouki Taniyama, *Total curvature of graphs in Euclidean spaces*, Differential Geom. Appl. **8**:2 (1998), 135–155.

[Ton21] Leonida Tonelli, *Fondamenti di calcolo delle variazione*, Zanichelli, Bologna, 1921.

[Vos55] Konrad Voss, *Eine Bemerkung über die Totalkrümmung geschlossener Raumkurven*, Arch. Math. **6** (1955), 259–263.

[Wor08] Max Wardetzky, *Convergence of the cotangent formula: An overview*, Discrete Differential Geometry (A.I. Bobenko, P. Schröder, J.M. Sullivan, G.M. Ziegler, eds.), Oberwolfach Seminars, vol. 38, Birkhäuser, 2008, this volume, pp. 275–286.

[Zie89] William P. Ziemer, *Weakly differentiable functions*, GTM 120, Springer, 1989.

John M. Sullivan
Institut für Mathematik, MA 3–2
Technische Universität Berlin
Str. des 17. Juni 136
10623 Berlin
Germany
e-mail: sullivan@math.tu-berlin.de

Discrete Differential Geometry, A.I. Bobenko, P. Schröder, J.M. Sullivan and G.M. Ziegler, eds.
Oberwolfach Seminars, Vol. 38, 163–174

Convergence and Isotopy Type
for Graphs of Finite Total Curvature

Elizabeth Denne and John M. Sullivan

Abstract. Generalizing Milnor's result that an FTC (finite total curvature) knot has
an isotopic inscribed polygon, we show that any two nearby knotted FTC graphs are
isotopic by a small isotopy. We also show how to obtain sharper constants when the
starting curve is smooth. We apply our main theorem to prove a limiting result for
essential subarcs of a knot.

Keywords. Knots, knotted graphs, isotopy, convergence.

1. Introduction

A *tame* knot or link type is one that can be represented by a polygonal space curve. It
is clear that the corners of a polygon can be rounded off, so that a tame link can also be
represented by a smooth, even C^∞, curve. Conversely, given a C^2 curve, it is relatively
easy to find an isotopic inscribed polygon. With less smoothness, the arguments become
more intricate but are by now standard. Crowell and Fox [CF63, App. I] gave a detailed
proof for C^1 curves, while Milnor [Mil50] introduced the class of *finite total curvature
(FTC) curves*, and showed they also have isotopic inscribed polygons.

Thus any embedded FTC curve represents a tame link type; but tame links (even the
unknot) can of course also be represented by curves of infinite total curvature or even infinite length. However, our work in geometric knot theory has suggested to us that FTC links
are a very useful class, which might be considered as "geometrically tame". (See [Sul08],
in this volume, for a survey on FTC curves. We show, for instance, that this is the natural
class for standard results like Schur's comparison theorem.)

Here, with a view towards convergence results in geometric knot theory, we examine the question of how close two space curves must be in order to be (ambient) isotopic.
Of course, the phenomenon of local knotting means that our notion of distance between
space curves must control tangent directions as well as position. (Otherwise, within any
distance ε of any knot there would be infinitely many composite knots—with extra summands contained in ε-balls—and even certain prime satellites of the original knot.) We

also show that, when the initial curves are close enough, the ambient isotopy can be made arbitrarily small.

The arguments of [CF63, Mil50] not only produce one isotopic inscribed polygon, but in fact show that all sufficiently fine inscribed polygons are isotopic to a given C^1 or FTC curve. We recover this result, since finely inscribed polygons are nearby in our sense (with one caveat, discussed later).

We prove three versions of our theorem. The first is restricted curves of positive thickness, that is, to $C^{1,1}$ links, but it allows us to get nearly optimal bounds (in terms of the thickness) for how close the two curves must be before they are guaranteed to be isotopic. The second version, our main theorem, applies not just to links but to arbitrary knotted FTC graphs; it gives us control over the distance points are moved by the isotopy. The final version applies to C^1 links and allows us to prove that the isotopy is small even in a stronger C^1 sense.

As an application of the main theorem, we consider the notion of essential arcs and secants of a knot, defined [DDS06] in terms of certain knotted Θ-graphs. We show that essential secants remain essential in limits, which is important for results [Den04] on quadrisecants.

2. Definitions

Definition. A *(knotted) rectifiable graph* is a (multi-)graph—of fixed finite combinatorial type—embedded in space such that each edge is a rectifiable arc. An *FTC graph* is a rectifiable graph where each arc has finite total curvature.

We note that Taniyama [Tan98] has considered minimization of the total curvature of knotted graphs, including analogs of the Fáry/Milnor theorem.

For an introduction to the theory of knotted graphs, see [Kau89]: in particular, we are interested in graphs with what Kauffman calls topological (rather than rigid) vertices. The k edges incident to a vertex v can be braided arbitrarily near v without affecting the knot type of the graph. We do not require our graphs to be connected, so our results apply to links as well as knots.

As mentioned above, in order to conclude that nearby knots are isotopic, we need to use a notion of distance that controls tangent vectors. A standard notion would be, for instance, C^1 convergence of C^1 curves. Our goal, however, is to consider curves which need not be C^1. Recalling that a rectifiable curve has a well-defined tangent vector almost everywhere, we can make the following definition.

Definition. Given two rectifiable embeddings Γ and Γ' of the same combinatorial graph, we say they are (δ, θ)-*close* if there exists a homeomorphism between them such that corresponding points are within distance δ of each other, and corresponding tangent vectors are within angle θ of each other almost everywhere.

Remember [Sul08] that any curve or graph Γ of finite total curvature has well-defined one-sided tangent vectors everywhere; these are equal and opposite except at countably many *corners* of Γ (including of course the vertices of the graph). In the definition of (δ, θ)-close it would be equivalent—in the case of FTC or piecewise C^1 graphs—to require that the one-sided tangent vectors be *everywhere* within the angle θ.

3. Isotopy for thick knots

Our first isotopy result looks at thick curves (that is, embedded $C^{1,1}$ curves). Here we can get close to optimal bounds on the δ and θ needed to conclude that nearby curves are isotopic.

The *thickness* $\tau(K)$ of a space curve K is defined [GM99] to be twice the infimal radius of circles through any three distinct points of K. A link is $C^{1,1}$ (that is, C^1 with Lipschitz tangent vector, or, equivalently, with a weak curvature bound) if and only if it has positive thickness [CKS02]. Of course, when K is C^1, we can define normal tubes around K, and then $\tau(K)$ is the supremal diameter of such a tube that remains embedded. (We note that in the existing literature thickness is sometimes defined to be the radius rather than diameter of this thick tube.) Points inside the tube have a unique nearest neighbor on K: the tube's radius is also the *reach* of K in the sense of Federer [Fed59].

Fix a link K of thickness $\tau > 0$. Any other link which follows K once around within its thick tube, transverse to the normal disks, will be isotopic to K. This might seem to say that any link $(\tau/2, \pi/2)$-close to K is isotopic, but this would only work if we knew the correspondence used in the definition of closeness were the closest-point projection to K. We do see that (δ, θ)-closeness is no guarantee of isotopy if $\theta > \pi/2$ (allowing local knotting) or if $\delta > \tau/2$ (allowing global strand passage). So the following proposition (which extends Lemma 6.1 from [DDS06], with essentially the same proof) is certainly close to optimal.

(We note, however, that $\delta > \tau/2$ only allows global strand passage in the case where the thickness of K is controlled by self-distance rather than curvature. This caveat explains why we can later get some analogs of the following proposition even for curves whose thickness is zero because they have corners.)

Proposition 3.1. *Fix a $C^{1,1}$ link K of thickness $\tau > 0$. For any $\delta < \tau/4$, set $\theta := \pi/2 - 2\arcsin(2\delta/\tau)$. Then any (rectifiable) link K' which is (δ, θ)-close to K is ambient isotopic; the isotopy can be chosen to move no point a distance more than δ.*

Proof. For simplicity, rescale so that $\tau = 1$, meaning the curvature of K is bounded by 2. Suppose p and p' are corresponding points on K and K', so $|p' - p| < \delta$. Let p_0 be the (unique) closest point on K to p'. Then of course $|p' - p_0| \leq |p' - p| < \delta$, so $|p - p_0| < 2\delta$. Standard results on the geometry of thick curves [DDS06, Lem. 3.1] show that the arc length of K between p and p_0 is at most $\arcsin(2\delta)$, so by the curvature bound, the angle between the tangents T_p and T_{p_0} at these two points is at most $2\arcsin(2\delta)$. The definition of (δ, θ)-close means that the tangent vector (if any) to K' at p' makes an angle less than θ with T_p, so by definition of θ it makes an angle less than $\pi/2$ with T_{p_0}.

This shows that K' is transverse to the foliation of the thick tube around K by normal disks. We now construct the isotopy from K' to K as the union of isotopies in these disks; on each disk we move p' to p_0, coning this outwards to the fixed boundary. No point in the disk moves further than p' does, and this is less than δ.   □

4. Isotopy for graphs of finite total curvature

When generalizing our isotopy results to knotted graphs, we must allow corners at the vertices of the graph, so it is natural to consider FTC graphs with possibly additional corners along the arcs. Our main theorem will again show that sufficiently close curves are isotopic.

Any curve with a corner has zero thickness $\tau = 0$, so it is perhaps surprising that an analog of Proposition 3.1 still holds. Perhaps the maximum possible value of δ here could be taken as a different notion of thickness of the curve; we do not explore this idea, but content ourselves with merely proving the existence of some positive δ.

A preliminary combing lemma allows the main theorem to treat vertices on knotted graphs as well as sharp corners along a single strand. It was inspired by Alexander and Bishop's correction [AB98] to Milnor's treatment of corners of angle π. This case—where two or more strands leave a corner in the same direction, allowing infinite winding or braiding—is the most difficult, and seems to be the obstacle to creating an isotopy which is (δ, θ)-small in the sense described in Section 7. But this case does not need to be treated separately in our proof.

This combing lemma is an example of the "Alexander trick" (compare [Thu97, Prob. 3.2.10]). In general the trick shows that any homeomorphism of the unit d-ball which fixes the boundary is isotopic to the identity.

Lemma 4.1. *Let B be a round ball centered at p, and suppose graphs Γ and Γ' each consist of n labeled arcs starting at p and proceeding out to ∂B transverse to the nested spheres around p. Then Γ and Γ' are ambient isotopic. In fact, any isotopy of ∂B which takes the n points of Γ to those of Γ' can be extended to an isotopy of the whole ball.*

Proof. It suffices to show that Γ is isotopic to a configuration with n straight radial segments to the same n points on ∂B. For then the same is true of Γ', and any two straight configurations are clearly isotopic (for instance, by extending radially any given boundary isotopy).

To straighten Γ, we comb the n strands inwards from ∂B, which we identify with the unit sphere. For $0 < \lambda \leq 1$, let $p_i(\lambda)$ be the n points of intersection of Γ with the sphere $\lambda \partial B$. Start with f_1 being the identity map, and define a family of maps f_λ from the unit sphere to itself, continuous in λ, such that f_λ takes $p_i(1)$ to $p_i(\lambda)/\lambda$ for each i. The combing isotopy, at each time t, maps each concentric sphere $\lambda \partial B$ to itself, using the map $f_{\lambda+(1-\lambda)t} \circ f_\lambda^{-1}$. Thus at time t on the sphere of size λ, we see the same picture that was initially on the sphere of size $\lambda + (1 - \lambda)t$; see Figure 1. It follows that at time $t = 1$ the strands are all straight.   □

As we have mentioned, if two (or more) strands of Γ leave p with a common tangent, then they might twist (or braid) around each other infinitely often. But outside any

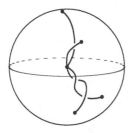 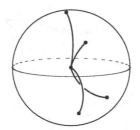 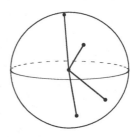

FIGURE 1. The isotopy of Lemma 4.1 straightens strands in a ball by combing them inwards. Only the outer half of the time 0 pattern (left) is still seen at time $\frac{1}{2}$ (center); at time 1 (right) we have straight radii.

sphere $\lambda \partial B$ they can twist only finitely many times, explaining why the isotopy constructed in the lemma has no trouble combing this out.

Theorem 4.2. *Suppose Γ is a knotted graph of finite total curvature and $\varepsilon > 0$ is given. Then there exists $\delta > 0$ such that any (rectifiable) graph Γ' which is $(\delta, \pi/8)$-close to Γ is ambient isotopic to Γ, via an isotopy which moves no point by more than ε.*

Proof. We begin by selecting a finite number of points p_j on Γ (including all its vertices) such that these points divide Γ into arcs α_k each of total curvature less than $\pi/8$. (Note that any corner in Γ of turning angle at least $\pi/8$ must be included among the p_j.)

Let r_1 be the minimum distance between any two arcs α_k which are not incident to a common p_j (or the minimum distance between points p_j, if this is smaller). Set $r_2 := \min(r_1/2, \varepsilon/2)$. Consider disjoint open balls B_j of radius r_2 centered at the p_j. Each arc α_k leaving p_j proceeds monotonically outwards to the boundary of B_j (since its curvature is too small to double back). Also, B_j contains no other arcs (since r_2 was chosen small enough); indeed no other arcs come within distance r_2 of B_j.

Note that $\Gamma \smallsetminus \bigcup B_j$ is a compact union of disjoint arcs $\beta_k \subset \alpha_k$. Let r_3 be the minimum distance between any two of these arcs β_k. Considering end points of various β_k on the boundary of a single B_j, we note that $r_3 \le 2r_2 < \varepsilon$.

For each β_k, we construct a tube T_k around it as follows. Suppose p_j and $p_{j'}$ are the two end points of $\alpha_k \supset \beta_k$. Foliate $\mathbb{R}^3 \smallsetminus (B_j \cup B_{j'})$ by spheres of radius at least r_2 (and one plane, bisecting $\overline{p_j p_{j'}}$) in the obvious smooth way. (Their signed curvatures vary linearly along $\overline{p_j p_{j'}}$.) Note that because the arc α_k has total curvature less than $\pi/8$, by [Sul08, Lem. 2.3] it is contained in a spindle of revolution from p to p', of curvature $\pi/4$. Within this spindle, the normal vectors to the foliating spheres stay within angle $\pi/8$ of $\overline{p_j p_{j'}}$, as do the (one-sided) tangent vectors to α_k. Thus β_k is within angle $\pi/4$ of being normal to the foliation; in particular it is transverse. Finally, we set $r_4 := r_3/6$ and, for each point $q \in \beta_k$, we consider the foliating sphere through q and in particular its intersection with $B_{r_4}(q)$, a (slightly curved) disk D_q. The tubular neighborhood T_k is the union of these disks over all $q \in \beta_k$. Since $r_4 < r_2/3$, the normal vector along each D_q stays within angle $2 \arcsin \frac{1}{6} < \pi/8$ of that at its center q, hence within $\pi/4$ of $\overline{p_j p_{j'}}$.

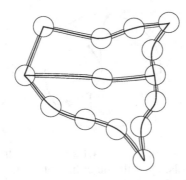

FIGURE 2. An FTC graph has a neighborhood consisting of balls and tubes, inside of which we perform our isotopy to any nearby graph.

We have thus found a neighborhood N of Γ which, as shown in Figure 2, is foliated by round spheres near the p_j and almost flat disks along the β_k; the graph Γ is transverse to this foliation F.

Now take $\delta := r_4/3 < r_2/9$ and consider some Γ' which is $(\delta, \pi/8)$-close to Γ. We will see that Γ' lies within our neighborhood N and is also transverse to the foliation F, except perhaps near the p_j. To find an ambient isotopy from Γ' to Γ, we proceed in two stages.

First we find the points $p'_j \in \Gamma'$ corresponding to $p_j \in \Gamma$, and construct an isotopy I supported in the B_j as follows: the map moves each point along a straight line; p'_j moves to p_j, each boundary point $q \in \partial B_j$ is fixed, and the map is linear on each segment $\overline{p_j q}$. Because Γ' was $(\delta, \pi/8)$-close to Γ, and because p'_j was so close to p_j compared to the radius r_2 of B_j, the resulting $\Gamma'' := I_1(\Gamma')$ at time 1 has radial strands within each B_j, so we will be able to use the combing lemma.

In fact, we claim that Γ'' is contained in N and is transverse to the foliation F. The original correspondence between Γ' and Γ (showing they were $(\delta, \pi/8)$-close) did not pair points in the same leaf of the foliation. But outside the B_j, for each point q' on the arc of $\Gamma'' = \Gamma'$ corresponding to β_k, consider the leaf of the foliation it lies in and the unique point $q_0 \in \beta_k$ on the same leaf. The distance $|q' - q_0|$ might be slightly bigger than the distance $|q' - q| < \delta$ in the original pairing. However, since the angles of the curves and the foliation are so well controlled, it is certainly well less than $2\delta < r_4$; the tangent directions at q and q_0 are certainly within $\pi/8$ of each other. That is, even under the new correspondence (q' to q_0) these arcs of Γ' and Γ are $(2\delta, \pi/4)$-close. This proves the claim.

Second, we construct an isotopy J which is supported in N and which preserves each leaf of the foliation F. On the disks in the tubes T_k this is easy: we have to move one given point in the disk to the center, and can choose to do this in a continuous way. On the boundary ∂B_j we define J to match these motions in the disks and fix the rest of the sphere. Finally, we use Lemma 4.1 to fill in the isotopy J in the interior of the balls.

By construction, the overall isotopy I followed by J moves points less than ε, as desired: within each ball B_j we have little control over the details but certainly each point moves less than the diameter $2r_2 < \varepsilon$; within T_k each point moves less than $r_4 \ll \varepsilon$.   □

Even though we have emphasized the class of FTC curves, we note that a very similar proof could be given for the case when Γ is piecewise C^1. We would simply choose the points p_j such that the intervening arcs α_k, while potentially of infinite total curvature, had tangent vectors staying within angle $\pi/8$ of the vector between their end points. Compare the beginning of the proof of Proposition 7.3 below.

As we noted in the introduction, we can recover the following result: given any FTC (or C^1) link, any sufficiently fine inscribed polygon is isotopic. When the link is C^1, this is an immediate corollary, since the polygon will be (δ, θ)-close. For an FTC link K with corners, even very finely inscribed polygons will typically cut those corners and thus deviate by more than angle $\pi/8$ from the tangent directions of K. But, using Lemma 4.1 near those sharp corners, we see immediately that the polygon is isotopic to one that does use the corner as a vertex, and thus to K.

5. Tame and locally flat links and graphs

Remember that a tame link is one isotopic to a polygon. Tame links clearly satisfy the following local condition.

Definition. A link K is *locally flat* if each point $p \in K$ has a neighborhood U such that $(U, U \cap K)$ is an unknotted ball–arc pair. (Unknotted means that the pair is homeomorphic to a round ball with its diameter.)

We can generalize the definitions to knotted graphs.

Definition. A knotted graph Γ is *tame* if it is isotopic to a polygonal embedding; it is *locally flat* if each point $p \in \Gamma$ has a neighborhood U such that $(U, U \cap \Gamma)$ is homeomorphic to a standard model. At a k-fold vertex this standard model is a round ball with k of its radii; along an arc of Γ it is an unknotted ball–arc pair.

Again, tame graphs are clearly locally flat. In the mid-1950s, Bing [Bin54] and Moise [Moi54] independently proved the much more difficult converse: any locally flat graph is tame.

Our main theorem showed that an FTC graph is isotopic to any other nearby graph. This implies that FTC graphs are tame: as long as we include sharp corners among the polygon vertices, sufficiently fine inscribed polygons will be $(\delta, \pi/8)$-close.

Here we note that it is much easier to prove directly that FTC graphs are locally flat.

Proposition 5.1. *Graphs of finite total curvature are locally flat.*

Proof. Let $p_0 \in \Gamma$ be any point along a graph of finite total curvature. Repeat the construction at the beginning of the proof of Theorem 4.2, but include p_0 among the p_j. This gives a ball B_0 around p_0 containing only k radial arcs of Γ. By Lemma 4.1 this ball is homeomorphic to any other such ball, in particular to the standard model.   □

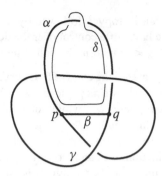

FIGURE 3. In this knotted Θ-graph the ordered triple (α, β, γ) is essential. The curve δ is the parallel to $\alpha \cup \beta$ having linking number zero with $\alpha \cup \gamma$. The fact that it is homotopically nontrivial in the knot complement $\mathbb{R}^3 \smallsetminus (\alpha \cup \gamma)$ is the obstruction to the existence of a spanning disk. In this illustration, β is the straight segment $\overline{pq}$, so we equally say that the arc α of the knot $\alpha \cup \gamma$ is essential.

6. Applications to essential arcs

One strand of work in geometric knot theory attempts to show that knotted curves are geometrically more complex than unknots. One of the earliest results in this direction is the Fáry/Milnor theorem, which says a knotted curve has more than twice the total curvature of a round circle. Interesting measures of geometric complexity include rope length [CKS02] and distortion [KS97].

We have recently obtained new lower bounds for both rope length [DDS06] and distortion [DS04] of knotted curves using the notion of essential arcs. Generically, a knot K together with one of its chords $\overline{pq}$ forms a Θ-graph in space; being essential is a topological feature of this knotted graph. The following definition, introduced in [Den04, DDS06] and illustrated in Figure 3, is an extension of ideas of Kuperberg [Kup94].

Definition. Suppose α, β and γ are three disjoint simple arcs from p to q, forming a knotted Θ-graph in $\mathbb{R}^3$. We say that (α, β, γ) is *inessential* if there is a disk D bounded by the knot $\alpha \cup \beta$ and having no interior intersections with the knot $\alpha \cup \gamma$. (We allow self-intersections of D, and interior intersections with β; the latter are necessary if $\alpha \cup \beta$ is knotted.)

Now suppose p and q are two points along a knot K, dividing it into complementary subarcs γ_{pq} and γ_{qp}. We say γ_{pq} is *essential* in K if for every $\varepsilon > 0$ there exists some ε-perturbation S of $\overline{pq}$ (with end points fixed) such that $K \cup S$ is an embedded Θ-graph in which $(\gamma_{pq}, S, \gamma_{qp})$ is essential.

Allowing the ε-perturbation ensures that the set of essential arcs is closed in the set of all subarcs of K; it also lets us handle the case when S intersects K. We could allow the perturbation only in case of such intersections; applying an argument like Lemma 4.1 in the case of an embedded Θ (no intersections) shows this definition would be equivalent, at least when the knot K is FTC. We require only that the perturbation be small in a pointwise sense. Thus S could be locally knotted, but since the disk in the definition can intersect $\beta = S$ this is irrelevant.

This definition of essential in terms of Θ-graphs was what led us to consider knotted graphs and to look for a result like Theorem 4.2. We now apply that theorem to show that essential arcs are preserved in limits. (In [DDS06, Prop. 6.2] we gave a similar result, but since we did not yet have Theorem 4.2, we had to restrict to $C^{1,1}$ knots only. The proof below does closely follow the proof there, Theorem 4.2 being the main new ingredient.)

Proposition 6.1. *Suppose a sequence of rectifiable knots K_i converge to an FTC limit K in the sense that K_i is $(\delta_i, \pi/10)$-close to K with $\delta_i \to 0$. Suppose that the K_i have essential subarcs $\gamma_{p_i q_i}$ with $p_i \to p$ and $q_i \to q$. Then γ_{pq} is an essential subarc of K.*

Proof. We can reduce to the case $p_i = p, q_i = q$ (but with $\pi/10$ above replaced by $\pi/8$) by appyling euclidean similarities (approaching the identity) to the K_i.

Given any $\varepsilon > 0$, we prove that there is an 2ε-perturbation S of $\overline{pq}$ for which $(\gamma_{pq}, S, \gamma_{qp})$ is essential. Then by definition γ_{pq} is essential.

We want to find an ambient isotopy I from some K_i to K which moves points less than ε. So apply Theorem 4.2 to K to determine $\delta > 0$ and then pick i large enough so that the knot K_i is $(\delta, \pi/8)$-close to K. The ambient isotopy I guaranteed by Theorem 4.2, with $K = I(K_i)$, moves points less than ε, as desired.

Since $\gamma_{p_i q_i} \subset K_i$ is essential, by definition, we can find an ε-perturbation S_i of $\overline{p_i q_i}$ such that $(\gamma_{p_i q_i}, S_i, \gamma_{q_i p_i})$ is essential. Setting $S := I(S_i)$, this is the desired 2ε-perturbation of $\overline{pq}$. By definition, the Θ-graph $K_i \cup S_i$ is isotopic via I to $K \cup S$, so in the latter $(\gamma_{pq}, S, \gamma_{qp})$ is also essential. $\square$

7. Small isotopies in a stronger sense

The isotopy constructed in Proposition 3.1 is small in a pointwise sense. But we have required our curves to be close in a stronger sense. We will say that a C^1 isotopy I, or more precisely its time-1 map I_1, is (δ, θ)-small if I_1 moves no point distance more than δ and its derivative turns no tangent vector more than angle θ. Note that under such a small isotopy, every rectifiable graph Γ is (δ, θ)-close to its image $I_1(\Gamma)$. (Although we could extend the definition of "small" for certain non-smooth isotopies, we do not pursue this idea here. It remains unclear if the ball isotopy of Lemma 4.1 could be made small in some such sense.)

Lemma 7.1. *Suppose an ambient isotopy is given for $t \in [0, 1]$ by $I_t(p) = p + tf(p)$, where $f : \mathbb{R}^3 \to \mathbb{R}^3$ is smooth, bounded in norm by δ, and λ-Lipschitz. Then I is $(\delta, \arctan \lambda)$-small.*

Proof. Clearly each point $p \in \Gamma$ is within distance δ of $p' := I_1(p)$. If v is the unit tangent vector to Γ at p, then its image, the tangent vector to Γ' at p', is $v + \partial_v f$, where $|\partial_v f| < \lambda$ by the Lipschitz condition. This differs in angle from v by at most $\arctan \lambda$. $\square$

The isotopy we built disk-by-disk in Proposition 3.1 is not necessarily smooth, and does not necessarily satisfy the hypotheses of this lemma. In particular, since K might not be piecewise C^2, its normal disks might not form a piecewise C^1 foliation, and the resulting isotopy would not even have one-sided derivatives everywhere.

So we will use a different construction, not using the thick tube. Here, as in Theorem 4.2, we abandon any attempt to get optimal constants, but the starting curve can be any C^1 curve, as in Crowell and Fox's construction [CF63, App. I] of an isotopic inscribed polygon. We begin with a lemma which captures what we feel is the essential ingredient of that argument.

Lemma 7.2. *Given a C^1 link K and an angle $\theta > 0$, there is some $\ell > 0$ such that along any subarc α of K of length at most ℓ, the tangent vector stays within angle θ of the chord vector v connecting the end points of α.*

Proof. Because K is compact, its continuous tangent vector is in fact uniformly continuous. Noting that v is an average of the tangent vectors along α, the result follows immediately. $\qquad\square$

This implies that the distortion (the supremal arc/chord ratio) of any C^1 curve is bounded: by the lemma the ratio approaches 1 for short arcs, so the supremum is achieved.

Proposition 7.3. *Given a C^1 link K, we can find $\varepsilon > 0$ such that the following holds. For any $\delta < \varepsilon$ and $\theta < \pi/6$, and any C^1 link K' which is (δ, θ)-close to K, there is a $(2\delta, 2\theta)$-small ambient isotopy taking K' to K.*

Sketch of proof. Fix $\varphi < 5°$, and find $\ell > 0$ as in Lemma 7.2. Now set $\tau := \min|x - y|$, where the minimum is taken over all points $x, y \in K$ not connected by a subarc of length less than ℓ. (Note that $\tau < \ell$.)

Along each component of K, pick $r \in (\tau/50, \tau/40)$ and place points p_j spaced equally at arc length r. The resulting inscribed polygon P has edge lengths in the interval $[r\cos\varphi, r]$ and is $(r\sin\varphi, \varphi)$-close to K, with turning angles at the p_j less than 2φ. Rounding off the corners in a suitable fashion, we obtain a C^2 curve L whose radius of curvature is never less than $5r$, and which is (d, φ)-close to P for $d \leq (\sec\varphi - 1)r/2 < r/500$. We see that L is $(\tau/400, 2\varphi)$-close to K, and has radius of curvature at least $\tau/10$.

We claim that the normal tube around L of diameter $\tau/5$ is embedded. This is true locally by the curvature bound. By standard results on thickness (cf. [CKS02, DDS06]) it can then only fail if there is a doubly critical pair on L at distance less than $\tau/5$: a pair whose chord is perpendicular to L at both ends. Then there would be a nearby pair on K, still at distance less than $\tau/4$. Since an arc of L connecting the doubly critical pair must turn at least π, the pair on X cannot be connected by an arc of length less than ℓ. Thus we have a contradiction to the choice of τ.

Now choose $\varepsilon := \tau/100$. For any $\delta < \varepsilon$ and $\theta < \pi/6$, if K' is (δ, θ)-close to K, then it is easy to check that both are C^1, transverse sections in the tube of radius $\tau/50$ around L. We construct an isotopy between them, supported in the embedded tube of radius $\tau/10$ around L. On each normal disk, we move a subdisk of radius $\tau/25$ rigidly, to move its center from K to K'. In the outer part of the thick tube, we extend in such a way that the isotopy is C^1 throughout space. The isotopy moves no points more than the distance between K and K' in the normal disk. This may be somewhat more than δ, since the original pairing between the two links may have been different, but is certainly less than 2δ.

To check that this isotopy turns tangent vectors at most 2θ, it is easiest to apply Lemma 7.1. The Lipschitz constant λ is basically determined by needing to turn K' by approximately angle θ to match K. Because the normal disks get slightly closer to one another as we move off L in the direction of its curvature, and since we have smoothed outside the thinner tube, the constant goes up slightly, but not enough to violate the 2θ. $\qquad\square$

Acknowledgments

We extend our thanks to Stephanie Alexander, Dick Bishop and Dylan Thurston for their interest and helpful conversations.

References

[AB98] Stephanie B. Alexander and Richard L. Bishop, *The Fáry Milnor theorem in Hadamard manifolds*, Proc. Amer. Math. Soc. **126**:11 (1998), 3427–3436.

[Bin54] RH Bing, *Locally tame sets are tame*, Ann. of Math. (2) **59**:1 (1954), 145–158.

[CF63] Richard H. Crowell and Ralph H. Fox, *Introduction to knot theory*, GTM 57, Springer, 1963.

[CKS02] Jason Cantarella, Robert B. Kusner, and John M. Sullivan, *On the minimum ropelength of knots and links*, Invent. Math. **150** (2002), 257–286; arXiv.org/math.GT/0103224.

[DDS06] Elizabeth Denne, Yuanan Diao, and John M. Sullivan, *Quadrisecants give new lower bounds for the ropelength of a knot*, Geometry and Topology **10** (2006), 1–26; arXiv.org/math.DG/0408026.

[Den04] Elizabeth Denne, *Alternating quadrisecants of knots*, Ph.D. thesis, Univ. Illinois, Urbana, 2004, arXiv.org/math.GT/0510561.

[DS04] Elizabeth Denne and John M. Sullivan, *The distortion of a knotted curve*, preprint, 2004, arXiv.org/math.GT/0409438.

[Fed59] Herbert Federer, *Curvature measures*, Trans. Amer. Math. Soc. **93** (1959), 418–491.

[GM99] Oscar Gonzalez and John H. Maddocks, *Global curvature, thickness, and the ideal shapes of knots*, Proc. Nat. Acad. Sci. (USA) **96** (1999), 4769–4773.

[Kau89] Louis H. Kauffman, *Invariants of graphs in three-space*, Trans. Amer. Math. Soc. **3** (1989), 697–710.

[KS97] Robert B. Kusner and John M. Sullivan, *On distortion and thickness of knots*, Topology and Geometry in Polymer Science (S.G. Whittington, D.W. Sumners, and T. Lodge, eds.), IMA Vol. 103, Springer, 1997, pp. 67–78; arXiv.org/dg-ga/9702001.

[Kup94] Greg Kuperberg, *Quadrisecants of knots and links*, J. Knot Theory Ramifications **3** (1994), 41–50; arXiv.org/math.GT/9712205.

[Mil50] John W. Milnor, *On the total curvature of knots*, Ann. of Math. **52** (1950), 248–257.

[Moi54] Edwin E. Moise, *Affine structures in 3-manifolds. VIII. Invariance of the knot-types; local tame imbedding*, Ann. of Math. **59** (1954), 159–170.

[Sul08] John M. Sullivan, *Curves of finite total curvature*, Discrete Differential Geometry (A.I. Bobenko, P. Schröder, J.M. Sullivan, G.M. Ziegler, eds.), Oberwolfach Seminars, vol. 38, Birkhäuser, 2008, this volume, pp. 137–161; arXiv.org/math.GT/0606007.

[Tan98] Kouki Taniyama, *Total curvature of graphs in Euclidean spaces*, Differential Geom. Appl. **8**:2 (1998), 135–155.

[Thu97] William P. Thurston, *Three-dimensional geometry and topology*, Princeton, 1997, edited by Silvio Levy.

Elizabeth Denne
Department of Mathematics & Statistics
Smith College
Northampton, MA 01063
USA
e-mail: edenne@email.smith.edu

John M. Sullivan
Institut für Mathematik, MA 3–2
Technische Universität Berlin
Str. des 17. Juni 136
10623 Berlin
Germany
e-mail: sullivan@math.tu-berlin.de

Discrete Differential Geometry, A.I. Bobenko, P. Schröder, J.M. Sullivan and G.M. Ziegler, eds.
Oberwolfach Seminars, Vol. 38, 175–188

Curvatures of Smooth and Discrete Surfaces

John M. Sullivan

Abstract. We discuss notions of Gauss curvature and mean curvature for polyhedral surfaces. The discretizations are guided by the principle of preserving integral relations for curvatures, like the Gauss–Bonnet theorem and the mean-curvature force balance equation.

Keywords. Discrete Gauss curvature, discrete mean curvature, integral curvature relations.

The curvatures of a smooth surface are local measures of its shape. Here we consider analogous quantities for discrete surfaces, meaning triangulated polyhedral surfaces. Often the most useful analogs are those which preserve integral relations for curvature, like the Gauss–Bonnet theorem or the force balance equation for mean curvature. For simplicity, we usually restrict our attention to surfaces in euclidean three-space $\mathbb{E}^3$, although some of the results generalize to other ambient manifolds of arbitrary dimension.

This article is intended as background for some of the related contributions to this volume. Much of the material here is not new; some is even quite old. Although some references are given, no attempt has been made to give a comprehensive bibliography or a full picture of the historical development of the ideas.

1. Smooth curves, framings and integral curvature relations

A companion article [Sul08] in this volume investigates curves of finite total curvature. This class includes both smooth and polygonal curves, and allows a unified treatment of curvature. Here we briefly review the theory of smooth curves from the point of view we will later adopt for surfaces.

The curvatures of a smooth curve γ (which we usually assume is parametrized by its arc length s) are the local properties of its shape, invariant under euclidean motions. The only first-order information is the tangent line; since all lines in space are equivalent, there are no first-order invariants. Second-order information (again, independent of parametrization) is given by the osculating circle; the one corresponding invariant is its curvature $\kappa = 1/r$.

(For a plane curve given as a graph $y = f(x)$ let us contrast the notions of curvature κ and second derivative f''. At a point p on the curve, we can find either one by translating p to the origin, transforming so the curve is horizontal there, and then looking at the second-order behavior. The difference is that for curvature, the transformation is a euclidean rotation, while for second derivative, it is a shear $(x, y) \mapsto (x, y - ax)$. A parabola has constant second derivative f'' because it looks the same at any two points after a shear. A circle, on the other hand, has constant curvature because it looks the same at any two points after a rotation.)

A plane curve is completely determined (up to rigid motion) by its (signed) curvature $\kappa(s)$ as a function of arc length s. For a space curve, however, we need to look at the third-order invariants; these are the torsion τ and the derivative κ', but the latter of course gives no new information. Curvature and torsion now form a complete set of invariants: a space curve is determined by $\kappa(s)$ and $\tau(s)$.

Generically speaking, while second-order curvatures usually suffice to determine a hypersurface (of codimension 1), higher-order invariants are needed for higher codimension. For curves in $\mathbb{E}^d$, for instance, we need $d - 1$ generalized curvatures, of order up to d, to characterize the shape.

Let us examine the case of space curves $\gamma \subset \mathbb{E}^3$ in more detail. At every point $p \in \gamma$ we have a splitting of the tangent space $T_p\mathbb{E}^3$ into the tangent line $T_p\gamma$ and the normal plane. A framing along γ is a smooth choice of a unit normal vector N_1, which is then completed to the oriented orthonormal frame (T, N_1, N_2) for $\mathbb{E}^3$, where $N_2 = T \times N_1$. Taking the derivative with respect to arc length, we get a skew-symmetric matrix (an infinitesimal rotation) that describes how the frame changes:

$$\begin{pmatrix} T \\ N_1 \\ N_2 \end{pmatrix}' = \begin{pmatrix} 0 & \kappa_1 & \kappa_2 \\ -\kappa_1 & 0 & \tau \\ -\kappa_2 & -\tau & 0 \end{pmatrix} \begin{pmatrix} T \\ N_1 \\ N_2 \end{pmatrix}.$$

Here, $T'(s) = \sum \kappa_i N_i$ is the *curvature vector* of γ, while τ measures the twisting of the chosen orthonormal frame.

If we fix N_1 at some base point along γ, then one natural framing is the *parallel frame* or *Bishop frame* [Bis75] defined by the condition $\tau = 0$. Equivalently, the vectors N_i are parallel-transported along γ from the base point, using the Riemannian connection on the normal bundle induced by the immersion in $\mathbb{E}^3$. One should note that this is not necessarily a closed framing along a closed loop γ; when we return to the base point, the vector N_1 has been rotated through an angle called the *writhe* of γ.

Other framings are also often useful. For instance, if γ lies on a surface M with unit normal ν, it is natural to choose $N_1 = \nu$. Then $N_2 = \eta := T \times \nu$ is called the *conormal vector*, and (T, ν, η) is the *Darboux frame* (adapted to $\gamma \subset M \subset \mathbb{E}^3$). The curvature vector of γ decomposes into parts tangent and normal to M as $T' = \kappa_n \nu + \kappa_g \eta$. Here, $\kappa_n = -\nu' \cdot T$ measures the *normal curvature* of M in the direction T, and is independent of γ, while κ_g, the *geodesic curvature* of γ in M, is an intrinsic notion, showing how γ sits in M, and is unchanged if we isometrically deform the immersion of M into space.

When the curvature vector of γ never vanishes, we can write it as $T' = \kappa N$, where N is a unit vector, the *principal normal*, and $\kappa > 0$. This yields the orthonormal *Frenet frame* (T, N, B), whose twisting τ is the *torsion* of γ.

The *total curvature* of a smooth curve is $\int \kappa \, ds$. In [Sul08] we review a number of standard results: For closed curves, the total curvature is at least 2π (Fenchel) and for knotted space curves the total curvature is at least 4π (Fáry/Milnor). For plane curves, we can consider instead the signed curvature, and find that $\int \kappa \, ds$ is always an integral multiple of 2π. Suppose (following Milnor) we define the total curvature of a polygonal curve simply to be the sum of the turning angles at the vertices. Then, as we explain in [Sul08], all these theorems on total curvature remain true. Our goal, when defining curvatures for polyhedral surfaces, will be to ensure that similar integral relations remain true.

2. Curvatures of smooth surfaces

Given a (two-dimensional, oriented) surface M smoothly immersed in $\mathbb{E}^3$, we understand its local shape by looking at the Gauss map $\nu : M \to \mathbb{S}^2$ given by the unit normal vector $\nu = \nu_p$ at each point $p \in M$. The derivative of the Gauss map at p is a linear map from $T_p M$ to $T_{\nu_p} \mathbb{S}^2$. Since these spaces are naturally identified, being parallel planes in $\mathbb{E}^3$, we can view the derivative as an endomorphism $-S_p : T_p M \to T_p M$. The map S_p is called the shape operator (or Weingarten map).

The shape operator is the complete second-order invariant (or curvature) which determines the original surface M. (This statement has been left intentionally a bit vague, since without a standard parametrization like arc length, it is not quite clear how one should specify such an operator along an unknown surface.) Usually, however, it is more convenient to work not with the operator S_p but instead with scalar quantities. Its eigenvalues κ_1 and κ_2 are called the *principal curvatures*, and (since they cannot be globally distinguished) it is their symmetric functions which have the most geometric meaning.

We define the *Gauss curvature* $K := \kappa_1 \kappa_2$ as the determinant of S_p and the *mean curvature* $H := \kappa_1 + \kappa_2$ as its trace. Note that the sign of H depends on the choice of the unit normal ν, so often it is more natural to work with the *vector mean curvature* (or mean curvature vector) $\mathbf{H} := H\nu$. Furthermore, some authors use the opposite sign on S_p and thus $\mathbf{H}$, and many use $H = (\kappa_1 + \kappa_2)/2$, justifying the name "mean" curvature. Our conventions mean that the mean curvature vector for a convex surface points inwards (like the curvature vector for a circle). For a unit sphere oriented with inward normal, the Gauss map ν is the antipodal map, $S_p = I$, and $H \equiv 2$.

The Gauss curvature is an intrinsic notion, depending only on the pullback metric on the surface M, and not on the immersion into space. That is, K is unchanged by bending the surface without stretching it. For instance, a developable surface like a cylinder or cone has $K \equiv 0$ because it is obtained by bending a flat plane. One intrinsic characterization of $K(p)$ is obtained by comparing the circumferences C_ε of (intrinsic) ε-balls around p

to the value $2\pi\varepsilon$ in $\mathbb{E}^2$. We get

$$\frac{C_\varepsilon}{2\pi\varepsilon} = 1 - \frac{\varepsilon^2}{6} K(p) + \mathcal{O}(\varepsilon^3).$$

Mean curvature, on the other hand, is certainly not intrinsic, but it has a nice variational interpretation. Consider a variation vector field ξ on M; for simplicity assume ξ is compactly supported away from any boundary. Then $H = -\delta\,\text{Area}\,/\delta\,\text{Vol}$ in the sense that

$$\delta_\xi\,\text{Vol} = \int \xi \cdot \nu\,dA, \qquad \delta_\xi\,\text{Area} = -\int \xi \cdot H\nu\,dA.$$

With respect to the L^2 inner product $\langle \xi, \eta \rangle := \int \xi_p \cdot \eta_p\,dA$ on vector fields, the vector mean curvature is thus the negative gradient of the area functional, often called the first variation of area: $\mathbf{H} = -\nabla\,\text{Area}$. (Similarly, the negative gradient of length for a curve is its curvature vector κN.)

Just as κ is the geometric version of the second derivative for curves, mean curvature is the geometric version of the Laplacian Δ. Indeed, if a surface M is written locally near a point p as the graph of a height function f over its tangent plane $T_p M$, then $H(p) = \Delta f$. Alternatively, we can write $\mathbf{H} = \nabla_M \cdot \nu = \Delta_M \mathbf{x}$, where $\mathbf{x}$ is the position vector in $\mathbb{E}^3$ and Δ_M is the Laplace–Beltrami operator, the intrinsic surface Laplacian.

We can flow a curve or surface to reduce its length or area, by following the gradient vector field κN or $H\nu$; the resulting parabolic (heat) flow is slightly nonlinear in a natural geometric way. This so-called *mean-curvature flow* has been extensively studied as a geometric smoothing flow. (See, among many others, [GH86, Gra87] for the curve-shortening flow and [Bra78, Hui84, Ilm94, Eck04, Whi05] for higher dimensions.)

3. Integral curvature relations for surfaces

For surfaces, we will consider various integral curvature relations that relate area integrals over a region $D \subset M$ to arc length integrals over the boundary $\gamma := \partial D$. First, the Gauss–Bonnet theorem says that, when D is a disk,

$$2\pi - \iint_D K\,dA = \oint_\gamma \kappa_g\,ds = \oint_\gamma T' \cdot \eta\,ds = -\oint_\gamma \eta' \cdot d\mathbf{x}.$$

Here, $d\mathbf{x} = T\,ds$ is the vector line element along γ, and $\eta = T \times \nu$ is again the conormal. In particular, this theorem implies that the total Gauss curvature of D depends only on a collar neighborhood of γ: if we make any modification to D supported away from the boundary, the total curvature is unchanged (as long as D remains topologically a disk). We will extend the notion of (total) Gauss curvature from smooth surfaces to more general surfaces (in particular polyhedral surfaces) by requiring that this property remain true.

Our other integral relations are all proved by Stokes's theorem, and thus require only that γ be the boundary of D in a homological sense; for these D need not be a disk. First consider the *vector area*

$$\mathbf{A}_\gamma := \frac{1}{2}\oint_\gamma \mathbf{x} \times d\mathbf{x} = \frac{1}{2}\oint_\gamma \mathbf{x} \times T\,ds = \iint_D \nu\,dA.$$

The right-hand side represents the total vector area of any surface spanning γ, and the relation shows this to depend only on γ (and this time not even on a collar neighborhood). The integrand on the left-hand side depends on a choice of the origin for the coordinates, but because we integrate over a closed loop, the integral is independent of this choice. Both sides of this vector area formula can be interpreted directly for a polyhedral surface, and the equation remains true in that case. We note also that this vector area $\mathbf{A}_\gamma$ is one of the quantities preserved when γ evolves under the Hasimoto or smoke-ring flow $\dot{\gamma} = \kappa B$. (Compare [LP94, PSW07, Hof08].)

A simple application of the fundamental theorem of calculus to the tangent vector of a curve γ from p to q shows that

$$T(q) - T(p) = \int_p^q T'(s)\,ds = \int_p^q \kappa N\,ds.$$

This can be viewed as a balance between elastic tension forces trying to shrink the curve, and sideways forces holding it in place. It is the key step in verifying that the vector curvature κN is the first variation of length.

The analog for a surface patch D is the mean-curvature force balance equation

$$\oint_\gamma \eta\,ds = -\oint_\gamma \nu \times d\mathbf{x} = \iint_D H\nu\,dA = \iint_D \mathbf{H}\,dA.$$

Again this represents a balance between surface tension forces acting in the conormal direction along the boundary of D and what can be considered as pressure forces (especially in the case of constant H) acting normally across D. We will use this equation to develop our analog of mean curvature for discrete surfaces.

The force balance equation can be seen to arise from the translational invariance of curvatures. It has been important for studying surfaces of constant mean curvature; see, for instance, [KKS89, Kus91, GKS03]. The rotational analog is the following torque balance:

$$\oint_\gamma \mathbf{x} \times \eta\,ds = \oint_\gamma \mathbf{x} \times (\nu \times d\mathbf{x}) = \iint_D H(\mathbf{x} \times \nu)\,dA = \iint_D \mathbf{x} \times \mathbf{H}\,dA.$$

Somewhat related is the following equation:

$$\oint_\gamma \mathbf{x} \cdot \eta\,ds = \oint_\gamma \mathbf{x} \cdot (\nu \times d\mathbf{x}) = \iint_D (\mathbf{H} \cdot \mathbf{x} - 2)\,dA.$$

It gives, for example, an interesting expression for the area of a minimal ($H \equiv 0$) surface.

4. Discrete surfaces

For us, a discrete or polyhedral surface $M \subset \mathbb{E}^3$ will mean a triangulated surface with a continuous map into space, linear on each triangle. In more detail, we start with an abstract combinatorial triangulation—a simplicial complex—representing a 2-manifold with boundary. We then pick positions $p \in \mathbb{E}^3$ for all the vertices, which uniquely determine a linear map on each triangle; these maps fit together to form the polyhedral surface.

The union of all triangles containing a vertex p is called Star(p), the *star* of p. Similarly, the union of the two triangles containing an edge e is Star(e).

4.1. Gauss curvature

It is well known how the notion of Gauss curvature extends to such discrete surfaces M. (Banchoff [Ban67, Ban70] was probably the first to discuss this in detail, though he notes that Hilbert and Cohn-Vossen [HCV32, §29] had already used a polyhedral analog to motivate the intrinsic nature of Gauss curvature.) Any two adjacent triangles (or, more generally, any simply connected region in M not including any vertices) can be flattened—developed isometrically into the plane. Thus the Gauss curvature is supported on the vertices $p \in M$. In fact, to keep the Gauss–Bonnet theorem true, we must take

$$\iint_D K \, dA := \sum_{p \in D} K_p, \qquad \text{with} \quad K_p := 2\pi - \sum_i \theta_i.$$

Here, the angles θ_i are the interior angles at p of the triangles meeting there, and K_p is often known as the angle defect at p. If D is any neighborhood of p contained in Star(p), then $\oint_{\partial D} \kappa_g \, ds = \sum \theta_i$; when the triangles are acute, this is most easily seen by letting ∂D be the path connecting their circumcenters and crossing each edge perpendicularly. Analogous to our intrinsic characterization of Gauss curvature in the smooth case, note that the circumference of a small ε ball around p here is exactly $2\pi\varepsilon - \varepsilon K_p$.

This version of discrete Gauss curvature is quite natural, and seems to be the correct analog when Gauss curvature is used intrinsically. But the Gauss curvature of a smooth surface in $\mathbb{E}^3$ also has extrinsic meaning; for instance, the total absolute Gauss curvature is proportional to the average number of critical points of different height functions. For such considerations, Brehm and Kühnel [BK82] suggest the following: when a vertex p is extreme on the convex hull of its star, but the star itself is not a convex cone, then we should think of p as having both positive and negative curvature. We let K_p^+ be the curvature at p of the convex hull, and set $K_p^- := K_p^+ - K_p \geq 0$. Then the absolute curvature at p is $K_p^+ + K_p^-$, which is greater than $|K_p|$. This discretization is of course also based on preserving an integral curvature relation—a different one. (See also [BK97, vDA95].)

We can use the same principle as before—preserving the Gauss–Bonnet theorem— to define Gauss curvature, as a measure, for much more general surfaces. For instance, on a piecewise smooth surface, we have ordinary K within each face, a point mass (again the angle defect) at each vertex, and a linear density along each edge, equal to the difference in the geodesic curvatures of that edge within its two incident faces. Indeed, clothes are often designed from pieces of (intrinsically flat) cloth, joined so that each vertex is intrinsically flat and thus all the curvature is along the edges; corners would be unsightly in clothes.

Returning to polyhedral surfaces, we note that K_p is clearly an intrinsic notion (as it should be) depending only on the angles of each triangle and not on the precise embedding into $\mathbb{E}^3$. Sometimes it is useful to have a notion of combinatorial curvature, independent of *all* geometric information. Given just a combinatorial triangulation, we can pretend that each triangle is equilateral with angles $\theta = 60°$. (Such a euclidean metric with cone points at certain vertices exists on the abstract surface, independent of whether or not it could be embedded in space. See the survey [Tro07].)

The curvature of this metric, $K_p = \frac{\pi}{3}(6 - \deg p)$, can be called the *combinatorial (Gauss) curvature* of the triangulation. (See [Thu98, IK$^+$08] for combinatorial applications of this notion.) In this context, the global form $\sum K_p = 2\pi\chi(M)$ of Gauss–Bonnet amounts to nothing more than Euler's formula $\chi = V - E + F$. (We note that Forman has proposed a combinatorial Ricci curvature [For03]; although for smooth surfaces, Ricci curvature is Gauss curvature, for discrete surfaces Forman's combinatorial curvature does not agree with ours, so he fails to recover the Gauss–Bonnet theorem.)

Our discrete Gauss curvature K_p is of course an integrated quantity. Sometimes it is desirable to have instead a curvature density, dividing K_p by the surface area associated to the vertex p. One natural choice is $A_p := \frac{1}{3}\,\mathrm{Area}\big(\mathrm{Star}(p)\big)$, but this does not always behave nicely for irregular triangulations. One problem is that, while K_p is intrinsic, depending only on the cone metric of the surface, A_p depends also on the choice of which pairs of cone points are connected by triangle edges. One fully intrinsic notion of the area associated to p would be the area of its intrinsic Voronoi cell in the sense of Bobenko and Springborn [BS05]; perhaps this would be the best choice for computing Gauss curvature density.

4.2. Vector area

The vector area formula

$$\mathbf{A}_\gamma := \frac{1}{2}\oint_\gamma \mathbf{x} \times d\mathbf{x} = \iint_D \nu\, dA$$

needs no special interpretation for discrete surfaces: both sides of the equation make sense directly, since the surface normal ν is well defined almost everywhere. However, it is worth interpreting this formula for the case when D is the star of a vertex p. More generally, suppose γ is any closed curve (smooth or polygonal), and D is the cone from p to γ (the union of all line segments pq for $q \in \gamma$). Fixing γ and letting p vary, we find that the volume enclosed by this cone is an affine linear function of $p \in \mathbb{E}^3$, and thus

$$\mathbf{A}_p := \nabla_p \mathrm{Vol}\, D = \frac{\mathbf{A}_\gamma}{3} = \frac{1}{6}\oint_\gamma \mathbf{x} \times d\mathbf{x}$$

is independent of the position of p. We also note that any such cone D is intrinsically flat except at the cone point p, and that $2\pi - K_p$ is the cone angle at p.

4.3. Mean curvature

The mean curvature of a discrete surface M is supported along the edges. If e is an edge, and $e \subset D \subset \mathrm{Star}(e) = T_1 \cup T_2$, then we set

$$\mathbf{H}_e := \iint_D \mathbf{H}\, dA = \oint_{\partial D} \eta\, ds = e \times \nu_1 - e \times \nu_2 = J_1 e - J_2 e.$$

Here ν_i is the normal vector to the triangle T_i, and the operator J_i rotates by $90°$ in the plane of that triangle. Note that $|\mathbf{H}_e| = 2\sin(\theta_e/2)\,|e|$, where θ_e is the exterior dihedral angle along the edge, defined by $\cos\theta_e = \nu_1 \cdot \nu_2$.

No nonplanar discrete surface has $\mathbf{H}_e = 0$ along every edge. But this discrete mean curvature can cancel out around vertices. We set

$$2\mathbf{H}_p := \sum_{e \ni p} \mathbf{H}_e = \iint_{\text{Star}(p)} \mathbf{H}\, dA = \oint_{\partial \text{Star}(p)} \eta\, ds.$$

The area of the discrete surface is a function of the vertex positions; if we vary only one vertex p, we find that $\nabla_p \text{Area}(M) = -\mathbf{H}_p$. This mirrors the variational characterization of mean curvature for smooth surfaces, and we see that a natural notion of discrete minimal surfaces is to require $\mathbf{H}_p \equiv 0$ for all vertices [Bra92, PP93].

Suppose that the vertices adjacent to p, in cyclic order, are $p_1, \ldots, p_n$. Then we can express $\mathbf{A}_p$ and $\mathbf{H}_p$ explicitly in terms of these neighbors. We get

$$3\mathbf{A}_p = 3\nabla_p \text{Vol} = \iint_{\text{Star}(p)} \nu\, dA = \frac{1}{2}\oint_{\partial \text{Star}(p)} \mathbf{x} \times d\mathbf{x} = \frac{1}{2}\sum_i p_i \times p_{i+1}$$

and similarly

$$2\mathbf{H}_p = \sum \mathbf{H}_{pp_i} = -2\nabla_p \text{Area} = \sum J_i(p_{i+1} - p_i)$$
$$= \sum_i (\cot \alpha_i + \cot \beta_i)(p - p_i),$$

where α_i and β_i are the angles opposite the edge pp_i in the two incident triangles. This latter equation is the famous "cotangent formula" [PP93, War08] which also arises naturally in a finite-element discretization of the Laplacian.

Suppose we change the combinatorics of a discrete surface M by introducing a new vertex p along an existing edge e and subdividing the two incident triangles. Then $\mathbf{H}_p$ in the new surface equals the original $\mathbf{H}_e$, independent of where along e we place p. This allows a variational interpretation of $\mathbf{H}_e$.

4.4. Minkowski-mixed-volumes

A somewhat different interpretation of mean curvature for convex polyhedra is found in the context of Minkowski's theory of mixed volumes. (In this simple form, it dates back well before Minkowski, to Steiner [Ste40].) If X is a smooth convex body in $\mathbb{E}^3$ and $B_t(X)$ denotes its t-neighborhood, then its Steiner polynomial is:

$$\text{Vol}(B_t(X)) = \text{Vol}\, X + t\, \text{Area}\, \partial X + \frac{t^2}{2}\int_{\partial X} H\, dA + \frac{t^3}{3}\int_{\partial X} K\, dA.$$

Here, by Gauss–Bonnet, the last integral is always 4π.

When X is instead a convex polyhedron, we already understand how to interpret each term except $\int_{\partial X} H\, dA$. The correct replacement for this term, as Steiner discovered, is $\sum_e \theta_e |e|$. This suggests $H_e := \theta_e |e|$ as a notion of total mean curvature for the edge e.

Note the difference between this formula and our earlier $|\mathbf{H}_e| = 2\sin(\theta_e/2)\, |e|$. Either one can be derived by replacing the edge e with a sector of a cylinder of length $|e|$ and arbitrary (small) radius r. We have $\iint \mathbf{H}\, dA = \mathbf{H}_e$, so that

$$\left|\iint \mathbf{H}\, dA\right| = |\mathbf{H}_e| = 2\sin(\theta_e/2)\, |e| < \theta_e |e| = H_e = \iint H\, dA.$$

The difference is explained by the fact that one formula integrates the scalar mean curvature while the other integrates the vector mean curvature. Again, these two discretizations both arise through preservation of (different) integral relations for mean curvature.

See [Sul08] for a more extensive discussion of the analogous situation for curves: although, as we have mentioned, the sum of the turning angles ψ_i is often the best notion of total curvature for a polygon, in certain situations the "right" discretization is instead the sum of $2\sin\psi_i/2$ or $2\tan\psi_i/2$.

The interpretation of curvatures in terms of the mixed volumes or Steiner polynomial actually works for arbitrary convex surfaces. (Compare [Sch08] in this volume.) Using geometric measure theory—and a generalized normal bundle called the normal cycle—one can extend both Gauss and mean curvature in a similar way to quite general surfaces. See [Fed59, Fu94, CSM03, CSM06].

4.5. Constant mean curvature and Willmore surfaces

A smooth surface which minimizes area under a volume constraint has constant mean curvature; the constant H can be understood as the Lagrange multiplier for the constrained minimization problem. A discrete surface which minimizes area among surfaces of fixed combinatorial type and fixed volume will have constant discrete mean curvature H in the sense that at every vertex, $\mathbf{H}_p = H\mathbf{A}_p$, or, equivalently, ∇_p Area $= -H\nabla_p$Vol.

In general, of course, the vectors $\mathbf{H}_p$ and $\mathbf{A}_p$ are not even parallel: they give two competing notions of a normal vector to the discrete surface at the vertex p. Still,

$$h_p := \frac{|\nabla_p\,\text{Area}\,|}{|\nabla_p\,\text{Vol}\,|} = \frac{|\mathbf{H}_p|}{|\mathbf{A}_p|} = \frac{|\iint_{\text{Star}(p)}\mathbf{H}\,dA|}{|\iint_{\text{Star}(p)}\nu\,dA|}$$

gives a better notion of mean curvature density near p than, say, the smaller quantity

$$\frac{|\mathbf{H}_p|}{A_p} = \frac{|\iint\mathbf{H}\,dA|}{\iint 1\,dA},$$

where $A_p := \frac{1}{3}\,\text{Area}(\text{Star}(p))$.

Suppose we want to discretize the elastic bending energy for surfaces, $\iint H^2\,dA$, known as the Willmore energy. The discussion above shows why $\sum_p h_p^2 A_p$ (which was used in [FS$^+$97]) is a better discretization than $\sum_p |\mathbf{H}_p|^2/A_p$ (used fifteen years ago in [HKS92]). Several related discretizations are by now built into Brakke's Evolver; see the discusion in [Bra07]. Recently, Bobenko [Bob05, Bob08] has described a completely different approach to discretizing the Willmore energy, which respects the Möbius invariance of the smooth energy.

4.6. Relation to discrete harmonic maps

As mentioned above, we can define a *discrete minimal surface* to be a polyhedral surface with $\mathbf{H}_p \equiv 0$. An early impetus to the field of discrete differential geometry was the realization (starting with [PP93]) that discrete minimal surfaces are not only critical points for area (fixing the combinatorics), but also have other properties similar to those of smooth minimal surfaces.

For instance, in a conformal parameterization of a smooth minimal surface, the co-ordinate functions are harmonic. To interpret this for discrete surfaces, we are led to the question of when a discrete map should be considered conformal. In general this is still open. (Interesting suggestions come from the theory of circle packings, and this is an area of active research. See, for instance, [Ste05, BH03, Bob08, KSS06, Spr06].)

However, we should certainly agree that the identity map is conformal. A polyhedral surface M comes with an embedding $\mathrm{Id}_M : M \to \mathbb{E}^3$ which we consider as the identity map. Indeed, we then find (following [PP93]) that M is discrete minimal if and only if Id_M is discrete harmonic. Here a polyhedral map $f : M \to \mathbb{E}^3$ is called *discrete harmonic* if it is a critical point for the Dirichlet energy, written as the following sum over the triangles T of M:

$$E(f) := \sum_T |\nabla f_T|^2 \, \mathrm{Area}_M(T).$$

We can view $E(f) - \mathrm{Area}\, f(M)$ as a measure of non-conformality. For the identity map, $E(\mathrm{Id}_M) = \mathrm{Area}(M)$ and $\nabla_p E(\mathrm{Id}_M) = \nabla_p \mathrm{Area}(M)$, confirming that M is minimal if and only if Id_M is harmonic.

5. Vector bundles on polyhedral manifolds

We now give a general definition of vector bundles and connections on polyhedral mani-folds; this leads to another interpretation of the Gauss curvature for a polyhedral surface.

A polyhedral n-manifold P^n means a CW-complex which is homeomorphic to an n-dimensional manifold, and which is regular and satisfies the intersection condition. (Compare [Zie08] in this volume.) That is, each n-cell (called a *facet*) is embedded in P^n with no identifications on its boundary, and the intersection of any two cells (of any dimension) is a single cell (if non-empty). Because P^n is topologically a manifold, each $(n-1)$-cell (called a *ridge*) is contained in exactly two facets.

Definition. A *discrete vector bundle* V^k of rank k over P^n consists of a vector space $V_f \cong \mathbb{E}^k$ for each facet f of P. A *connection* on V^k is a choice of isomorphism ϕ_r between V_f and $V_{f'}$ for each ridge $r = f \cap f'$ of P. We are most interested in the case where the vector spaces V_f have inner products, and the isomorphisms ϕ_r are orthogonal.

Consider first the case $n = 1$, where P is a polygonal curve. On an arc (an open curve) any vector bundle is trivial. On a loop (a closed curve), a vector bundle of rank k is determined (up to isomorphism) simply by its *holonomy* around the loop, an automorphism $\phi : \mathbb{E}^k \to \mathbb{E}^k$.

Now suppose P^n is linearly immersed in $\mathbb{E}^d$ for some d. That is, each k-face of P is mapped homeomorphically to a convex polytope in an affine k-plane in $\mathbb{E}^d$, and the star of each vertex is embedded. Then it is clear how to define the discrete tangent bundle T (of rank n) and normal bundle N (of rank $d-n$). Namely, each T_f is the n-plane parallel to the affine hull of the facet f, and N_f is the orthogonal $(d-n)$-plane. These inherit inner products from the euclidean structure of $\mathbb{E}^d$.

There are also natural analogs of the Levi-Civita connections on these bundles. Namely, for each ridge r, let $\alpha_r \in [0, \pi)$ be the exterior dihedral angle bewteen the facets f_i meeting along r. Then let $\phi_r : \mathbb{E}^d \to \mathbb{E}^d$ be the simple rotation by this angle, fixing the affine hull of r (and the space orthogonal to the affine hull of the f_i). We see that ϕ_r restricts to give maps $T_f \to T_{f'}$ and $N_f \to N_{f'}$; these form the connections we want. (Note that $T \oplus N = \mathbb{E}^d$ is a trivial vector bundle over P^n, but the maps ϕ_r give a nontrivial connection on it.)

Consider again the example of a closed polygonal curve $P^1 \subset \mathbb{E}^d$. The tangent bundle has rank 1 and trivial holonomy. The holonomy of the normal bundle is some rotation of $\mathbb{E}^{d-1}$. For $d = 3$ this rotation of the plane $\mathbb{E}^2$ is specified by an angle equal (modulo 2π) to the writhe of P^1. (To define the writhe of a curve as a real number, rather than just modulo 2π, requires a bit more care, and requires the curve P to be embedded.)

Next consider a two-dimensional polyhedral surface P^2 and its tangent bundle. Around a vertex p we can compose the cycle of isomorphisms ϕ_e across the edges incident to p. This gives a self-map $\phi_p : T_f \to T_f$. This is a rotation of the tangent plane by an angle which—it is easy to check—equals the discrete Gauss curvature K_p.

Now consider the general case of the tangent bundle to a polyhedral manifold $P^n \subset \mathbb{E}^d$. Suppose p is a codimension-two face of P. Then composing the ring of isomorphisms across the ridges incident to p gives an automorphism of $\mathbb{E}^n$ which is the local holonomy, or *curvature*, of the Levi-Civita connection around p. We see that this is a rotation fixing the affine hull of p. To define this curvature (which can be interpreted as a sectional curvature in the two-plane normal to p) as a real number and not just modulo 2π, we should look again at the angle defect around p, which is $2\pi - \sum \beta_i$ where the β_i are the interior dihedral angles along p of the facets f incident to p.

In the case of a hypersurface P^{d-1} in $\mathbb{E}^d$, the one-dimensional normal bundle is locally trivial: there is no curvature or local holonomy around any p. Globally, the normal bundle is of course trivial exactly when P is orientable.

References

[Ban67] Thomas F. Banchoff, *Critical points and curvature for embedded polyhedra*, J. Differential Geometry **1** (1967), 245–256.

[Ban70] _____ , *Critical points and curvature for embedded polyhedral surfaces*, Amer. Math. Monthly **77** (1970), 475–485.

[BH03] Philip L. Bowers and Monica K. Hurdal, *Planar conformal mappings of piecewise flat surfaces*, Visualization and Mathematics III (H.-C. Hege and K. Polthier, eds.), Springer, 2003, pp. 3–34.

[Bis75] Richard L. Bishop, *There is more than one way to frame a curve*, Amer. Math. Monthly **82** (1975), 246–251.

[BK82] Ulrich Brehm and Wolfgang Kühnel, *Smooth approximation of polyhedral surfaces regarding curvatures*, Geom. Dedicata **12**:4 (1982), 435–461.

[BK97] Thomas F. Banchoff and Wolfgang Kühnel, *Tight submanifolds, smooth and polyhedral*, Tight and taut submanifolds (Berkeley, CA, 1994), Math. Sci. Res. Inst. Publ., vol. 32, Cambridge Univ. Press, Cambridge, 1997, pp. 51–118.

[Bob05]   Alexander I. Bobenko, *A conformal energy for simplicial surfaces*, Combinatorial and computational geometry, Math. Sci. Res. Inst. Publ., vol. 52, Cambridge Univ. Press, Cambridge, 2005, pp. 135–145.

[Bob08]   _____, *Surfaces from circles*, Discrete Differential Geometry (A.I. Bobenko, P. Schröder, J.M. Sullivan, G.M. Ziegler, eds.), Oberwolfach Seminars, vol. 38, Birkhäuser, 2008, this volume, pp. 3–35; arXiv.org/0707.1318.

[Bra78]   Kenneth A. Brakke, *The motion of a surface by its mean curvature*, Mathematical Notes, vol. 20, Princeton University Press, Princeton, N.J., 1978.

[Bra92]   _____, *The Surface Evolver*, Experiment. Math. **1**:2 (1992), 141–165.

[Bra07]   _____, *The Surface Evolver*, www.susqu.edu/brakke/evolver, online documentation, accessed September 2007.

[BS05]   Alexander I. Bobenko and Boris A. Springborn, *A discrete Laplace–Beltrami operator for simplicial surfaces*, preprint, 2005, arXiv:math.DG/0503219.

[CSM03]   David Cohen-Steiner and Jean-Marie Morvan, *Restricted Delaunay triangulations and normal cycle*, SoCG '03: Proc. 19th Sympos. Comput. Geom., ACM Press, 2003, pp. 312–321.

[CSM06]   _____, *Second fundamental measure of geometric sets and local approximation of curvatures*, J. Differential Geom. **74**:3 (2006), 363–394.

[Eck04]   Klaus Ecker, *Regularity theory for mean curvature flow*, Progress in Nonlinear Differential Equations and their Applications, 57, Birkhäuser, Boston, MA, 2004.

[Fed59]   Herbert Federer, *Curvature measures*, Trans. Amer. Math. Soc. **93** (1959), 418–491.

[For03]   Robin Forman, *Bochner's method for cell complexes and combinatorial Ricci curvature*, Discrete Comput. Geom. **29**:3 (2003), 323–374.

[FS⁺97]   George Francis, John M. Sullivan, Robert B. Kusner, Kenneth A. Brakke, Chris Hartman, and Glenn Chappell, *The minimax sphere eversion*, Visualization and Mathematics (H.-C. Hege and K. Polthier, eds.), Springer, Heidelberg, 1997, pp. 3–20.

[Fu94]   Joseph H.G. Fu, *Curvature measures of subanalytic sets*, Amer. J. Math. **116**:4 (1994), 819–880.

[GH86]   Michael E. Gage and Richard S. Hamilton, *The heat equation shrinking convex plane curves*, J. Differential Geom. **23**:1 (1986), 69–96.

[GKS03]   Karsten Große-Brauckmann, Robert B. Kusner, and John M. Sullivan, *Triunduloids: Embedded constant mean curvature surfaces with three ends and genus zero*, J. reine angew. Math. **564** (2003), 35–61; arXiv:math.DG/0102183.

[Gra87]   Matthew A. Grayson, *The heat equation shrinks embedded plane curves to round points*, J. Differential Geom. **26**:2 (1987), 285–314.

[HCV32]   David Hilbert and Stephan Cohn-Vossen, *Anschauliche Geometrie*, Springer, Berlin, 1932, second printing 1996.

[HKS92]   Lucas Hsu, Robert B. Kusner, and John M. Sullivan, *Minimizing the squared mean curvature integral for surfaces in space forms*, Experimental Mathematics **1**:3 (1992), 191–207.

[Hof08]   Tim Hoffmann, *Discrete Hashimoto surfaces and a doubly discrete smoke-ring flow*, Discrete Differential Geometry (A.I. Bobenko, P. Schröder, J.M. Sullivan, G.M. Ziegler, eds.), Oberwolfach Seminars, vol. 38, Birkhäuser, 2008, this volume, pp. 95–115; arXiv.org/math.DG/0007150.

[Hui84]   Gerhard Huisken, *Flow by mean curvature of convex surfaces into spheres*, J. Differential Geom. **20**:1 (1984), 237–266.

[IK⁺08]   Ivan Izmestiev, Robert B. Kusner, Günter Rote, Boris A. Springborn, and John M. Sullivan, *Torus triangulations . . .* , in preparation.

[Ilm94]   Tom Ilmanen, *Elliptic regularization and partial regularity for motion by mean curvature*, Mem. Amer. Math. Soc. **108**:520 (1994), x+90 pp.

[KKS89]   Nicholas J. Korevaar, Robert B. Kusner, and Bruce Solomon, *The structure of complete embedded surfaces with constant mean curvature*, J. Differential Geom. **30**:2 (1989), 465–503.

[KSS06]   Liliya Kharevych, Boris A. Springborn, and Peter Schröder, *Discrete conformal maps via circle patterns*, ACM Trans. Graphics **25**:2 (2006), 412–438.

[Kus91]   Robert B. Kusner, *Bubbles, conservation laws, and balanced diagrams*, Geometric analysis and computer graphics, Math. Sci. Res. Inst. Publ., vol. 17, Springer, New York, 1991, pp. 103–108.

[LP94]   Joel Langer and Ron Perline, *Local geometric invariants of integrable evolution equations*, J. Math. Phys. **35**:4 (1994), 1732–1737.

[MD⁺03]   Mark Meyer, Mathieu Desbrun, Peter Schröder, and Alan H. Barr, *Discrete differential-geometry operators for triangulated 2-manifolds*, Visualization and Mathematics III (H.-C. Hege and K. Polthier, eds.), Springer, Berlin, 2003, pp. 35–57.

[PP93]   Ulrich Pinkall and Konrad Polthier, *Computing discrete minimal surfaces and their conjugates*, Experiment. Math. **2**:1 (1993), 15–36.

[PSW07]   Ulrich Pinkall, Boris A. Springborn and Steffen Weißmann, *A new doubly discrete analogue of smoke ring flow and the real time simulation of fluid flow*, J. Phys. A **40** (2007), 12563–12576; arXiv.org/0708.0979.

[Sch08]   Peter Schröder, *What can we measure?*, Discrete Differential Geometry (A.I. Bobenko, P. Schröder, J.M. Sullivan, G.M. Ziegler, eds.), Oberwolfach Seminars, vol. 38, Birkhäuser, 2008, this volume, pp. 263–273.

[Spr06]   Boris A. Springborn, *A variational principle for weighted Delaunay triangulations and hyperideal polyhedra*, preprint, 2006, arXiv.org/math.GT/0603097.

[Ste40]   Jakob Steiner, *Über parallele Flächen*, (Monats-) Bericht Akad. Wiss. Berlin (1840), 114–118; Ges. Werke II, 2nd ed. (AMS/Chelsea, 1971), 171–176; bibliothek.bbaw.de/bbaw/bibliothek-digital/digitalequellen/schriften/anzeige/index_html?band=08-verh/1840&seite:int=114.

[Ste05]   Kenneth Stephenson, *Introduction to circle packing*, Cambridge Univ. Press, Cambridge, 2005.

[Sul08]   John M. Sullivan, *Curves of finite total curvature*, Discrete Differential Geometry (A.I. Bobenko, P. Schröder, J.M. Sullivan, G.M. Ziegler, eds.), Oberwolfach Seminars, vol. 38, Birkhäuser, 2008, this volume, pp. 137–161; arXiv.org/math/0606007.

[Thu98]   William P. Thurston, *Shapes of polyhedra and triangulations of the sphere*, The Epstein birthday schrift, Geom. Topol. Monogr., vol. 1, Geom. Topol. Publ., Coventry, 1998, pp. 511–549.

[Tro07]   Marc Troyanov, *On the moduli space of singular euclidean surfaces*, preprint, 2007, arXiv:math/0702666.

[vDA95]   Ruud van Damme and Lyuba Alboul, *Tight triangulations*, Mathematical methods for curves and surfaces (Ulvik, 1994), Vanderbilt Univ. Press, Nashville, TN, 1995, pp. 517–526.

[War08]   Max Wardetzky, *Convergence of the cotangent formula: An overview*, Discrete Differential Geometry (A.I. Bobenko, P. Schröder, J.M. Sullivan, G.M. Ziegler, eds.), Oberwolfach Seminars, vol. 38, Birkhäuser, 2008, this volume, pp. 275–286.

[Whi05]   Brian White, *A local regularity theorem for mean curvature flow*, Ann. of Math. (2) **161**:3 (2005), 1487–1519.

[Zie08]   Günter M. Ziegler, *Polyhedral surfaces of high genus*, Discrete Differential Geometry (A.I. Bobenko, P. Schröder, J.M. Sullivan, G.M. Ziegler, eds.), Oberwolfach Seminars, vol. 38, Birkhäuser, 2008, this volume, pp. 191–213; arXiv.org/math/0412093.

John M. Sullivan
Institut für Mathematik, MA 3–2
Technische Universität Berlin
Str. des 17. Juni 136
10623 Berlin
Germany
e-mail: sullivan@math.tu-berlin.de

Part III

Geometric Realizations of Combinatorial Surfaces

Discrete Differential Geometry, A.I. Bobenko, P. Schröder, J.M. Sullivan and G.M. Ziegler, eds.
Oberwolfach Seminars, Vol. 38, 191–213

Polyhedral Surfaces of High Genus

Günter M. Ziegler

Abstract. The construction of the *combinatorial* data for a surface of maximal genus with n vertices is a classical problem: The maximal genus $g = \lfloor \frac{1}{12}(n-3)(n-4) \rfloor$ was achieved in the famous "Map Color Theorem" by Ringel et al. (1968). We present the nicest one of Ringel's constructions, for the case $n \equiv 7 \mod 12$, but also an alternative construction, essentially due to Heffter (1898), which easily and explicitly yields surfaces of genus $g \sim \frac{1}{16} n^2$.

For *geometric* (polyhedral) surfaces in $\mathbb{R}^3$ with n vertices the maximal genus is not known. The current record is $g \sim \frac{1}{8} n \log_2 n$, due to McMullen, Schulz & Wills (1983). We present these surfaces with a new construction: We find them in Schlegel diagrams of "neighborly cubical 4-polytopes," as constructed by Joswig & Ziegler (2000).

Keywords. Polyhedral surfaces, high genus, neighborly surfaces, geometric construction, projected deformed cubes.

1. Introduction

In the following we present constructions for surfaces that have extremely and perhaps surprisingly high topological complexity (genus, Euler characteristic) compared to their number of vertices. We believe that not only the resulting surfaces, but also the constructions themselves are interesting and worth studying—also in the hope that they can be substantially improved.

1.1. What is a surface?

What do we mean by "a surface"? This is not a stupid question, since combinatorialists, geometers, and topologists work with quite different frameworks, definitions and concepts of surfaces. As we will see, in the high-genus case it is not clear that the same results are valid for the different concepts.

A *topological* surface may be defined as a closed (compact, without boundary), connected, orientable, Hausdorff, 2-dimensional manifold. By adopting this model, we already indicate that one could have worked in much greater generality: Here we do not consider the non-orientable case, we do not worry about manifolds with boundary, etc.

The *combinatorial* version of a surface may be presented by listing the *faces* (vertices, edges, and 2-cells), and giving the necessary incidence information (for example, by specifying for each face the vertices in its boundary, in clockwise order according to the orientation). Such combinatorial data must, of course, satisfy some consistency conditions if we are to be guaranteed that they do correspond to a surface. Such conditions are easy to derive.

The vertices and edges of a combinatorial surface together make up its *graph* or 1-skeleton.

In the following, we will insist throughout that the combinatorial surface data we look at are regular (no identifications on the boundary of any cell), and they must satisfy the *intersection condition*: The intersection of any two faces is again a face (which may be empty). This condition implies that any two vertices are connected by at most one edge, and that any two 2-faces have at most two vertices in common (which must then be connected by an edge).[1]

Geometric surfaces consist of flat convex polygons. We require that all these faces are simultaneously realized in $\mathbb{R}^3$, without intersections. It is customary to also assume that adjacent faces are not coplanar. Any such geometric surface yields a combinatorial surface, which in turn determines a topological manifold.

1.2. The f-vector

The f-vector of a combinatorial or geometric surface S is the triple

$$f(S) := (f_0, f_1, f_2),$$

where f_0 denotes the number of vertices, f_1 is the number of edges, and f_2 is the number of 2-dimensional cells.

The f-vector contains a lot of information. For example, we can tell from the f-vector whether the surface is simplicial. Indeed, one always has $3f_2 \le 2f_1$, by double-counting: Every 2-face has at least three edges; every edge lies in two 2-faces. Equality $3f_2 = 2f_1$ holds if and only if every face is bounded by exactly three edges, that is, for a triangulated (simplicial) surface.

Similarly, we have $f_1 \le \binom{f_0}{2}$, with equality for a *neighborly* surface, meaning one whose 1-skeleton is a complete graph. A neighborly surface is necessarily simplicial.

1.3. The genus

The classification of (orientable, closed, connected—the generality outlined above) surfaces up to homeomorphism is well-known: For each integer $g \ge 0$, there is exactly one topological type, "the surface S_g of genus g," which may be obtained by attaching g handles to the 2-sphere S^2.

The genus of a surface may be defined, viewed, and computed in various different ways, also depending on the model in which the surface is presented.

Topologically, the genus may, for example, be obtained from a homology group, as $g = \frac{1}{2} \dim H_1(S_g; \mathbb{Q})$. Alternatively, the genus may be expressed as the maximal number

---

[1] A combinatorial surface with the intersection condition is called a "polyhedral map" in some of the discrete geometry literature; see Brehm & Wills [10].

of disjoint, non-separating, closed loops (this is the definition given by Heffter [16]). It is also half the maximal number of non-separating loops that are disjoint except for a common base point.

Combinatorially, we can compute the genus in terms of the Euler characteristic, $\chi(S_g) = 2 - 2g = f_0 - f_1 + f_2$. So combinatorially the genus is given by the formula $g = 1 - \frac{1}{2}(f_0 - f_1 + f_2) \geq 0$.

1.4. The construction and realization problems

Any combinatorial surface describes a topological space. Conversely, any 2-manifold can be triangulated, but it is not at all clear a priori how many vertices would be needed for that. Thus we have the construction problem for combinatorial surfaces:

Combinatorial construction problem: For which parameters (f_0, f_1, f_2) are there combinatorial surfaces?

This is not an easy problem; in the triangulated case of $2f_1 = 3f_2$ it is solved by Ringel's Map Color Theorem, discussed below.

Any geometric surface yields a combinatorial surface, but in the passage from combinatorial to geometric surfaces, there are substantial open problems:

Geometric construction problem: For which parameters (f_0, f_1, f_2) are there geometric surfaces?

This problem is much harder. It may be factored into two steps, where the first one asks for a classification or enumeration of the combinatorial surfaces with the given parameters, and the second one tries to solve the following realization problem for all the combinatorial types:

Realization problem: Which combinatorially given surfaces have geometric realizations?

In general the answer to the geometric construction problem *does not* coincide with the answer for the combinatorial construction problem, that is, the second step may fail even if the first one succeeds. Let's look at some special cases:

- In the case of genus 0, that is, $f_1 = f_0 + f_2 - 2$, the construction problem was solved by Steinitz [34]: The necessary and sufficient conditions both for combinatorial and for geometric surfaces are $3f_2 \leq 2f_1$ (as discussed before) and $3f_1 \leq 2f_0$, or, equivalently, $f_2 \leq 2f_0 - 4$ and $f_0 \leq 2f_2 - 4$.
  By a second, much deeper, theorem by Steinitz [35] [36] [38, Lect. 4], every combinatorial surface of genus 0 has a geometric realization in $\mathbb{R}^3$, as the boundary of a convex polytope. This solves the realization problem for the case $g = 0$.
- In the case of a simplicial torus, the case of genus 1, the combinatorially possible f-vectors are easily seen to be $(n, 3n, 2n)$, for $n \geq 7$. Apparently, Archdeacon et al. [1] have just proved an old conjecture of Grünbaum [15, Exercise 13.2.3, p. 253] that *every* triangulated torus (surface of genus 1, with f-vector $(n, 3n, 2n)$) has a geometric realization in $\mathbb{R}^3$. Certainly for each $n \geq 7$ there is at least one simplicial torus with f-vector $(n, 3n, 2n)$ that is realizable in $\mathbb{R}^3$, so for simplicial tori the set of f-vectors for the combinatorial and for the geometric model coincide.

- On the other hand, there are combinatorial tori with f-vector $(2n, 3n, n)$, but *none* of them has a geometric realization. Indeed, the condition $3f_0 = 2f_1$ means that the surface in question has a cubic graph (all vertices have degree 3). Thus we are looking at the dual cell decompositions of the simplicial tori. But any geometric surface with a cubic graph is necessarily convex—and thus a 2-sphere (cf. [15, Exercise 11.1.7, p. 206]).

- Rather little is known about geometric surfaces of genus $g \geq 2$: Lutz [23] enumerated that there are 865 triangulated surfaces of genus 2 and 20 surfaces of genus 3 on 10 vertices. Together with J. Bokowski he showed that all of these have geometric realizations; Hougardy, Lutz & Zelke [19] even obtained small integer vertex coordinates for all of these. Specific examples of geometric surfaces of genus $g = 3$ and $g = 4$ with a (combinatorially) minimal number of vertices were constructed by Bokowski & Brehm [7] [8].

- There are also triangulated combinatorial surfaces that have no geometric realization: Let's look at the f-vector $(12, 66, 44)$, which corresponds to a neighborly surface of genus 6 with 12 vertices. Altshuler [3] has enumerated that there are exactly 59 types of such triangulations. One single one, number 54, which is particularly symmetric, was shown not to be geometrically realizable by Bokowski & Guedes de Oliveira [9]. Thus, 58 possible types remain; recently Schewe [33] has proved that none of them can be realized geometrically in $\mathbb{R}^3$.

In general, it seems difficult to show for any given triangulated surface that no geometric realization exists. Besides the oriented matroid methods of Bokowski and Guedes de Oliveira, the obstruction theory criteria of Novik [27] and a linking-number approach of Timmreck [37] have been developed in an attempt to do such non-realizability proofs.

In these lectures we look at families of combinatorial surfaces whose genus grows quadratically in the number of vertices, such as the neighborly triangulated surfaces on $n > 7$ vertices. We *think* that no geometric realizations exist for these, but no methods to prove such a general result seem to be available yet. On the geometric side we present a construction for surfaces of genus $n \log n$, which may be considered "high genus" in the category of geometric surfaces. We hope that someone will be able to show that this is good, or even best possible, or to improve upon it.

2. Two combinatorial constructions

Let us now look at a combinatorial surface with $f_0 = n$ vertices. The following upper bound is quite elementary—the challenge is in the construction of examples that meet it, or at least get close.

Lemma 2.1. *A combinatorial surface with n vertices has genus at most*

$$g \leq \tfrac{1}{12}(n-3)(n-4). \tag{2.1}$$

Equality can hold only if n is congruent to $0, 3, 4$ or $7 \bmod 12$, for a triangulated surface that is neighborly.

Proof. Due to the intersection condition, any two vertices are connected by at most one edge, and thus $f_1 \leq \binom{n}{2}$.

In the case of a triangulated/simplicial surface, we have $3f_2 = 2f_1$. With this, a simple calculation yields

$$g = 1 - \tfrac{1}{2}(f_0 - f_1 + f_2) = 1 - \tfrac{1}{2}f_0 + \tfrac{1}{6}f_1 \leq 1 - \tfrac{1}{2}n + \tfrac{1}{6}\binom{n}{2} = \tfrac{1}{12}(n-3)(n-4).$$

This holds with equality only if the surface is neighborly. The genus $\frac{1}{12}(n-3)(n-4)$ then is an integer, that is, $n \equiv 0, 3, 4$ or $7 \bmod 12$.

If the surface is not simplicial, then it can be triangulated by introducing diagonals, without new vertices, and without changing the genus. However, this always results in triangulated surfaces that are not neighborly (diagonals that have not been chosen are missing), and thus in surfaces that do not achieve equality in (2.1).   □

The case of neighborly surfaces is indeed very interesting, and has received a lot of attention. In particular, it occurred first in connection with (a generalization of) the four color problem: Its analog on surfaces of genus $g > 0$, known as the "Problem der Nachbargebiete," the problem of neighboring countries, is solved by constructing a maximal configuration of "countries" that are pairwise adjacent. If one draws the dual graph to such a configuration, then this will yield a (neighborly) triangulation of the surface (Kempe 1879 [22]; Heffter 1891 [16]). As the "thread problem" ("Fadenproblem") the question was presented in the famous book by Hilbert & Cohn-Vossen [18].

The case $n = 4$ is trivial (realized by the tetrahedron); the first interesting case is $n = 7$, where a combinatorially-unique configuration exists, the simplicial "Möbius Torus" on 7 vertices [26]. We will look at it below. Möbius's triangulation was rediscovered a number of times, realized by Császár [12], and finally exhibited in the Schlegel diagram of a cyclic 4-polytope on 7 vertices, by Duke [13] and Altshuler [2]. For the other neighborly cases, $n \geq 12$, no realizations are known.

When n is not congruent to $0, 3, 4$, or 7, the maximal genus of a surface on n vertices is of course smaller than the bound given above, but it could be just the bound rounded down, and indeed it is.

Theorem 2.2 (Ringel et al. (1968); see [31]). *For each $n \geq 4$, $n \neq 9$, there is a (combinatorial) n-vertex surface of genus*

$$g_{\max} = \left\lfloor \frac{(n-3)(n-4)}{12} \right\rfloor.$$

In his 1891 paper, Heffter [16, §3] proved this theorem for $n \leq 12$; in particular, in doing this he introduced some of the basic concepts and notation, and thus "set the stage." From then, it needed another 77 years to complete the proof of Theorem 2.2. The full proof is complicated, with intricate combinatorial arguments divided into twelve cases (according to $n \bmod 12$) and a number of ad-hoc constructions needed for sporadic cases of "small n." In the following we will sketch Ringel's construction for the nicest of the twelve cases, the case of $n \equiv 7 \bmod 12$. This is the only case where we can get a triangulated surface with an orientation-preserving cyclic symmetry, according to Heffter, and in fact we do! (This special case was first solved by Ringel in 1961, but our exposition follows his book [31] from 1974, to which we also refer for the other eleven cases.) Then we also present a second construction, based on a paper by Heffter [17] from 1898: This produces surfaces that are not quite neighborly, but they still do have genus that grows quadratically with the number of vertices. Moreover, this construction is very conceptual and explicit. We will give a simple combinatorial description. However, the surface has a $\mathbb{Z}_q$-action, whose quotient is the "perfect" cellulation of the surface of genus g with just one vertex and one 2-cell that arises if one identifies the opposite edges of a $4g$-gon. Heffter's surface thus arises as an abelian covering of this perfect cellulation.

2.1. A neighborly triangulation for $n \equiv 7 \bmod 12$

It was observed already by Heffter that a combinatorial surface is completely determined if we label the vertices and for each vertex describe the cycle of its neighbors (in counter-clockwise/orientation order).

For example, Figure 1 shows a "square pyramid" in top view (a 2-sphere with 5 vertices, consisting of one quadrilateral and four triangles). It is given by a *rotation scheme* of the form

$$
\begin{array}{ccl}
0 & : & (1,2,3,4) \\
1 & : & (0,4,2) \\
2 & : & (0,1,3) \\
3 & : & (0,2,4) \\
4 & : & (0,3,1),
\end{array}
$$

which says that $1, 2, 3, 4$ are the neighboring vertices, in cyclic order, for vertex 0, etc. In particular, we could have written $(2, 3, 4, 1)$ instead of $(1, 2, 3, 4)$, since this denotes the same cyclic permutation. Some checking is needed, of course, to see whether a scheme of this form actually describes a surface that satisfies the intersection condition.

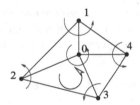

FIGURE 1. Top view of a polyhedral surface (boundary of a square pyramid), with the data that yield the rotation scheme.

In the case of a triangulated surface, the corresponding consistency conditions are rather easy to describe. Indeed, if j, k appear adjacent in the cyclic list of neighbors of a vertex i, then this means that $[i, j, k]$ is an oriented triangle of the surface—and thus k, i have to be adjacent in this order in the cycle of neighbors of j, and similarly i, j have to appear in the list for k.

Thus in terms of the rotation scheme, the *triangulation condition* (which Ringel calls the "rule Δ^*") says that if the row for vertex i reads

$$
i \quad : \quad (\ldots \ldots \ldots, j, k, \ldots \ldots \ldots),
$$

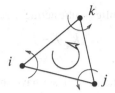

FIGURE 2. Reading off data of the rotation scheme from a triangle in an oriented surface.

then in the rows for j and k we have to get

$$j \quad : \quad (\ldots \ldots, k, i, \ldots \ldots \ldots \ldots)$$
$$k \quad : \quad (\ldots \ldots \ldots \ldots, i, j, \ldots \ldots).$$

We want to construct neighborly orientable triangulated surfaces with n vertices and a cyclic automorphism group $\mathbb{Z}_n$—so the scheme for one vertex should yield each of the others by addition of a constant, modulo n. For example, for $n = 7$ there is such a surface, the Möbius torus [26] of Figure 3. Unfortunately, this is possible *only* for $n \equiv 7 \bmod 12$, according to Heffter [16, §4]. (You may observe that in the case $n = 4$ the S_4 automorphism group of the tetrahedron has a $\mathbb{Z}_4$ subgroup, but that does not preserve the orientation of the surface; for $n = 5$ there is the torus solution of Figure 8 that is not triangulated, and does not satisfy the intersection condition.)

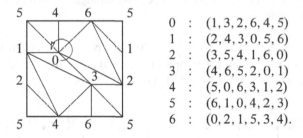

FIGURE 3. The Möbius torus, and its rotation scheme. Read off the rotation scheme from the sketch! Observe how the first row determines all others by addition modulo 7.

Now let's assume we have a rotation scheme for a triangulated surface with automorphism group $\mathbb{Z}_n$. If the row for vertex 0 reads

$$0 \quad : \quad (\ldots \ldots \ldots, j, k, \ldots \ldots \ldots),$$

then the triangulation condition, rule Δ^*, yields that

$$j \quad : \quad (\ldots \ldots, k, 0, \ldots \ldots \ldots \ldots)$$
$$k \quad : \quad (\ldots \ldots \ldots \ldots, 0, j, \ldots \ldots)$$

and then the $\mathbb{Z}_n$-automorphism implies (subtracting j resp. k) that

$$0 \; : \; (\ldots \ldots , k-j, -j, \ldots \ldots \ldots)$$
$$0 \; : \; (\ldots \ldots \ldots , -k, j-k, \ldots \ldots).$$

In other words, if in the neighborhood of 0 we have that "k follows j", then also "$-j$ follows $k-j$", and "$j-k$ follows $-k$" (where all vertex labels are interpreted in $\mathbb{Z}_n$, that is, modulo n).

The condition that we have thus obtained can be viewed as a flow condition (a "Kirchhoff law") in a cubic graph. Indeed, the cyclic order in the neighborhood of 0 can be derived from a walk in a plane drawing of an edge-labelled directed cubic graph, whose edge labels satisfy a flow condition—see Figure 4.

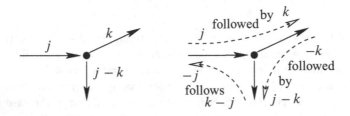

FIGURE 4. The flow condition, "Kirchhoff's law." The left figure shows how a flow of size j is split into two parts. The right figure shows how to read the labels when passing through the vertex, turning left in each case.

Thus in order to obtain a valid "row 0" we have to produce a cyclic permutation of $1, 2, \ldots, n-1$ that can be read off from a flow (circulation) in a cubic graph. Ringel's solution for this in the case $n \equiv 7 \bmod 12$ is given by Figure 5.

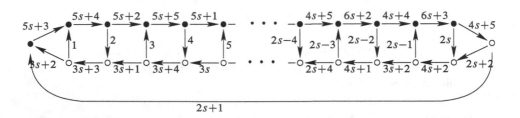

FIGURE 5. A network for a neighborly surface with $n = 12s + 7$ vertices.

It is based on writing $\mathbb{Z}_n = \{0, \pm 1, \pm 2, \ldots, \pm(6s + 3)\}$. The figure encodes the full construction: It describes a cubic graph with $2s + 4$ vertices and $6s + 3$ edges, where
 – each edge label from $\{1, 2, \ldots, 6s + 3\}$ occurs exactly once,
 – at each vertex, the flow condition is satisfied (modulo n).

Now the construction rule is the following: Travel on this graph,

- at each black vertex • turn left, to the next arc in clockwise ordering of the arcs that are adjacent to the vertex, at each white vertex ○ turn right, and
- record the label of each edge traversed in arrow direction, resp. the negative of the label if traversed against arrow direction,

The main claim to be checked is that this prescription leads to a single cycle in which each edge is traversed in each direction exactly once, so each value in $\{\pm 1, \pm 2, \ldots, \pm(6s+3)\}$ occurs exactly once. For example, if we start at the arrow labelled "1," then the sequence we follow will start

$$1, \; -(5s+3), \; -(3s+2), \; -(3s+3), \; -(3s+1), \; -(3s+4), \; -3s, \; -(3s+5), \; \ldots$$

This is the first line of the rotation diagram for Ringel's neighborly surface with $n = 12s + 7$ vertices.

Note: From *any* cyclic order one can derive a neighborly cellular surface with $\mathbb{Z}_n$ symmetry, but subtle combinatorics is needed if we want a triangulated surface, satisfying the intersection property, or if we only want to control the genus. It is interesting to study the random surfaces of random genus that arise this way from a random cyclic order of the neighbors of 0, or from other random combinatorial constructions. See Pippenger & Schleich [29] for a current discussion of such models.

2.2. Heffter's surface and a triangulation

Here is a much simpler construction, which yields a not-quite neighborly surface. The underlying remarkable cellular surface was first discovered by Heffter [17], much later rediscovered by Eppstein et al. [14]. (See also Pfeifle & Ziegler [28].)

Let $q = 4g + 1$ be a prime power with $g \geq 1$ (one can find suitable primes q, or simply take $q = 5^r$). The one algebraic fact we need is that there is a finite field $\mathbb{F}_q$ with q elements, and that the multiplicative group $\mathbb{F}_q^* = \mathbb{F}_q \setminus \{0\}$ is cyclic (of order $q - 1 = 4g$), that is, there is a generator $\alpha \in \mathbb{F}_q^*$ such that $\mathbb{F}_q^* = \{\alpha, \alpha^2, \ldots, \alpha^{q-1}\}$, with $\alpha^{q-1} = \alpha^{4g} = 1$, and $\alpha^{2g} = -1$. For example, for $g = 3$ and $q = 13$ we may take $\alpha = 2$, with $(1, \alpha, \alpha^2, \ldots) = (1, 2, 4, 8, 3, 6, 12, 11, 9, 5, 10, 7)$.

For any $g \geq 1$, a cellulation of S_g is obtained from a $4g$-gon by identifying opposite edges in parallel. In the prime power case, a combinatorial description for this is as follows: Label the directed edges of the $4g$-gon by $1, \alpha, \alpha^2, \alpha^3, \ldots$ in cyclic order, and identify the antiparallel edges labelled s and $-s$. (Compare Figure 6.)

The resulting cell complex has the f-vector $f = (1, 2g, 1)$. It is *perfect* in the view of Morse theory, since this is also the sequence of Betti numbers (ranks of the homology groups). However, this cell decomposition is not "regular" in the terminology of cell complexes, since there are identifications on the boundaries of cells: All the ends of the edges are identified, so all the edges are loops, and there are lots of identifications on the boundary of the 2-cell.

Now we explicitly write down a q-fold "abelian covering" of this perfect cellulation. It has both its vertices and its 2-cells indexed by $\mathbb{F}_q$: For simplicity we identify the vertices with the q elements of $\mathbb{F}_q$. The surface has q 2-faces F_s, indexed by $s \in \mathbb{F}_q$. The face F_s

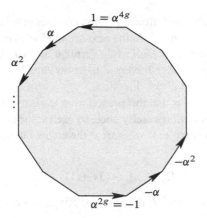

FIGURE 6. Identifying the opposite edges of a $4g$-gon we obtain a perfect cellulation of S_g: All vertices are identified.

has the vertices

$$s, \ s+1, \ s+1+\alpha, \ s+1+\alpha+\alpha^2, \ \ldots \ , s+1+\alpha+\cdots+\alpha^{4g-2},$$

in cyclic order (as indicated by Figure 7), that is,

$$v_s^k := s + \sum_{i=0}^{k-1} \alpha^i \ = \ s + \frac{\alpha^k - 1}{\alpha - 1} \qquad \text{for} \quad 0 \le k < 4g - 1.$$

For each face F_s this yields $q - 1$ distinct values/vertices: α^k takes on every value except for 0, and thus $s + \frac{\alpha^k - 1}{\alpha - 1}$ yields all elements of $\mathbb{F}_q$ except for $s + \frac{-1}{\alpha - 1}$. (See Figure 8 for the case $q = 5$. The case $q = 9$ appears in [28].)

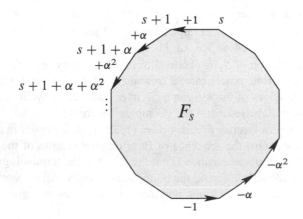

FIGURE 7. One of the q 2-cells, and the labelling of its $q - 1 = 4g$ vertices (by elements of $\mathbb{F}_q$).

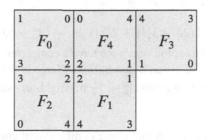

FIGURE 8. The Heffter surface for $q = 5, g = 1, \mathbb{F}_5 = \mathbb{Z}_5$, and $\alpha = 5$ is glued from five squares, as indicated.

Now we have to verify that this prescription does indeed give a surface: For this, check that each vertex comes to lie in a cyclic family of $q - 1$ faces.

We thus have a quite remarkable combinatorial structure: The cellular surface $\widetilde{S}_g$ has q vertices and q faces; the vertices have degree $q - 1$, the faces have $q - 1$ neighbors. Thus the 1-skeleton of the surface is a complete graph K_q (each vertex is adjacent to every other vertex), and so is the dual graph (each 2-face is adjacent to every other one). So the surface is neighborly and dual neighborly! Moreover, the surface is self-dual, that is, isomorphic to its dual cell decomposition.

With all the combinatorial facts just mentioned, we have in particular computed the f-vector of the surface: It is

$$ f(\widetilde{S}_g) = (q, 2gq, q) = (q, \tbinom{q}{2}, q). $$

Thus we have an orientable surface with Euler characteristic $2q - \binom{q}{2} = 2 - 2g$, and genus $g = \frac{1}{2}\binom{q}{2} - q + 1$.

Moreover, "by construction" the surface is very symmetric: First, there clearly is an $\mathbb{F}_q$-action by addition; and if we mod out by this action, we recover the original, "perfect" cell decomposition $\widetilde{S}_g / \mathbb{Z}_q \cong S_g$ with one face. But also the multiplicative group $\mathbb{F}_q^*$ acts by multiplication, with $\alpha \cdot (s + 1 + \cdots + \alpha^{k-1}) = \alpha s + \alpha + \cdots + \alpha^k = (\alpha s - 1) + 1 + \alpha + \cdots + \alpha^k$. Thus the action on the faces is given by

$$ F_s \mapsto F_{\alpha s - 1}, \qquad v_s^k \mapsto v_{\alpha s - 1}^{k+1}. $$

The full symmetry group of S_g is a "metacyclic group" with $q(q - 1)$ elements.

The surface $\widetilde{S}_g$ is a regular cell complex, in that its 2-cells have no identifications on the boundary, but it does not satisfy the intersection condition: Any two 2-cells intersect in $q - 2$ vertices (since each facet includes all but one of the q vertices).

Thus we triangulate $\widetilde{S}_g$, by stellar subdivision of the 2-cells: Then we have $n = 2q$ vertices, q of degree $q - 1$, and q of degree $2q - 2$. Furthermore, there are $f_1 = 3\binom{q}{2}$ edges, namely $\binom{q}{2}$ "old" ones and $q(q - 1)$ "new ones" introduced by the q stellar subdivisions. Furthermore, we have now $q(q - 1)$ triangle faces, which yields an f-vector

$$ f = \left(2q, 3\tbinom{q}{2}, 2\tbinom{q}{2} \right), $$

and hence
$$g = 1 - \tfrac{1}{2}(2q - 3\binom{q}{2} + 2\binom{q}{2})) = \tfrac{1}{16}(n^2 - 10n + 16).$$
So for these simplicial surfaces, for which we have a completely explicit and very simple combinatorial description, the genus is quadratically large in the number of vertices, $g \sim \frac{n^2}{16}$, but they don't quite reach the value $g \sim \frac{n^2}{12}$ of neighborly surfaces.

Conclusion

So what is the moral? The moral is that using combinatorial constructions, we do obtain triangulated surfaces whose genus grows quadratically with the number of vertices. To find the constructions for surfaces with the exact maximal genus is very tricky, and certainly one would hope for simpler and more conceptual descriptions/constructions, but combinatorial surfaces whose genus grows quadratically with the number of vertices are quite easy to get.

3. A geometric construction

If a surface is smoothly embedded or immersed in $\mathbb{R}^3$, then any generic linear function on $\mathbb{R}^3$ is a Morse function for the surface: All critical points are isolated, and a small neighborhood looks like a quadratic surface: a minimum, a maximum, or a saddle point. Morse theory then tells us that the topological complexity of the surface is bounded by the number of critical points. In particular, we get a chain of inequalities

$$2g = \dim H_1(S) < \dim H_*(S) \leq \# \text{ critical points.}$$

If we think of a simplicial/polyhedral surface in $\mathbb{R}^3$ as an approximation to a smooth surface, then we might use a general-position linear function as a Morse function. We might say that all the critical points should certainly be at the vertices, and thus the genus g cannot be larger than the number of vertices for an embedded (or immersed) surface.

   However, in the case of high genus, the approximation of a smooth surface by a simplicial surface is not good, it is very coarse, and the critical points induced by a linear function on a simplicial surface certainly do not satisfy the Morse condition of looking like quadratic surfaces. (Barnette, Gritzmann & Höhne [6] analyzed the local combinatorics of the critical points for a linear function.) And indeed, the result suggested by our argument is far from being true. It was disproved by McMullen, Schulz & Wills [25], who in 1983 presented "polyhedral 2-manifolds in E^3 with unusually high genus": They produced sequences both of simplicial and of quad-surfaces on n vertices whose genus grows like $n \log n$.

   In the following we will give a simple combinatorial description of their quad-surfaces Q_m, and describe an explicit, new geometric construction for them, which is non-inductive, yields explicit coordinates, and "for free" even yields a cubification of the convex hull of the surface without additional vertices. We obtain this by putting together (simplified versions of) several recent constructions: Based on intuition from Amenta & Ziegler [4], a simplified construction of the neighborly cubical polytopes of Joswig & Ziegler [21], which are connected to the construction of high genus surfaces via Babson,

Billera & Chan [5] and observations by Joswig & Rörig [20]. The constructions as presented here can be generalized and extended quite a bit, which constitutes both recent work as well as promising and exciting directions for further research. See, e.g., [39].

The construction in the following will be in five parts:

1. combinatorial description of the surface as the mirror-surface of the n-gon, embedded into the n-dimensional standard cube,
2. construction of a deformed n-cube,
3. general definition and characterization of faces that are "strictly preserved" under a polytope projection,
4. determining some faces of our deformed n-cube that are strictly preserved under projection to $\mathbb{R}^4$, and
5. realizing the desired surfaces in $\mathbb{R}^3$ via Schlegel diagrams.

3.1. Combinatorial description

The surface Q_m is most easily described as a subcomplex of the m-dimensional cube $C_m = [0, 1]^m$.

Any non-empty face of C_m consists of those points in C_m for which some coordinates are fixed to be 0, others are fixed to be 1, and the rest are left free to vary in $[0, 1]$. Thus there is a bijection of the non-empty faces with $\{0, 1, *\}^m$.

Definition 3.1. For $m \geq 3$, the quad-surface Q_m is given by all the faces of C_m for which only two, cyclically-successive coordinates may be left free.

Thus the subset $|Q_m| \subset [0, 1]^m$ consists of all points that have at most two fractional coordinates—and if there are two, they have to be either adjacent, or they have to be the first and the last coordinate. (This description perhaps first appeared in Ringel [30]. Compare Coxeter [11, p. 57].) In particular, Q_3 is just the boundary of the unit 3-cube.

Let's list the faces of Q_m: These are *all* the $f_0(Q_m) = 2^m$ vertices of the 0/1-cube, encoded by $\{0, 1\}^m$; then Q_m contains *all* the $f_1(Q_m) = m2^{m-1}$ edges of the m-cube, corresponding to strings with exactly one $*$ and 0/1-entries otherwise. And finally we have $f_2(Q_m) = m2^{m-2}$ quad faces, corresponding to strings with two cyclically-adjacent $*$s and 0/1s otherwise.

Why is this a surface? This is since all the vertex links are circles. Indeed, if we look at any vertex, then we see in its star the m edges emanating from the vertex, and the m square faces between them, which connect the edges in the cyclic order, as in Figure 9.

The surface is indeed orientable: An explicit orientation is obtained by dictating that in the boundary of any 2-face for which the fractional coordinates are $k - 1$ and k (modulo m), the edges with a fractional $(k - 1)$-coordinate should be oriented from the even-sum vertex to the odd-sum vertex, while the edges corresponding to a fractional k-coordinate are oriented from the odd-sum vertex to the even-sum vertex (cf. Figure 10). Thus, Q_m is an orientable polyhedral surface, realized geometrically in $\mathbb{R}^m$ as a subcomplex of C_m. Its Euler characteristic is

$$\chi(Q_m) = 2^m - m2^{m-1} + m2^{m-2} = (4 - 2m + m)2^{m-2}$$

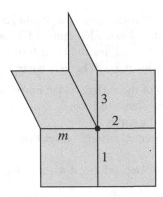

FIGURE 9. The star of a vertex in Q_m.

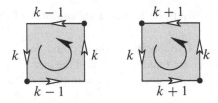

FIGURE 10. The orientation of Q_m described in the text; here the black vertices are the ones with an even sum of coordinates.

and thus with $g = 1 - \frac{1}{2}\chi$ and $n := f_0(Q_m) = 2^m$ the genus is

$$g(Q_m) = 1 + (m-4)2^{m-3} = 1 + \frac{n}{8}\log_2\frac{n}{16} = \Theta(n\log n).$$

So we are dealing with a 2-sphere for $m = 3$, with a torus for $m = 4$, while for $m = 5$ we already get a surface of genus 5. There are also simple recursive ways to describe the surface Q_m, as given by McMullen, Schulz & Wills in their original paper [25]. The combinatorial description here is a special case of the "mirror complex" construction of Babson, Billera & Chan [5], which from any simplicial d-complex on n vertices produces a cubical $(d + 1)$–dimensional subcomplex of the n-cube on 2^n vertices with the given complex as the link of each vertex: Here we are dealing with the case of $d = 1$, where the simplicial complex is a cycle on m vertices.

3.2. Construction of a deformed m-cube

In the last section, we have described a surface Q_m as a subcomplex of the standard orthogonal m-cube $C_m = [0, 1]^m$, and thus as a polyhedral complex in $\mathbb{R}^m$. If we take any other realization of the m-cube, then this yields a corresponding realization of our surface as a subcomplex. The object of this section is to describe an entirely explicit "deformed" cube realization D_m^ε, which contains the surface Q_m^ε as a subcomplex. Here it is.

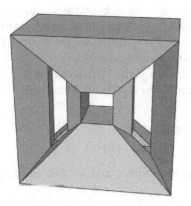 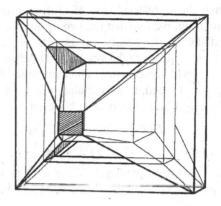

FIGURE 11. Realizations of the surface Q_5 of genus 5 in $\mathbb{R}^3$. On the left we show `polymake/javaview` graphics by Thilo Rörig; on the right a figure from McMullen, Schulte & Wills [24, p. 300].

Definition 3.2. For $m \geq 4$ and $\varepsilon > 0$ let D_m^ε be the set of all points $x \in \mathbb{R}^m$ that satisfy the linear system of $2m$ linear inequalities

$$
\begin{pmatrix}
\pm\varepsilon \\
2 & \pm\varepsilon \\
-7 & 2 & \pm\varepsilon \\
7 & -7 & 2 & \pm\varepsilon \\
-2 & 7 & -7 & 2 & \pm\varepsilon \\
& -2 & 7 & -7 & \cdot & \cdot \\
& & -2 & \cdot & -7 & 2 & \pm\varepsilon \\
& & & \cdot & 7 & -7 & 2 & \pm\varepsilon \\
& & & & -2 & 7 & -7 & 2 & \pm\varepsilon \\
& & & & & -2 & 7 & -7 & 2 & \pm\varepsilon
\end{pmatrix}
\begin{pmatrix}
x_1 \\ x_2 \\ x_3 \\ x_4 \\ \cdot \\ \cdot \\ \cdot \\ x_{m-2} \\ x_{m-1} \\ x_m
\end{pmatrix}
\leq
\begin{pmatrix}
b_1 \\ b_2 \\ b_3 \\ b_4 \\ \cdot \\ \cdot \\ \cdot \\ b_{m-2} \\ b_{m-1} \\ b_m
\end{pmatrix}.
\tag{3.1}
$$

This defines a polytope with the combinatorics of C_m if $\varepsilon > 0$ is small enough and if the sequence of right-hand side entries $b_1, b_2, b_3, \ldots$ grows fast enough. The following lemma gives concrete values "that work."

Lemma 3.3. *For $0 < \varepsilon < \frac{1}{2}$ and $b_k = (\frac{6}{\varepsilon})^{k-1}$, the set D_m^ε is combinatorially equivalent to the m-cube.*

Proof. The k-th pair of inequalities from (3.1) may be written as

$$
|x_k| \leq \frac{1}{\varepsilon}(b_k - 2x_{k-1} + 7x_{k-2} - 7x_{k-3} + 2x_{k-4}),
\tag{3.2}
$$

with $x_0 \equiv x_{-1} \equiv x_{-2} \equiv x_{-3} := 0$. So if $x_{k-1}, x_{k-2}, \ldots$ are bounded, and if b_k is guaranteed to be larger than

$$
L_k := 2|x_{k-1}| + 7|x_{k-2}| + 7|x_{k-3}| + 2|x_{k-4}|,
$$

then the right-hand side of (3.2) is strictly positive, and x_k is bounded again. In this situation, we find that the two inequalities represented by (3.2) cannot be simultaneously satisfied with equality, but any one of them can. Thus, by induction we get that the first $2k$ inequalities of the system (3.1) define a k-cube (in the first k variables).

With the concrete values suggested by the lemma, we verify by induction that $|x_k| < \frac{1}{3}(\frac{6}{\varepsilon})^k$. Indeed, this is certainly true for $k \leq 0$, where we have $x_k \equiv 0$. Thus, with $\varepsilon < \frac{1}{2}$ for $k \geq 1$ and induction on k we get

$$
\begin{aligned}
L_k &= 2|x_{k-1}| + 7|x_{k-2}| + 7|x_{k-3}| + 2|x_{k-4}| \\
&< (\tfrac{6}{\varepsilon})^{k-1}\left[\tfrac{2}{3} + \tfrac{7}{3}\tfrac{\varepsilon}{6} + \tfrac{7}{3}(\tfrac{\varepsilon}{6})^2 + \tfrac{2}{3}(\tfrac{\varepsilon}{6})^3\right] < (\tfrac{6}{\varepsilon})^{k-1} = b_k
\end{aligned}
$$

and thus the right-hand side in (3.2) is always strictly positive, and we also get the inequality $|x_k| < \frac{1}{\varepsilon}(b_k + L_k) = \frac{2}{\varepsilon}(\frac{6}{\varepsilon})^{k-1} < \frac{1}{3}(\frac{6}{\varepsilon})^k$. $\square$

3.3. Strictly preserved faces

In the following, we consider an arbitrary convex m-dimensional polytope $P \subset \mathbb{R}^m$, but of course you should think of $P = D_m^\varepsilon$, the polytope that we will want to apply this to.

The nontrivial *faces* $G \subset P$ of such a polytope are defined by linear functions: A nonzero linear function $x \mapsto c^t x$ *defines* the face $G \subset P$ if G consists of the points of P for which the value $c^t x$ is maximal, that is, if

$$
G = \{x \in P : c^t x = c_0\} = P \cap H,
$$

where $c_0 = \max\{c^t x : x \in P\}$, and where $H = \{x \in \mathbb{R}^m : c^t x = c_0\}$ is a hyperplane.

Given a face G, how do we find a linear functional $c^t x$ that defines it? For *facets* (faces of dimension $m - 1$) $F \subset P$ the linear functional is unique up to taking positive multiples: $F = \{x \in P : n_F{}^t x = c_F\}$, where n_F is a normal vector to F, and c_F is the maximal value that the linear functional $x \mapsto n_F{}^t x$ takes on P. From this it is easy to check (see [38, Lect. 2] for proofs, and Figure 12 for intuition) that c defines G if and only if it is a linear combination, with positive coefficients, of normal vectors n_F of the facets $F \subset P$ that contain G.

In particular, the affine hull of G, aff G, is the intersection of all the hyperplanes spanned by the facets F that contain G:

$$
\operatorname{aff} G = \{x \in \mathbb{R}^m : n_F{}^t x = \max \text{ for all facets } F \supseteq G\}.
$$

Now we look at a projection of P, that is, we look at a surjective linear map $\pi : \mathbb{R}^m \to \mathbb{R}^k$. The image $\pi(P)$ is then a k-dimensional polytope, and the faces of $\pi(P)$ are all induced by faces of P: If $\bar{G} \subset \pi(P)$ is a face of $\pi(P)$, then $\pi^{-1}(\bar{G})$ is a unique face of P.

The faces $\bar{G}$ of the projected polytope $\pi(P)$ thus correspond to those faces of P that can be defined by hyperplanes that are parallel to the kernel of the projection. Equivalently, the faces of $\pi(P)$ correspond to those faces of P that can be defined by normal vectors that are orthogonal to (the kernel of) the projection.

However, in general the face $\pi^{-1}(\bar{G})$ is not the only face that projects to $\bar{G}$, and in general it will have a higher dimension than $\bar{G}$, and it will have faces that do not project to faces of $\pi(P)$. (See Figure 12 for easy examples.) Thus, we single out a very specific,

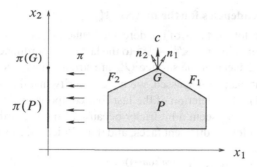

FIGURE 12. The normal vector c for the face G can be written as a positive combination of the normal vectors of the facets F_1, F_2 that contain G. The face $G \subset P$ is strictly preserved by the projection π of P to the last coordinate.

nice situation, where this does not happen: G will map to a face $\pi(G) \subseteq \pi(P)$ of the same dimension as G, and the faces of G map to the faces of $\pi(G)$.

Definition 3.4 (Strictly preserved faces). Let $\pi : P \to \pi(P)$ be a polytope projection. A nontrivial face $G \subset P$ is *strictly preserved* by the projection if $\pi(G)$ is a face of $\pi(P)$, with $G = \pi^{-1}(\pi(G))$, and such that the map $G \to \pi(G)$ is injective.

One can work out linear algebra conditions that characterize faces G that are strictly preserved by a projection (see [39]): We need that the normal vectors n_F to the facets F that contain G, after projection to the kernel (or to a fiber) of the projection do span this fiber positively, that is, the projected vectors have to span the fiber, and they have to be linearly dependent with positive coefficients.

Here we want to apply this only in a very specific situation, namely for an orthogonal projection to the last k coordinates, that is, for the projection $\pi : \mathbb{R}^m \to \mathbb{R}^k$ given by $x = (x', x'') \mapsto x''$, where x'' denotes the last k coordinates of x, and x' denotes the first $m - k$ coordinates. For this situation, the characterization of strictly preserved faces boils down to the following.

Lemma 3.5. *Let $P \subset \mathbb{R}^m$ be an m-dimensional polytope, and let $\pi : \mathbb{R}^m \to \mathbb{R}^k$, $(x', x'') \mapsto x''$ be the projection to the last k coordinates, which maps P to the k-polytope $\pi(P)$.*

Then a nontrivial face $G \subset P$ is strictly preserved by the projection if and only if the facet normals n_F to the facets $F \subset P$ that contain G satisfy the following two conditions: Their restrictions $n'_F \in \mathbb{R}^{m-k}$ to the first $m - k$ coordinates

- *must be positively dependent, that is, they must satisfy a linear relation of the form $\sum_{F \supset G} \lambda_F n'_F = 0$ with real coefficients $\lambda_F > 0$, and*
- *they have full rank, that is, the vectors n'_F span $\mathbb{R}^{m-k}$.*

3.4. Positive row dependencies for the matrix A'_m

Our aim is to prove that lots of faces of the deformed cube D^ε_m constructed in Section 3.2 are preserved by the projection $\pi : \mathbb{R}^m \to \mathbb{R}^4$ to the last four coordinates; in particular, we want to see that all the 2-faces of the surface Q^ε_m are strictly preserved by the projection.

In view of the criteria just discussed, we have to verify that the corresponding rows of the matrix from (3.1), after deletion of the last four components, are positively dependent and spanning. This may seem a bit tricky because of the ε coordinates around, and because we have to treat lots of different faces, and thus choices of rows. However, it turns out to be surprisingly easy.

We start with the matrix $A'_m \in \mathbb{R}^{m \times (m-4)}$,

$$
A'_m := \begin{pmatrix}
0 & & & & & & & \\
2 & 0 & & & & & & \\
-7 & 2 & 0 & & & & & \\
7 & -7 & 2 & 0 & & & & \\
-2 & 7 & -7 & 2 & \cdot & & & \\
& -2 & 7 & -7 & \cdot & 0 & & \\
& & -2 & \cdot & \cdot & -7 & 2 & \\
& & & \cdot & 7 & -7 & \\
& & & & -2 & 7 & \\
& & & & & -2 &
\end{pmatrix}.
$$

This is the matrix that you get from the left-hand side matrix of (3.1) if you put ε to zero, and if you delete the last four coordinates in each row.

The vectors

$$
\begin{pmatrix} 1 \\ 0 \\ 0 \\ \vdots \\ 0 \end{pmatrix}, \quad
\begin{pmatrix} 1 \\ 1 \\ 1 \\ \vdots \\ 1 \end{pmatrix}, \quad
\begin{pmatrix} 1 \\ 2 \\ 4 \\ \vdots \\ 2^{m-1} \end{pmatrix}, \quad
\begin{pmatrix} 1 \\ 1/2 \\ 1/4 \\ \vdots \\ 1/2^{m-1} \end{pmatrix}
$$

are orthogonal to the columns of this matrix, that is, they describe row dependencies. Indeed, the coefficients $(2, -7, 7, -2)$ that appear in the columns of A^ε_m, and hence of A'_m, have been chosen exactly to make this true.

In particular, for any $t \in \mathbb{Z}$ the rows of A'_m are dependent with the coefficient 0 for the first row, and coefficients

$$
(2^{i-t} - 1)(1 - 2^{t+1-i}) = 2^{-t}2^i + 2^{t+1}\tfrac{1}{2^i} - 3
$$

for the i-th row, $2 \le i \le m - 4$. These coefficients are positive, except for the coefficients for $i = 0, t, t + 1$, which are zero. Thus, if the first, t-th and $(t + 1)$-st row are deleted from A'_m, the remaining $m - 3$ rows are positively dependent. Moreover, the remaining $m - 3$ rows span $\mathbb{R}^{m-3}$, as one sees by inspection of A'_m: The rows $2, \ldots, t - 1$ have the same span as the first $t - 2$ unit vectors $e_1, \ldots, e_{t-2}$, since the corresponding submatrix has lower-triangular form with diagonal entries $+2$, and the rows numbered $t + 2, \ldots, m$

together have the same span as $e_{t-2}, \ldots, e_{m-4}$, due to a corresponding upper-triangular submatrix with diagonal entries -2.

So the $m - 3$ rows from A'_m corresponding to the index set $[n] \setminus \{1, t, t + 1\}$ are positively dependent and spanning, for $1 < t < n$. In fact, this is true for the rows with index set $[n] \setminus \{t, t + 1\}$ for $1 \leq t < n$ as well as for the rows given by $[n] \setminus \{1, n\}$. That is, if we delete any two cyclically-adjacent rows from A'_m, then the remaining rows are positively dependent and spanning. Moreover, the property of a vector configuration to be "positively dependent and spanning" is stable under sufficiently small perturbations: Thus if we delete the last four columns, the first row, and any two adjacent rows from A^ε_m, then the rows of the resulting matrix will be positively dependent, and spanning. Thus we have proved the following result.

Proposition 3.6. *For sufficiently small $\varepsilon > 0$, the projection $\pi : \mathbb{R}^m \to \mathbb{R}^4$ yields a polyhedral realization $\pi(D^\varepsilon_m)$ of the surface Q^ε_m in $\mathbb{R}^4$, as part of the boundary complex of the polytope $\pi(D^\varepsilon_m)$.*

3.5. Completion of the construction, via Schlegel diagrams

In the last section, we have constructed a 4-dimensional polytope

$$\bar{P}_m := \pi(D^\varepsilon_m) \subset \mathbb{R}^4$$

as the projection of an m-cube. One can quite easily prove that the projection is in sufficiently general position with respect to the D^ε_m, so the resulting 4-polytope is *cubical*: All its facets are combinatorial cubes.

Moreover, the 1-skeleton of this polytope is exactly that of the m-cube: We have constructed *neighborly cubical* 4-polytopes. (Indeed, our construction is very closely related to the original one by Joswig & Ziegler [21].)

The boundary complex of any 4-polytope may be visualized in terms of a Schlegel diagram (see [38, Lect. 5]): By radial projection from a point that is very close to a facet $F_0 \subset \bar{P}_m$, we obtain a polytopal complex $\mathcal{D}(\bar{P}_m, F_0)$ that faithfully represents all the faces of $\bar{P}_m$, except for F_0 and $\bar{P}_m$ itself. Hence we have arrived at the goal of our construction.

Theorem 3.7. *For $m \geq 3$, there is a polyhedral realization of the surface Q_m, the "mirror complex of an m-gon", in $\mathbb{R}^3$.*

For $m \geq 4$ such a realization may be found as a subcomplex of

$$\mathcal{D}(\pi(D^\varepsilon_m), F_0),$$

the Schlegel diagram (with respect to an arbitrary facet F_0) of a projection of the deformed m-cube $D^\varepsilon_m \subset \mathbb{R}^m$ (with sufficiently small ε) to the last 4 coordinates.

Thus we have obtained quadrilateral surfaces, polyhedrally realized in $\mathbb{R}^3$, of remarkably high genus. If you prefer to have triangulated surfaces, you may of course further triangulate the surfaces just obtained, without introduction of new vertices. This yields a simplicial surface embedded in $\mathbb{R}^3$, with f-vector

$$(2^m, 3m2^{m-2}, m2^{m-1}).$$

For even $m \geq 4$ this may be done in such a way that the resulting surface has all vertex degrees equal (to $\frac{3}{2}m$): To achieve this, triangulate the faces with fractional coordinates $k-1$ and k by using the diagonal between the even-sum vertices if k is even, and the diagonal between the odd-sum vertices if k is odd. (Figure 13 indicates how two adjacent quadrilateral faces are triangulated by this rule.)

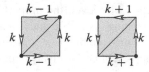

FIGURE 13. The triangulation of Q_m described above. Here we assume that k is even. The black vertices are the ones with an even sum of coordinates.

In other words, this yields *equivelar* triangulated surfaces of high genus, which is what McMullen et al. sought in [25].

Let's finally note that this construction has lots of interesting components that may be further analyzed, varied, and extended. Thus a lot remains to be done, and further questions abound. To note just a few aspects briefly:

- Give explicit bounds for some $\varepsilon > 0$ that is "small enough" for Proposition 3.6.
- There are higher-dimensional analogues of this: So, extend the construction as given here in order to get neighborly cubical d-polytopes, with the $(\frac{d}{2} - 1)$-skeleton of the N-cube, for $N \geq d \geq 2$. (Compare [21].)
- Extend this to surfaces that you get as "mirror complexes" in products of polygons, rather than just m-cubes (which are products of quadrilaterals, for even m).

See Ziegler [39], Joswig & Rörig [20] and Sanyal et al. [32] for further work and ideas related to these questions.

Acknowledgments

Thanks to everyone at the Oberwolfach Seminar and in particular to Michael Joswig, Frank Lutz, Raman Sanyal, and Thilo Rörig for interesting and helpful discussions, hints, and comments. I am especially grateful to John Sullivan for an extraordinarily careful reading and many detailed and insightful suggestions and improvements. Thanks to Torsten Heldmann for ∞.

References

[1] Dan Archdeacon and C. Paul Bonnington and Joanna A. Ellis-Monaghan, *How to exhibit toroidal maps in space*, Discrete Comput. Geometry, **38** (2007), 573–594.

[2] Amos Altshuler, *Polyhedral realizations in $\mathbb{R}^3$ of triangulations of the torus and 2-manifolds in cyclic 4-polytopes*, Discrete Math. **1** (1971), 211–238.

[3] Amos Altshuler, Jürgen Bokowski, and Peter Schuchert, *Neighborly 2-manifolds with 12 vertices*, J. Combin. Theory Ser. A **75** (1996), no. 1, 148–162.

[4] Nina Amenta and Günter M. Ziegler, *Deformed products and maximal shadows*, Advances in Discrete and Computational Geometry (South Hadley, MA, 1996) (Providence RI) (B. Chazelle, J.E. Goodman, and R. Pollack, eds.), Contemporary Mathematics, vol. 223, Amer. Math. Soc., 1998, pp. 57–90.

[5] Eric K. Babson, Louis J. Billera, and Clara S. Chan, *Neighborly cubical spheres and a cubical lower bound conjecture*, Israel J. Math. **102** (1997), 297–315.

[6] David W. Barnette, Peter Gritzmann, and Rainer Höhne, *On valences of polyhedra*, J. Combinatorial Theory, Ser. A **58** (1991), 279–300.

[7] Jürgen Bokowski and Ulrich Brehm, *A new polyhedron of genus 3 with 10 vertices*, Intuitive geometry (Siófok, 1985), Colloq. Math. Soc. János Bolyai, vol. 48, North-Holland, Amsterdam, 1987, pp. 105–116.

[8] ———, *A polyhedron of genus 4 with minimal number of vertices and maximal symmetry*, Geom. Dedicata **29** (1989), no. 1, 53–64.

[9] Jürgen Bokowski and Antonio Guedes de Oliveira, *On the generation of oriented matroids*, Discrete Comput. Geometry **24** (2000), 197–208.

[10] Ulrich Brehm and Jörg M. Wills, *Polyhedral manifolds*, Handbook of Convex Geometry (P. Gruber and J. Wills, eds.), North-Holland, Amsterdam, 1993, pp. 535–554.

[11] Harold Scott MacDonald Coxeter, *Regular skew polyhedra in 3 and 4 dimensions and their topological analogues*, Proc. London Math. Soc. (2) **43** (1937), 33–62; Reprinted in "Twelve Geometric Essays," Southern Illinois U. Press, Carbondale, 1968, 75–105.

[12] Akos Csaszar, *A polyhedron without diagonals*, Acta Sci. Math. (Szeged) **13** (1949/50), 140–142.

[13] Richard A. Duke, *Geometric embeddings of complexes*, Amer. Math. Monthly **77** (1970), 597–603.

[14] David Eppstein, Greg Kuperberg, and Günter M. Ziegler, *Fat 4-polytopes and fatter 3-spheres*, Discrete Geometry: In honor of W. Kuperberg's 60th birthday (A. Bezdek, ed.), Pure and Applied Mathematics, vol. 253, Marcel Dekker Inc., New York, 2003, pp. 239–265; arXiv. org/math/0204007.

[15] Branko Grünbaum, *Convex Polytopes*, Graduate Texts in Math., vol. 221, Springer-Verlag, New York, 2003, Second edition prepared by V. Kaibel, V. Klee and G.M. Ziegler (original edition: Interscience, London 1967).

[16] Lothar Heffter, *Ueber das Problem der Nachbargebiete*, Math. Annalen **38** (1891), 477–508; www-gdz.sub.uni-goettingen.de/cgi-bin/digbib.cgi?PPN235181684_ 0038.

[17] ———, *Ueber metacyklische Gruppen und Nachbarconfigurationen*, Math. Annalen **50** (1898), 261–268; www-gdz.sub.uni-goettingen.de/cgi-bin/digbib.cgi? PPN235181684_0050.

[18] David Hilbert and Stephan Cohn-Vossen, *Anschauliche Geometrie*, Springer-Verlag, Berlin Heidelberg, 1932, Second edition 1996. English translation: Geometry and the Imagination, Chelsea Publ., 1952.

[19] Stefan Hougardy, Frank H. Lutz, and Mariano Zelke, *Polyhedra of genus 2 with 10 vertices and minimal coordinates*, preprint, 3 pages, 2005; arXiv.org/math/0507592.

[20] Michael Joswig and Thilo Rörig, *Neighborly cubical polytopes and spheres*, Israel J. Math **159** (2007), 221–242; arxiv.org/math/0503213.

[21] Michael Joswig and Günter M. Ziegler, *Neighborly cubical polytopes*, Discrete & Computational Geometry (Grünbaum Festschrift: G. Kalai, V. Klee, eds.) **24**:2–3 (2000), 325–344; arXiv.org/math/9812033.

[22] Alfred B. Kempe, *On the geographical problem of the four colours*, American J. Math. **2** (1879), 193–200.

[23] Frank H. Lutz, *Enumeration and random realization of triangulated surfaces*, Discrete Differential Geometry (A.I. Bobenko, P. Schröder, J.M. Sullivan, G.M. Ziegler, eds.), Oberwolfach Seminars, vol. 38, Birkhäuser, 2008, this volume, pp. 235–253; arXiv.org/math.CO/0506316.

[24] Peter McMullen, Egon Schulte, and Jörg M. Wills, *Infinite series of combinatorially regular polyhedra in three-space*, Geom. Dedicata **26** (1988), 299–307.

[25] Peter McMullen, Christoph Schulz, and Jörg M. Wills, *Polyhedral 2-manifolds in E^3 with unusually large genus*, Israel J. Math. **46** (1983), 127–144.

[26] August F. Möbius, *Mittheilungen aus Möbius' Nachlass: I. Zur Theorie der Polyëder und der Elementarverwandtschaft*, Gesammelte Werke II (F. Klein, ed.), Verlag von S. Hirzel, Leipzig, 1886, pp. 515–559.

[27] Isabella Novik, *Upper bound theorems for simplicial manifolds*, Israel J. Math. **108** (1998), 45–82.

[28] Julian Pfeifle and Günter M. Ziegler, *Many triangulated 3-spheres*, Math. Annalen **330** (2004), 829–837; arXiv.org/math/0304492.

[29] Nicholas Pippenger and Kristin Schleich, *Topological characteristics of random triangulated surfaces*, Random Struct. Algorithms **28** (2006), 247–288; arXiv.org/gr-qc/0306049.

[30] Gerhard Ringel, *Über drei kombinatorische Probleme am n-dimensionalen Würfel und Würfelgitter*, Abh. Math. Sem. Univ. Hamburg **20** (1955), 10–19.

[31] ———, *Map color theorem*, Grundlehren Series, vol. 234, Springer-Verlag, New York, 1974.

[32] Raman Sanyal, Thilo Rörig, and Günter M. Ziegler, *Polytopes and polyhedral surfaces via projection*, in preparation, 2006.

[33] Lars Schewe, *Satisfiability Problems in Discrete Geometry*, Dissertation, Technische Universität Darmstadt, 2007, 101 pages.

[34] Ernst Steinitz, *Über die Eulerschen Polyederrelationen*, Archiv für Mathematik und Physik **11** (1906), 86–88.

[35] ———, *Polyeder und Raumeinteilungen*, Encyklopädie der mathematischen Wissenschaften, Dritter Band: Geometrie, III.1.2., Heft 9, Kapitel III A B 12 (W.Fr. Meyer and H. Mohrmann, eds.), B.G. Teubner, Leipzig, 1922, pp. 1–139.

[36] Ernst Steinitz and Hans Rademacher, *Vorlesungen über die Theorie der Polyeder*, Springer-Verlag, Berlin 1934; Reprint, Springer-Verlag 1976.

[37] Dagmar Timmreck, *Necessary conditions for geometric realizability of simplicial complexes*, Discrete Differential Geometry (A.I. Bobenko, P. Schröder, J.M. Sullivan, G.M. Ziegler, eds.), Oberwolfach Seminars, vol. 38, Birkhäuser, 2008, this volume, pp. 215–233.

[38] Günter M. Ziegler, *Lectures on Polytopes*, Graduate Texts in Mathematics, vol. 152, Springer-Verlag, New York, 1995, Revised edition, 1998; "Updates, corrections, and more" at www.math.tu-berlin.de/~ziegler.

[39] _____ , *Projected products of polygons*, Electronic Research Announcements AMS **10** (2004), 122–134; www.ams.org/era/2004-10-14/S1079-6762-04-00137-4/

Günter M. Ziegler
Institut für Mathematik, MA 6–2
Technische Universität Berlin
Str. des 17. Juni 136
10623 Berlin
Germany
e-mail: ziegler@math.tu-berlin.de

Discrete Differential Geometry, A.I. Bobenko, P. Schröder, J.M. Sullivan and G.M. Ziegler, eds.
Oberwolfach Seminars, Vol. 38, 215–233
© 2008 Birkhäuser Verlag Basel/Switzerland

Necessary Conditions for Geometric Realizability of Simplicial Complexes

Dagmar Timmreck

Abstract. We associate with any simplicial complex K and any integer m a system of linear equations and inequalities. If K has a simplicial embedding in $\mathbb{R}^m$, then the system has an integer solution. This result extends the work of Novik (2000).

Keywords. Geometric realizability, obstruction theory, integer programming.

1. Introduction

In general, it is difficult to prove for a simplicial complex K that it does *not* have a simplicial embedding (or not even a simplicial immersion) into $\mathbb{R}^m$.

For example, the question whether any neighborly simplicial surface on $n \geq 12$ vertices can be realized in $\mathbb{R}^3$ leads to problems of this type. Specifically, Altshuler [2] enumerated the 59 combinatorial types of neighborly simplicial 2-manifolds of genus 6. Bokowski & Guedes de Oliveira [4] employed oriented matroid enumeration methods to show that one specific instance, number 54 from Altshuler's list, does not have a simplicial embedding; the other 58 cases were shown not to have simplicial embeddings only recently by Lars Schewe [10].

For *piecewise linear* non-embeddability proofs there is a classical setup via obstruction classes, due to Shapiro [11] and Wu [12]. In 2000, Isabella Novik [9] refined these obstructions for simplicial embeddability: She showed that if a *simplicial embedding* of K in $\mathbb{R}^m$ exists, then a certain polytope in the cochain space $C^m(\mathsf{K}^2_\Delta; \mathbb{R})$ must contain an integral point. Thus, infeasibility of a certain integer program might prove that a complex K has no geometric realization.

In the following, we present Novik's approach (cf. parts 1 and 4 of Theorem 5.3) in a reorganized way, so that we can work out more details, which allow us to sharpen some inequalities defining the polytope in $C^m(\mathsf{K}^2_\Delta; \mathbb{R})$ (cf. Theorem 5.3.2c). Further we interpret this polytope as a projection of a polytope in $C^m(\mathsf{S}^2_\Delta; \mathbb{R})$, where S denotes the simplicial complex consisting of all faces of the N-simplex. The latter polytope is easier to analyze. This setup is the right framework to work out the relations between variables (cf. Theorem 5.3.2) and to express linking numbers (cf. Theorem 5.3.3b), which are intersection numbers of cycles and empty simplices of K (which are present in S and therefore need no extra treatment.) Using the extensions based on linking numbers we can reprove for a first example (Brehm's triangulated Möbius strip [5]) that it is not simplicially embeddable in $\mathbb{R}^3$.

2. A quick walk-through

Let K be a finite (abstract) simplicial complex on the vertex set V, and fix a geometric realization $|K|$ in some euclidean space. Further let $f : V \to \mathbb{R}^m$ be any general position map (that is, such that any $m + 1$ points from V are mapped to affinely independent points in $\mathbb{R}^m$). Any such general position map extends affinely on every simplex to a *simplicial map* $f : |K| \to \mathbb{R}^m$ which we also denote by f. Such a simplicial map is a special case of a piecewise linear map.

Every piecewise linear general position map f defines an *intersection cocycle*

$$\varphi_f \in C^m(K_\Delta^2; \mathbb{Z}). \tag{2.1}$$

Here K_Δ^2 denotes the *deleted product* complex, which consists of all faces $\sigma_1 \times \sigma_2$ of the product $K \times K$ such that σ_1 and σ_2 are disjoint simplices (in K). As the deleted product is a polytopal complex we have the usual notions of homology and cohomology. For a detailed treatment of the deleted product complex we refer to [8].

The values of the intersection cocycle are given by

$$\varphi_f(\sigma_1 \times \sigma_2) = (-1)^{\dim \sigma_1} \mathcal{I}(f(\sigma_1), f(\sigma_2)),$$

where $\mathcal{I}$ denotes the signed intersection number of the oriented simplicial chains $f(\sigma_1)$ and $f(\sigma_2)$ of complementary dimensions in $\mathbb{R}^m$. These intersection numbers (and thus the values of the intersection cocycle) have the following key properties:

1. In the case of a simplicial map, all values $(-1)^{\dim \sigma_1} \mathcal{I}(f(\sigma_1), f(\sigma_2))$ are ± 1 or 0. (In the greater generality of piecewise linear general position maps $f : K \to \mathbb{R}^m$, as considered by Shapiro and by Wu, $\mathcal{I}(f(\sigma_1), f(\sigma_2))$ is an integer.)

2. If f is an embedding, then $\mathcal{I}(f(\sigma_1), f(\sigma_2)) = 0$ holds for any two disjoint simplices $\sigma_1, \sigma_2 \in K$.

3. In the case of the "cyclic map" which maps V to the monomial curve of order m (the "moment curve"), the coefficients $(-1)^{\dim \sigma_1} \mathcal{I}(f(\sigma_1), f(\sigma_2))$ are given combinatorially.

The intersection cocycle is of interest since it defines a cohomology class $\Phi_K = [\varphi_f]$ that does not depend on the specific map f. Thus, if some piecewise linear map f is an embedding, then Φ_K is zero.

But a simplicial embedding is a special case of a piecewise linear embedding. So the information Φ_K is not strong enough to establish simplicial non-embeddability for complexes that admit a piecewise linear embedding—such as, for example, orientable closed surfaces in $\mathbb{R}^3$.

According to Novik we should therefore study the specific coboundaries $\delta\lambda_{f,c}$ that establish equivalence between different intersection cocycles.

So, *Novik's Ansatz* is to consider

$$\boxed{\varphi_f - \varphi_c = \delta\lambda_{f,c}} \tag{2.2}$$

where

- $\varphi_f \in C^m(\mathsf{K}^2_\Delta;\mathbb{Z})$ is an integral vector, representing the intersection cocycle of a hypothetical embedding $f : \mathsf{K} \to \mathbb{R}^m$, so $\varphi_f \equiv 0$. (i.e., for every pair $\sigma_1,\sigma_2 \in \mathsf{K}$ of disjoint simplices, that $\varphi_f(\sigma_1 \times \sigma_2) = 0$),
- $\varphi_c \in C^m(\mathsf{K}^2_\Delta;\mathbb{Z})$ is an integral vector, whose coefficients $\varphi_c(\sigma_1 \times \sigma_2)$ are known explicitly, representing the intersection cochain of the cyclic map $c : \mathsf{K} \to \mathbb{R}^m$,
- δ is a known integral matrix with entries from $\{1,-1,0\}$ that represents the coboundary map $\delta : C^{m-1}(\mathsf{K}^2_\Delta;\mathbb{Z}) \to C^m(\mathsf{K}^2_\Delta;\mathbb{Z})$, and finally
- $\lambda_{f,c} \in C^{m-1}(\mathsf{K}^2_\Delta;\mathbb{Z})$ is an integral vector, representing the *deformation cochain*, whose coefficients are determined by f and c, via

$$\lambda_{f,c}(\tau_1 \times \tau_2) = \mathcal{I}\big(h_{f,c}(\tau_1 \times I), h_{f,c}(\tau_2 \times I)\big),$$

where $h_{f,c}(x,t) = tf(x) + (1-t)c(x)$ interpolates between f and c, for $t \in I := [0,1]$.

Thus if K has a simplicial embedding, then the linear system (2.2) in the unknown vector $\lambda_{f,c}$ has an *integral* solution. Moreover, Novik derived explicit bounds on the coefficients of $\lambda_{f,c}$, that is, on the signed intersection numbers between the parametrised surfaces $h_{f,c}(\tau_1 \times I)$ and $h_{f,c}(\tau_2 \times I)$.

The intersection cocycles and deformation cochains induced by the general position maps $f, g : V \to \mathbb{R}^m$ on *different* simplicial complexes K and $\tilde{\mathsf{K}}$ on the *same* vertex set V coincide on $\mathsf{K}^2_\Delta \cap \tilde{\mathsf{K}}^2_\Delta$. They are projections of the same intersection cocycle or deformation cochain on S^2_Δ, where S denotes the full face lattice of the simplex with vertex set V. We therefore investigate these largest cochains and get Novik's results back as well as some stronger results even in the original setting; see Theorem 4.8 and Remark 4.2.2.

In the following, we

- derive the validity of the basic equation (2.2), in Section 3,
- examine deformation cochains induced by general position maps on the vertex set in Section 4, and
- exhibit an obstruction system to geometric realizability in Section 5.

Furthermore, in Section 6 we discuss subsystems and report about computational results.

3. Obstruction theory

We state and prove the results of this section for *simplicial* maps only. They hold in the more general framework of *piecewise linear* maps as well. For proofs and further details in this general setting we refer to Wu [12].

3.1. Intersections of simplices and simplicial chains

Definition 3.1. Let σ and τ be affine simplices of complementary dimensions $k + \ell = m$ in $\mathbb{R}^m$ with vertices $\sigma_0, \ldots, \sigma_k$ and $\tau_0, \ldots, \tau_\ell$, respectively. Suppose that $\sigma_0, \ldots, \sigma_k$ and $\tau_0, \ldots, \tau_\ell$ are in general position and the simplices are oriented according to the increasing order of the indices. Then σ and τ intersect in at most one point. The *intersection number* $\mathcal{I}(\sigma, \tau)$ is defined to be zero if σ and τ don't intersect and ± 1 according to the orientation of the full-dimensional simplex $(p, \sigma_1, \ldots, \sigma_k, \tau_1, \ldots, \tau_\ell)$ if σ and τ intersect in p. This definition extends bilinearly to simplicial chains in $\mathbb{R}^m$. (We consider integral chains, that is, formal combinations of affine simplices in $\mathbb{R}^m$ with integer coefficients.)

Lemma 3.2. *Let x, y be simplicial chains in $\mathbb{R}^m$ with $\dim x = k$ and $\dim y = \ell$.*

(a) *If $k + \ell = m$, then $\mathcal{I}(x, y) = (-1)^{k\ell} \mathcal{I}(y, x)$.*
(b) *If $k + \ell = m + 1$, then $\mathcal{I}(\partial x, y) = (-1)^k \mathcal{I}(x, \partial y)$.*

Now we use intersection numbers to associate a cocycle to each general position map.

Notation. For $N \in \mathbb{N}$, let $[N] := \{1, \ldots, N\}$ and $\langle N \rangle := [N] \cup \{0\}$.

Lemma and Definition 3.3. *Let $f : \langle N \rangle \to \mathbb{R}^m$ be a general position map. The cochain defined by*

$$\varphi_f(\sigma_1 \times \sigma_2) := (-1)^{\dim \sigma_1} \mathcal{I}(f(\sigma_1), f(\sigma_2)) \qquad \text{for m-cells $\sigma_1 \times \sigma_2 \in \mathsf{K}_\Delta^2$}$$

is a cocycle. It is called the intersection cocycle *of f.*

The intersection cocycle has the following symmetries. For every m-cell $\sigma_1 \times \sigma_2$, with $\dim \sigma_1 = k$ and $\dim \sigma_2 = \ell$,

$$\varphi_f(\sigma_1 \times \sigma_2) = (-1)^{(k+1)(\ell+1)+1} \varphi_f(\sigma_2 \times \sigma_1).$$

Remark. Wu calls this cocycle the *imbedding cocyle* [12, p. 183]. If f is a piecewise linear embedding, then $\varphi_f = 0$. When we look at simplicial maps, we even have an equivalence: A simplicial map f is an embedding of K if and only if the intersection cocycle is 0. So φ_f measures the deviation of f from a geometric realization. This makes the intersection cocycle quite powerful.

3.2. Intersections of parametrized surfaces

In this section we sort out definitions, fix orientations and establish the fundamental relation in Proposition 3.6 (cf. [12, pp. 180 and 183]). Wu uses a simplicial homology between two different piecewise linear maps to establish the independence of the homology class of the particular piecewise linear map. We use a straight line homotopy instead.

Definition 3.4. Let $U \subset \mathbb{R}^k$ and $V \subset \mathbb{R}^\ell$ be sets that are closures of their interiors and $\varphi : U \to \mathbb{R}^m$ and $\psi : V \to \mathbb{R}^m$ smooth parametrized surfaces. The surfaces φ and ψ *intersect transversally* at $p = \varphi(\alpha) = \psi(\beta)$ with $\alpha \in \overset{\circ}{U}$ and $\beta \in \overset{\circ}{V}$, if

$$T_p \mathbb{R}^m = d\varphi(T_\alpha U) \oplus d\psi(T_\beta V).$$

In other words, $k + \ell = m$, and the vectors

$$\frac{\partial \varphi}{\partial u_1}\Big|_\alpha, \ldots, \frac{\partial \varphi}{\partial u_k}\Big|_\alpha, \frac{\partial \psi}{\partial v_1}\Big|_\beta, \ldots, \frac{\partial \psi}{\partial v_\ell}\Big|_\beta$$

span $\mathbb{R}^m$. In this situation the index of intersection of φ and ψ in p is defined by

$$\mathcal{I}_p(\varphi, \psi) := \operatorname{sgn} \det \left(\frac{\partial \varphi}{\partial u_1}\Big|_\alpha, \ldots, \frac{\partial \varphi}{\partial u_k}\Big|_\alpha, \frac{\partial \psi}{\partial v_1}\Big|_\beta, \ldots, \frac{\partial \psi}{\partial v_\ell}\Big|_\beta \right).$$

The surfaces φ and ψ are in *general position* if they intersect transversally only. In particular there are no intersections at the boundary. Surfaces in general position intersect in finitely many points only, and the *intersection number* is defined by

$$\mathcal{I}(\varphi, \psi) := \sum_{p = \varphi(\alpha) = \psi(\beta)} \mathcal{I}_p(\varphi, \psi).$$

We also write $\mathcal{I}\big(\varphi(U), \psi(V)\big)$ for $\mathcal{I}(\varphi, \psi)$ when we want to emphasize the fact that the images intersect.

We now give parametrizations of simplices so that the two definitions coincide.

Notation. Denote by $(e_1, \ldots, e_m)$ the standard basis of $\mathbb{R}^m$ and let $e_0 := 0$. For $I \subseteq \langle m \rangle$ let Δ_I denote the simplex $\operatorname{conv}\{e_i \mid i \in I\}$. Finally let $J = \{j_0, \ldots, j_k\}_<$ denote the set $\{j_0, \ldots, j_k\}$ with $j_0 < \ldots < j_k$.

For a simplex $\sigma = \operatorname{conv}\{\sigma_0, \ldots, \sigma_k\}$ the parametrization $\varphi_\sigma : \mathbb{R}^k \supseteq \Delta_{[k]} \to \sigma \subseteq \mathbb{R}^m$, $(u_1, \ldots, u_k) \mapsto \sigma_0 + \sum_{i=1}^{k} u_i (\sigma_i - \sigma_0)$ induces the orientation corresponding to the increasing order of the indices. Now consider two simplices $\sigma = \operatorname{conv}\{\sigma_0, \ldots, \sigma_k\}$ and $\tau = \operatorname{conv}\{\tau_0, \ldots, \tau_\ell\}$. If $\{\sigma_0, \ldots, \sigma_k, \tau_0, \ldots, \tau_\ell\}$ is in general position, then also φ_σ and φ_τ are in general position. Let σ and τ intersect in

$$p = \sum_{i=0}^{k} \alpha_i \sigma_i = \sigma_0 + \sum_{i=1}^{k} \alpha_i (\sigma_i - \sigma_0) = \varphi_\sigma(\alpha)$$

$$= \sum_{i=0}^{\ell} \beta_i \tau_i = \tau_0 + \sum_{i=1}^{\ell} \beta_i (\tau_i - \tau_0) = \varphi_\tau(\beta).$$

Then we have by a straightforward calculation:

$$\mathcal{I}(\sigma, \tau) = \operatorname{sgn} \det \left(\begin{pmatrix} 1 \\ p \end{pmatrix}; \begin{pmatrix} 1 \\ \sigma_1 \end{pmatrix}, \ldots, \begin{pmatrix} 1 \\ \sigma_k \end{pmatrix}, \begin{pmatrix} 1 \\ \tau_1 \end{pmatrix}, \ldots, \begin{pmatrix} 1 \\ \tau_\ell \end{pmatrix} \right)$$

$$= \operatorname{sgn} \det \left(\frac{\partial \varphi_\sigma}{\partial u_1}\Big|_\alpha, \ldots, \frac{\partial \varphi_\sigma}{\partial u_k}\Big|_\alpha, \frac{\partial \varphi_\tau}{\partial v_1}\Big|_\beta, \ldots, \frac{\partial \varphi_\tau}{\partial v_\ell}\Big|_\beta \right)$$

$$= \mathcal{I}(\varphi_\sigma, \varphi_\tau).$$

In the following we use the parametrization $\varphi_{|J|} \times \operatorname{id}$ that induces the product orientation on $|J| \times \mathbb{R}$.

Definition 3.5. Let $f, g : \langle N \rangle \to \mathbb{R}^m$ be two general position maps such that $\{f(i) : i \in \langle N \rangle\} \cup \{g(i) : i \in \langle N \rangle\}$ is in general position, where $f(i) = g(j)$ is permitted only if $i = j$. Define the *deformation map*

$$h_{f,g} : |\mathsf{K}| \times \mathbb{R} \ \to \ \mathbb{R}^m \times \mathbb{R}$$

$$h_{f,g}(x,t) \ := \ (tf(x) + (1-t)g(x), t).$$

and the *deformation cochain* $\lambda_{f,g} \in C^{m-1}(\mathsf{K}_\Delta^2)$ of f and g by

$$\lambda_{f,g}(\tau_1 \times \tau_2) \ := \ \mathcal{I}\big(h_{f,g}(|\tau_1| \times [0,1]), h_{f,g}(|\tau_2| \times [0,1])\big)$$

for $(m-1)$-cells $\tau_1 \times \tau_2 \in \mathsf{K}_\Delta^2$.

Proposition 3.6. *The cohomology class of φ_f is independent of the general position map f: For two general position maps f and g we have*

$$\delta\lambda_{f,g} \ = \ \varphi_f - \varphi_g.$$

Therefore the cohomology class $\Phi_\mathsf{K} := [\varphi_f] \in H^m(\mathsf{K}_\Delta^2; \mathbb{Z})$ is an invariant of the complex K itself.

Proof. Let $\sigma \times \tau \in \mathsf{K}_\Delta^2$, $\dim \sigma \times \tau = m$. In the following we omit the index f, g from $\lambda_{f,g}$ and $h_{f,g}$. We get the boundary of $h(\sigma \times [0,1])$ by taking the boundary first and then applying h. The intersections $h(\partial\sigma \times [0,1]) \cap h(\tau \times [0,1])$ are inner intersections. We extend the surface patch $h(\tau \times [0,1])$ to $h(\tau \times [-\varepsilon, 1+\varepsilon])$ so that the intersections $h(\sigma \times \{0\}) \cap h(\tau \times \{0\})$ and $h(\sigma \times \{1\}) \cap h(\tau \times \{1\})$ become inner intersections of $h(\partial(\sigma \times [0,1])) \cap h(\tau \times [-\varepsilon, 1+\varepsilon])$ as well but no new intersections occur. Then

$$\begin{aligned}
\lambda(\partial\sigma \times \tau) &= \mathcal{I}\big(h(\partial\sigma \times [0,1]), h(\tau \times [0,1])\big) \\
&= \mathcal{I}\big(h(\partial\sigma \times [0,1]), h(\tau \times [-\varepsilon, 1+\varepsilon])\big) \\
&= \mathcal{I}\big(h(\partial(\sigma \times [0,1])), h(\tau \times [-\varepsilon, 1+\varepsilon])\big) \\
&\quad + (-1)^{\dim\sigma}\mathcal{I}\big(h(\sigma \times \{0\}), h(\tau \times [-\varepsilon, 1+\varepsilon])\big) \\
&\quad - (-1)^{\dim\sigma}\mathcal{I}\big(h(\sigma \times \{1\}), h(\tau \times [-\varepsilon, 1+\varepsilon])\big) \\
&= (-1)^{\dim\sigma+1}\mathcal{I}\big(h(\sigma \times [0,1]), h(\partial(\tau \times [-\varepsilon, 1+\varepsilon]))\big) \\
&\quad + (-1)^{\dim\sigma}\mathcal{I}\big(f(\sigma), f(\tau)\big) - (-1)^{\dim\sigma}\mathcal{I}\big(g(\sigma), g(\tau)\big) \\
&= (-1)^{\dim\sigma+1}\big[\mathcal{I}\big(h(\sigma \times [0,1]), h(\partial\tau \times [-\varepsilon, 1+\varepsilon])\big) \\
&\qquad + \mathcal{I}\big(h(\sigma \times [0,1]), h(\tau \times \{-\varepsilon\})\big) \\
&\qquad - \mathcal{I}\big(h(\sigma \times [0,1]), h(\tau \times \{1+\varepsilon\})\big)\big] \\
&\quad + (-1)^{\dim\sigma}\mathcal{I}\big(f(\sigma), f(\tau)\big) - (-1)^{\dim\sigma}\mathcal{I}\big(g(\sigma), g(\tau)\big) \\
&= (-1)^{\dim\sigma+1}\lambda(\sigma \times \partial\tau) + \varphi_f(\sigma \times \tau) - \varphi_g(\sigma \times \tau).
\end{aligned}$$

$\square$

The deformation cochain has symmetries as well:

Lemma 3.7. *If $\tau_1 \times \tau_2$ is an $(m-1)$-cell of K_Δ^2, then $\tau_2 \times \tau_1$ is also an $(m-1)$-cell of K_Δ^2 and*

$$\lambda_{f,g}(\tau_1 \times \tau_2) \;=\; (-1)^{(\dim \tau_1 + 1)(\dim \tau_2 + 1)} \lambda_{f,g}(\tau_2 \times \tau_1).$$

Proof.

$$
\begin{aligned}
\lambda_{f,g}(\tau_1 \times \tau_2) &= \mathcal{I}\big(h_{f,g}(\tau_1 \times [0,1]), h_{f,g}(\tau_2 \times [0,1])\big) \\
&= (-1)^{(\dim \tau_1 + 1)(\dim \tau_2 + 1)} \mathcal{I}\big(h_{f,g}(\tau_2 \times [0,1]), h_{f,g}(\tau_1 \times [0,1])\big) \\
&= (-1)^{(\dim \tau_1 + 1)(\dim \tau_2 + 1)} \lambda_{f,g}(\tau_2 \times \tau_1).
\end{aligned}
$$
$\square$

Remark. Intersection cocycle and deformation cochain can also be defined for *piecewise linear* general position maps maintaining the same properties [12]. So the cohomology class $\Phi_K := [\varphi_f] \in H^m(K_\Delta^2)$ where $f : |K| \to \mathbb{R}^m$ is any piecewise linear map, only serves as an obstruction to piecewise linear embeddability. It cannot distinguish between piecewise linear embeddability and geometric realizability.

4. Distinguishing between simplicial maps and PL maps

In this section, we collect properties of deformation cochains between *simplicial* maps that do not necessarily hold for deformation cochains between arbitrary *piecewise linear* maps. The values of the intersection cocycles φ_f, φ_g and the deformation cochain $\lambda_{f,g}$ of two simplicial maps f and g depend only on the values that f and g take on the vertex set $\langle N \rangle$ of the complex in question. The complex itself determines the products $\sigma \times \tau$ on which $\lambda_{f,g}$ may be evaluated. So we examine what values these cochains take on $C^{m-1}(S_\Delta^2)$, where S denotes the full face lattice of the N-simplex. In Section 5 we derive further properties for the case that we deform into a geometric realization.

4.1. Linking numbers

Definition 4.1. Let x, y be simplicial cycles in $\mathbb{R}^m$, $\dim x + \dim y = m+1$, with disjoint supports. As every cycle bounds in $\mathbb{R}^m$ we find a chain γ such that $\partial \gamma = x$. The *linking number* of x and y is defined as $\mathcal{L}(x, y) := \mathcal{I}(\gamma, y)$.

Lemma 4.2. *Let σ, τ be affine simplices in $\mathbb{R}^m$. Then:*

(a) $|\mathcal{I}(\sigma, \tau)| \le 1$ *if* $\dim \sigma + \dim \tau = m$.
(b) $|\mathcal{L}(\partial \sigma, \partial \tau)| \le 1$ *if* $\dim \sigma = 2$ *and* $\dim \tau = m - 1$.

The next two conditions follow from the estimates in Lemma 4.2 on the intersection numbers φ_f.

Proposition 4.3. *Let f, g be two general position maps of $\langle N \rangle$ into $\mathbb{R}^m$. Then*

(a) $-1 - \varphi_g(\sigma \times \tau) \le \delta\lambda_{f,g}(\sigma \times \tau) \le 1 - \varphi_g(\sigma \times \tau)$
   for all $\sigma \times \tau \in S_\Delta^2$, $\dim \sigma + \dim \tau = m$.
(b) $-1 - \varphi_g(\sigma \times \partial\tau) \le \lambda_{f,g}(\partial\sigma \times \partial\tau) \le 1 - \varphi_g(\sigma \times \partial\tau)$
   for all $\sigma \times \tau \in S_\Delta^2$, $\dim \sigma = m - 1$ and $\dim \tau = 2$.

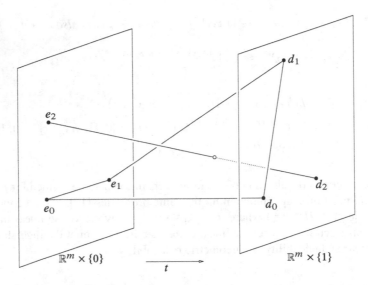

FIGURE 1. Intersecting surfaces $h(\Delta_{I_+} \times \mathbb{R})$ and $h(\Delta_{I_-} \times \mathbb{R})$ for the matrix (d_1, d_2) and the partition $I_+ = \{0, 1\}$ and $I_- = \{2\}$.

Proof. (a) As $\varphi_f(\sigma \times \tau) = (-1)^{\dim \sigma} \mathcal{I}(f(\sigma), f(\tau))$, we can bound φ_f in Proposition 3.6 by $|\varphi_f| \leq 1$.

(b) $\varphi_f(\sigma \times \partial \tau) = (-1)^{\dim \sigma} \mathcal{I}(f(\sigma), f(\partial \tau)) = (-1)^{\dim \sigma} \mathcal{L}(f(\partial \sigma), f(\partial \tau))$ and $\delta \lambda_{f,g}(\sigma \times \partial \tau) = \lambda_{f,g}(\partial \sigma \times \partial \tau)$. □

4.2. Deforming simplices

4.2.1. The simplest case.

For the following, homotopies between images of a simplicial complex under different general position maps play a crucial rôle. In this section we look at the simplest case: The homotopy from the standard simplex of $\mathbb{R}^m$ to an arbitrary one.

Let $D \in \mathbb{R}^{m \times m}$ be an arbitrary matrix with columns d_i, $i \in [m]$, and set $d_0 := 0$. Associate with D the map

$$h : \mathbb{R}^{m+1} \to \mathbb{R}^{m+1}, h(x, t) := ((tD + (1-t)E_m)x, t).$$

Then for every subset $I \subset \langle m \rangle$ the map $h|_{\Delta_I \times [0,1]}$ represents the homotopy of Δ_I into $\mathrm{conv}\{d_i \mid i \in I\}$, moving all points along straight line segments to corresponding points, i.e., it is a ruled m-dimensional surface.

We call an eigenvalue of a square matrix *general* if it is simple, its eigenvector v has no vanishing components, and $\sum v_i \neq 0$. This technical condition characterizes the situation where all pairs of ruled surfaces defined by disjoint subsets of the vertex set are transversal.

We begin by characterizing intersection points of pairs of surfaces in terms of eigenvalues of D.

Lemma 4.4. *Let $D \in \mathbb{R}^{m \times m}$ and $h : \mathbb{R}^{m+1} \to \mathbb{R}^{m+1}$ its associated map.*

(a) *Let $I_+, I_- \subset \langle m \rangle$ such that $I_+ \cap I_- = \varnothing$. If the surfaces $h(\Delta_{I_+} \times \mathbb{R})$ and $h(\Delta_{I_-} \times \mathbb{R})$ intersect at time t, then $1 - \frac{1}{t}$ is an eigenvalue of D.*

(b) *Let $u \neq 1$ be a general eigenvalue of D. Then u uniquely determines disjoint subsets I_+^u and $I_-^u \subset \langle m \rangle$ with $0 \in I_+^u$ such that $h(\Delta_{I_+^u} \times \mathbb{R})$ and $h(\Delta_{I_-^u} \times \mathbb{R})$ intersect at time*

$$t = \frac{1}{1-u}.$$

Another point of view: If $u \neq 1$ is an eigenvalue of D, then $h(\Delta_{\langle m \rangle} \times \{t\})$ fails to span $\mathbb{R}^m \times \{t\}$. So we get a Radon partition in some lower-dimensional subspace of $\mathbb{R}^m \times \{t\}$. If the eigenvector is general, then we get a unique Radon partition.

Proof. (a) Let $(p, t) \in h(\Delta_{I_+} \times \mathbb{R}) \cap h(\Delta_{I_-} \times \mathbb{R})$ be an intersection point. Then p has the representation

$$p = h(\alpha, t) = h(\beta, t),$$

that is,

$$p = \sum_{i \in I_+} \alpha_i (t d_i + (1-t) e_i) = \sum_{j \in I_-} \beta_j (t d_j + (1-t) e_j)$$

with

$$\sum_{i \in I_+} \alpha_i = \sum_{j \in I_-} \beta_j = 1,$$

$\alpha_i, \beta_j > 0$ for all $i \in I_+$, $j \in I_-$. Because of $t \neq 0$ and $e_0 = d_0 = 0$ we can rewrite this as

$$t \left(\sum_{i \in I_+} \alpha_i \left(d_i - (1 - \tfrac{1}{t}) e_i \right) + \sum_{j \in I_-} (-\beta_j) \left(d_j - (1 - \tfrac{1}{t}) e_j \right) \right) = 0.$$

Therefore $\sum_{i \in I_+} \alpha_i e_i + \sum_{j \in I_-} (-\beta_j) e_j$ is an eigenvector of D with eigenvalue $1 - \frac{1}{t}$.

(b) Let $u \neq 1$ be a general eigenvalue of D and v its eigenvector. Consider the sets $\tilde{I}_+^u := \{i \in [m] \mid v_i > 0\}$ and $I_-^u := \{i \in [m] \mid v_i < 0\}$ of positive and negative coefficients, respectively. Without loss of generality assume that $V := -\sum_{i \in I_-^u} v_i > \sum_{i \in \tilde{I}_+^u} v_i$. Denote $I_+^u := \tilde{I}_+^u \cup \{0\}$, $\alpha_i := \frac{v_i}{V}$ for $i \in \tilde{I}_+^u$ and $\alpha_0 := 1 - \sum_{i \in \tilde{I}_+^u} \alpha_i$, $\beta_j := -\frac{v_j}{V}$ for $j \in I_-$ and $t := \frac{1}{1-u}$. Then (p, t) with

$$p = \sum_{i \in I_+^u} \alpha_i (t d_i + (1-t) e_i) = \sum_{j \in I_-^u} \beta_j (t d_j + (1-t) e_j)$$

is an intersection point of the two simplices $h(\Delta_{I_+^u} \times \{t\})$ and $h(\Delta_{I_-^u} \times \{t\})$. So the surfaces $h(\Delta_{I_+^u} \times \mathbb{R})$ and $h(\Delta_{I_-^u} \times \mathbb{R})$ intersect at time t. $\qquad \square$

Remark. In the case of a general eigenvalue $u = 1$ we can still find the sets I_+^u and I_-^u. Then the surfaces $h(\Delta_{I_+^u} \times \mathbb{R})$ and $h(\Delta_{I_-^u} \times \mathbb{R})$ have parallel ends. This complements the preceding lemma because they then 'meet at time $t = \infty$'.

Denote by

$$\mathcal{P} := \{\{I_+, I_-\} \mid I_+ \cup I_- = \langle m \rangle, I_+ \cap I_- = \varnothing, 0 \in I_+\}$$

the set of all bipartitions of $\langle m \rangle$.

Corollary 4.5. *Let $D \in \mathbb{R}^{m \times m}$ and ℓ be the multiplicity of the eigenvalue 1 of D. Then*

$$\sum_{\{I_+, I_-\} \in \mathcal{P}} \# \left(h(\Delta_{I_+} \times \mathbb{R}) \cap h(\Delta_{I_-} \times \mathbb{R}) \right) \ \leq \ m - \ell,$$

that is, the total number of intersection points of pairs of surfaces of the form $h(\Delta_I \times \mathbb{R})$ and $h(\Delta_{\langle m \rangle \setminus I} \times \mathbb{R})$ can not exceed $m - \ell$.

Now we calculate intersection numbers of the surfaces found in Lemma 4.4. To this end we impose orientations on the surfaces in question. In the following let $h(\Delta_I \times \mathbb{R})$ carry the orientation induced by the parametrization $\psi := h \circ (\varphi_I \times \mathrm{id})$. Further let $I_+^u = \{i_0, \ldots, i_k\}_<$, $I_-^u = \{i_{k+1}, \ldots, i_m\}_<$ with $i_0 = 0$ and denote by (I_+^u, I_-^u) the 'shuffle' permutation $(i_0, \ldots, i_m) \mapsto (0, \ldots, m)$.

Lemma 4.6. *Let $D \in \mathbb{R}^{m \times m}$ and $u = 1 - \frac{1}{t}$ be an eigenvalue of D. Denote by (p, t) the intersection point of the surfaces $h(\Delta_{I_+^u} \times \mathbb{R})$ and $h(\Delta_{I_-^u} \times \mathbb{R})$. The surfaces intersect transversally and (p, t) is an inner point if and only if u is general. In this case we have*

$$\mathcal{I}\left(h(\Delta_{I_+^u} \times \mathbb{R}), h(\Delta_{I_-^u} \times \mathbb{R}) \right)\Big|_{(p,t)} = \mathrm{sgn}(I_+^u, I_-^u)\, \mathrm{sgn}(t^m \chi_D'(u)),$$

where χ_D' is the derivative of the characteristic polynomial $\chi_D(u) = \det(D - u E_m)$ of D.

Proof. Our calculations differ in so far from those in [9, Proof of Lemma 3.2] as we have to deal with the permutation $(I_+^u, I_-^u) : j \mapsto i_j$. Denote the intersection point of the surfaces in question by (p, t) where

$$p = t \sum_{j=0}^{k} \alpha_{i_j} d_{i_j}^u = t \sum_{j=k+1}^{m} \beta_{i_j} d_{i_j}^u \tag{4.1}$$

with $d_i^u := d_i - u e_i$.

For checking transversality as well as for the index of intersection at (p, t) we examine

$$\mathcal{D} := \det\left(\left(\frac{\partial \psi_+}{\partial \xi_1} \right), \ldots, \left(\frac{\partial \psi_+}{\partial \xi_k} \right), \left(\frac{\partial \psi_+}{\partial t} \right), \left(\frac{\partial \psi_-}{\partial \xi_1} \right), \ldots, \left(\frac{\partial \psi_-}{\partial \xi_{m-k}} \right), \left(\frac{\partial \psi_-}{\partial t} \right) \right),$$

where the first k derivatives are calculated at $(\alpha_{i_1}, \ldots, \alpha_{i_k}, t)$ and the last $m - k$ at $(\beta_{i_{k+2}}, \ldots, \beta_{i_m}, t)$. We therefore get

$$\mathcal{D} = \det\left(t\begin{pmatrix} d_{i_1}^u \\ 0 \end{pmatrix}, \ldots, t\begin{pmatrix} d_{i_k}^u \\ 0 \end{pmatrix}, \begin{pmatrix} \sum_{j=1}^k \alpha_{ij}(d_{ij} - e_{ij}) \\ 1 \end{pmatrix}, \right.$$
$$\left. t\begin{pmatrix} d_{i_{k+2}}^u - d_{i_{k+1}}^u \\ 0 \end{pmatrix}, \ldots, t\begin{pmatrix} d_{i_m}^u - d_{i_{k+1}}^u \\ 0 \end{pmatrix}, \begin{pmatrix} \sum_{j=k+1}^m \beta_{ij}(d_{ij} - e_{ij}) \\ 1 \end{pmatrix} \right).$$

With $i_0 = 0$, $d_0 = e_0 = 0$ we have

$$\begin{pmatrix} \sum_{j=1}^k \alpha_{ij}(d_{ij} - e_{ij}) \\ 1 \end{pmatrix} - \begin{pmatrix} \sum_{j=k+1}^m \beta_{ij}(d_{ij} - e_{ij}) \\ 1 \end{pmatrix} = \tfrac{1}{t} v,$$

as $v = \sum_{j=1}^k \alpha_{ij} e_{ij} - \sum_{j=k+1}^m \beta_{ij} e_{ij}$ is also an eigenvector of $D - E$ with eigenvalue $\frac{1}{t}$. Subtracting the last column from the $(k+1)$-st and using Laplace expansion with respect to the last row we get

$$\mathcal{D} = t^{m-2} \det(d_{i_1}^u, \ldots, d_{i_k}^u, v, d_{i_{k+2}}^u - d_{i_{k+1}}^u, \ldots, d_{i_m}^u - d_{i_{k+1}}^u)$$
$$= t^{m-2} \sum_{j=1}^m v_j \det(d_{i_1}^u, \ldots, d_{i_k}^u, e_j, d_{i_{k+2}}^u - d_{i_{k+1}}^u, \ldots, d_{i_m}^u - d_{i_{k+1}}^u). \qquad (4.2)$$

From (4.1) we have

$$0 = \sum_{j=1}^k \alpha_{ij} d_{ij}^u - d_{i_{k+1}}^u - \sum_{j=k+2}^m \beta_{ij}(d_{ij}^u - d_{i_{k+1}}^u).$$

Now we examine the summands of the last expression of $\mathcal{D}$ in three groups. In the first case, $j < k + 1$, we have $v_{ij} = \alpha_{ij}$. We substitute $\alpha_{ij} d_{ij}^u$, cancel all terms except $d_{i_{k+1}}^u$ and exchange $d_{i_{k+1}}^u$ and e_{ij}:

$$\alpha_{ij} \det(d_{i_1}^u, \ldots, d_{i_k}^u, e_{ij}, d_{i_{k+2}}^u - d_{i_{k+1}}^u, \ldots, d_{i_m}^u - d_{i_{k+1}}^u)$$
$$= \det(d_{i_1}^u, \ldots, \alpha_{ij} d_{ij}^u, \ldots, d_{i_k}^u, e_{ij}, d_{i_{k+2}}^u - d_{i_{k+1}}^u, \ldots, d_{i_m}^u - d_{i_{k+1}}^u)$$
$$= \det(d_{i_1}^u, \ldots, d_{i_{k+1}}^u, \ldots, d_{i_k}^u, e_{ij}, d_{i_{k+2}}^u - d_{i_{k+1}}^u, \ldots, d_{i_m}^u - d_{i_{k+1}}^u)$$
$$= -\det(d_{i_1}^u, \ldots, e_{ij}, \ldots, d_{i_m}^u).$$

By an analogous calculation the second case, $j > k + 1$, yields

$$-\beta_{ij} \det(d_{i_1}^u, \ldots, d_{i_k}^u, e_{ij}, d_{i_{k+2}}^u - d_{i_{k+1}}^u, \ldots, d_{i_m}^u - d_{i_{k+1}}^u)$$
$$= -\det(d_{i_1}^u, \ldots, e_{ij}, \ldots, d_{i_m}^u).$$

For the remaining term, $j = k$, we use the same procedure on each of the summands after the first step and evaluate the telescope sum in the last step. Thus we get:

$$
-\beta_{i_{k+1}} \det(d_{i_1}^u, \ldots, d_{i_k}^u, e_{i_{k+1}}, d_{i_{k+2}}^u - d_{i_{k+1}}^u, \ldots, d_{i_m}^u - d_{i_{k+1}}^u)
$$
$$
= -\det(d_{i_1}^u, \ldots, d_{i_k}^u, e_{i_{k+1}}, d_{i_{k+2}}^u - d_{i_{k+1}}^u, \ldots, d_{i_m}^u - d_{i_{k+1}}^u)
$$
$$
+ \sum_{j=k+2}^m \det(d_{i_1}^u, \ldots, d_{i_k}^u, e_{i_{k+1}}, d_{i_{k+2}}^u - d_{i_{k+1}}^u, \ldots, \beta_{ij}(d_{ij}^u - d_{i_{k+1}}^u), \ldots, d_{i_m}^u - d_{i_{k+1}}^u)
$$
$$
= -\det(d_{i_1}^u, \ldots, d_{i_k}^u, e_{i_{k+1}}, d_{i_{k+2}}^u - d_{i_{k+1}}^u, \ldots, d_{i_m}^u - d_{i_{k+1}}^u)
$$
$$
+ \sum_{j=k+2}^m \det(d_{i_1}^u, \ldots, d_{i_k}^u, e_{i_{k+1}}, d_{i_{k+2}}^u - d_{i_{k+1}}^u, \ldots, -d_{i_{k+1}}^u, \ldots, d_{i_m}^u - d_{i_{k+1}}^u)
$$
$$
= -\det(d_{i_1}^u, \ldots, d_{i_k}^u, e_{i_{k+1}}, d_{i_{k+2}}^u - d_{i_{k+1}}^u, \ldots, d_{i_m}^u - d_{i_{k+1}}^u)
$$
$$
- \sum_{j=k+2}^m \det(d_{i_1}^u, \ldots, d_{i_k}^u, e_{i_{k+1}}, d_{i_{k+2}}^u, \ldots, d_{i_{j-1}}^u, d_{i_{k+1}}^u, d_{i_{j+1}}^u - d_{i_{k+1}}^u \ldots, d_{i_m}^u - d_{i_{k+1}}^u)
$$
$$
= -\det(d_{i_1}^u, \ldots, d_{i_k}^u, e_{i_{k+1}}, d_{i_{k+2}}^u, \ldots, d_{i_m}^u) .
$$

So in every single case we have

$$
v_{ij} \det(d_{i_1}^u, \ldots, d_{i_k}^u, e_{ij}, d_{i_{k+2}}^u - d_{i_{k+1}}^u, \ldots, d_{i_m}^u - d_{i_{k+1}}^u) = -\det(d_{i_1}^u, \ldots, e_{ij}, \ldots, d_{i_m}^u) .
$$

To complete the calculation we insert these results into (4.2):

$$
\mathcal{D} = -t^{m-2} \sum_{j=1}^m \det(d_{i_1}^u, \ldots, e_{ij}, \ldots, d_{i_m}^u)
$$
$$
= \operatorname{sgn}(I_+^u, I_-^u) t^{m-2} \sum_{j=1}^m \det(d_1^u, \ldots, -e_j, \ldots, d_m^u)
$$
$$
= \operatorname{sgn}(I_+^u, I_-^u) t^{m-2} \chi_{\mathcal{D}}'(u).
$$

We have $\chi'(u) \neq 0$ since u is simple. Therefore the intersection is transversal and the index of intersection at the point under consideration is $\operatorname{sgn} \mathcal{D}$. $\qquad \square$

Corollary 4.7. *Let D be nonsingular, all its negative eigenvalues be general and ℓ_- the number of negative eigenvalues. Denote $\tilde{h}(J) := h(\Delta_J \times [0,1])$.*
Then we have

$$
\sum_{\{I_+, I_-\} \in \mathcal{P}} \operatorname{sgn}(I_+, I_-) \, \mathcal{I}(\tilde{h}(I_+), \tilde{h}(I_-)) = \begin{cases} 0 & \text{if } \det D > 0, \\ -1 & \text{if } \det D < 0. \end{cases} \tag{4.3}
$$

For every subset $S \subset \mathcal{P}$ we have

$$
-\left\lceil \frac{\ell_-}{2} \right\rceil \leq \sum_{\{I_+, I_-\} \in S} \operatorname{sgn}(I_+, I_-) \, \mathcal{I}(\tilde{h}(I_+), \tilde{h}(I_-)) \leq \left\lfloor \frac{\ell_-}{2} \right\rfloor . \tag{4.4}
$$

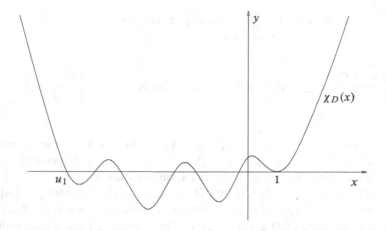

FIGURE 2. A characteristic polynomial with simple negative roots.

As a special case we have for every individual pair $\{I_+, I_-\} \in \mathcal{P}$ the estimates

$$-\left\lceil \frac{\ell_-}{2} \right\rceil \;\leq\; \mathrm{sgn}(I_+, I_-)\, \mathcal{I}\big(\tilde{h}(I_+), \tilde{h}(I_-)\big) \;\leq\; \left\lfloor \frac{\ell_-}{2} \right\rfloor. \tag{4.5}$$

Proof. Intersection times $t \in [0, 1]$ correspond to eigenvalues $u < 0$ of D. The first root u_1 of χ_D satisfies $\chi'_D(u_1) < 0$ and two consecutive roots $u, \hat{u}$ of χ_D satisfy $\mathrm{sgn}\, \chi'_D(u) = -\mathrm{sgn}\, \chi'_D(\hat{u})$. So χ_D has at most $\left\lceil \frac{\ell_-}{2} \right\rceil$ negative roots u with $\chi'_D(u) < 0$ and at most $\left\lfloor \frac{\ell_-}{2} \right\rfloor$ negative roots $\tilde{u}$ with $\chi'_D(\tilde{u}) > 0$. These are exactly the terms in the sums above.   □

4.2.2. Application to the deformation cochain.

The relations between coefficients of λ we develop here are local in the sense that we only look at few vertices at the same time. We restrict to subcomplexes of S consisting of $m + 1$ points. For a subset $J := \{j_0, \dots j_m\}_< \subset [N]$ and $k \in \mathbb{N}$ denote $\ell_J^k := \#\big((J \setminus \{j_0\}) \cap [k]\big)$ and

$$\mathcal{P}_J := \{\tau_+ \times \tau_- \in S_\Delta^2 \mid \dim(\tau_+ \times \tau_-) = m - 1, \tau_+ \cup \tau_- = J, j_0 \in \tau_+\}.$$

These are the products of simplices with vertices in J that we may insert into the deformation cochain.

Theorem 4.8 (Related coefficients of the deformation cochain). *Let f and g be general position maps of the vertex set $\langle N \rangle$ of S into $\mathbb{R}^m$ and $k \in \langle N \rangle$. Assume further that $f(i) = g(i)$ for $i \in \langle k \rangle$ and that the set $\{f(0), \dots, f(N), g(k + 1), \dots, g(N)\}$ is in general position.*

For every subset $J \subset \langle N \rangle$ with $|J| = m + 1$ denote by $\varepsilon_g(J)$ the orientation of the simplex $g(J)$. Then the deformation cochain $\lambda_{f,g} \in C^{m-1}(S_\Delta^2)$ has the following properties:

$$\sum_{\tau_+ \times \tau_- \in \mathcal{P}_J} |\lambda_{f,g}(\tau_+ \times \tau_-)| \;\leq\; m - \ell_J^k, \tag{4.6}$$

$$-1 \leq \varepsilon_g(J) \sum_{\tau_+ \times \tau_- \in \mathcal{P}_J} \mathrm{sgn}(\tau_+, \tau_-) \lambda_{f,g}(\tau_+ \times \tau_-) \leq 0, \tag{4.7}$$

and

$$-\left\lceil \frac{m - \ell_J^k}{2} \right\rceil \leq \varepsilon_g(J)\,\mathrm{sgn}(\tau_+, \tau_-)\lambda_{f,g}(\tau_+ \times \tau_-) \leq \left\lfloor \frac{m - \ell_J^k}{2} \right\rfloor \tag{4.8}$$

for every $\tau_+ \times \tau_- \in \mathcal{P}_J$.

Proof. Fix a subset $J := \{j_0, \dots j_m\}_< \subset \langle N \rangle$. Perform a basis transformation A_J that takes $(g(j_0), 0), \dots, (g(j_m), 0), (f(j_0), 1)$ to $e_0, \dots, e_{m+1}$, respectively. $\varepsilon_g(J)$ is the sign of the determinant of this basis transformation. Let $(d_1, 1), \dots, (d_m, 1)$ be the images of $(f(j_1), 1), \dots, (f(j_m), 1)$ and $D := (d_1, \dots, d_m)$. Denote $j : [m] \to J$, $i \mapsto j_i$. Then $h := A_J \circ h_{f,g} \circ j$ is of the form we considered in Subsection 4.2.1. Moreover the first ℓ_J^k columns of D are $e_1, \dots, e_{\ell_J^k}$. The eigenvalue 1 has at least multiplicity ℓ_J^k. Thus $m - \ell_J^k$ is an upper bound for the number of negative eigenvalues. Now

$$\lambda_{f,g}(\tau_+ \times \tau_-) = \mathcal{I}\big(h_{f,g}(\tau_+ \times [0,1]), h_{f,g}(\tau_- \times [0,1])\big)$$
$$= \varepsilon_g(J)\mathcal{I}\big(h(j^{-1}(\tau_+) \times [0,1]), h(j^{-1}(\tau_-) \times [0,1])\big)$$

and therefore

$$|\lambda_{f,g}(\tau_+ \times \tau_-)| \leq \#\big(h(j^{-1}(\tau_+) \times [0,1] \cap h(j^{-1}(\tau_-) \times [0,1])\big).$$

So we immediately get equation (4.6) from Corollary 4.5 and equations (4.7) and (4.8) from Corollary 4.7. $\qquad\square$

Remark. Condition (4.8) implies

$$-\left\lceil \frac{m}{2} \right\rceil \leq \lambda_{f,g}(\tau_+ \times \tau_-) \leq \left\lceil \frac{m}{2} \right\rceil$$

which are the restrictions on the values of $\lambda_{f,g}$ that Novik derived (cf. [9, Theorem 3.1]).

5. Geometric realizability and beyond

Up to now we have looked at arbitrary general position maps. In this section we compare a map with special properties such as a geometric realization with a reference map whose intersection cocycle can be easily computed.

5.1. The reference map

We start by defining our reference map:

Denote by $c : \langle N \rangle \to \mathbb{R}^m$ the *cyclic map* which maps vertex i to the point $c(i) = (i, i^2, \dots, i^m)^t$ on the moment curve.

Proposition 5.1 ([11, Lemma 4.2]). *Let* $k + \ell = m$, $k \geq \ell$, $s_0 < s_1 < \dots < s_k$, $t_0 < t_1 < \dots < t_\ell$. *If* $k = \ell$, *assume further that* $s_0 < t_0$. *The two simplices* $\sigma = \mathrm{conv}\{c(s_0), \dots, c(s_k)\}$ *and* $\tau = \mathrm{conv}\{c(t_0), \dots, c(t_\ell)\}$ *of complementary dimensions* *intersect if and only if their dimensions differ at most by one and their vertices alternate along the curve:*

$$k = \left\lceil \frac{m}{2} \right\rceil \quad and \quad s_0 < t_0 < s_1 < \ldots < s_{\lfloor \frac{m}{2} \rfloor} < t_{\lfloor \frac{m}{2} \rfloor} (< s_{\lceil \frac{m}{2} \rceil}).$$

In the case of intersection we have

$$\mathcal{I}(\sigma, \tau) = (-1)^{\frac{(k-1)k}{2}}.$$

Proof. For every set $\{c_0, \ldots, c_{m+1}\}$ consisting of $m + 2$ points $c_i = c(u_i)$ with $u_0 < u_1 < \ldots < u_{m+1}$ there is a unique affine dependence

$$\sum_{i=0}^{m+1} \alpha_i c_i = 0 \quad with \quad \sum_{i=0}^{m+1} \alpha_i = 0 \text{ and } \alpha_0 = 1.$$

We calculate the sign of the coefficients α_k.

$$\det\left(\binom{1}{c_0}, \ldots, \widehat{\binom{1}{c_k}}, \ldots, \binom{1}{c_{m+1}}\right) = -\alpha_k \det\left(\binom{1}{c_k}, \binom{1}{c_1}, \ldots, \widehat{\binom{1}{c_k}}, \ldots, \binom{1}{c_{m+1}}\right)$$

$$= (-1)^k \alpha_k \det\left(\binom{1}{c_1}, \ldots, \binom{1}{c_{m+1}}\right).$$

Since $\det\left(\binom{1}{c_0}, \ldots, \widehat{\binom{1}{c_k}}, \ldots, \binom{1}{c_{m+1}}\right)$ and $\det\left(\binom{1}{c_1}, \ldots, \binom{1}{c_{m+1}}\right)$ are both positive, we get

$$(-1)^k \alpha_k > 0,$$

that is,

$$\mathrm{sgn}\, \alpha_k = \begin{cases} +1 & \text{if } k \text{ is even} \\ -1 & \text{if } k \text{ is odd.} \end{cases}$$

The proposition follows.   □

5.2. Deformation cochains of geometric realizations

If a simplicial map defining the deformation cochain is a simplicial *embedding*, we know that the images of certain simplices don't intersect. The following trivial observation about the coefficients of deformation cochains is the key to bring in the combinatorics of the complex K.

Lemma 5.2. *Let $f, g : \langle N \rangle \to \mathbb{R}^m$ be general position maps. If $f(\sigma) \cap f(\tau) = \varnothing$ and $\dim \sigma + \dim \tau = m$, then*

$$\delta \lambda_{f,g}(\sigma \times \tau) = -\varphi_g(\sigma \times \tau).$$

Proof. $\varphi_f(\sigma \times \tau) = (-1)^{\dim \sigma} \mathcal{I}(f(\sigma), f(\tau)) = 0.$ □

Remark. The expression

$$\delta \lambda(\sigma \times \tau) = \sum_{i=0}^{\dim \sigma} (-1)^i \lambda(\sigma^i \times \tau) + \sum_{j=0}^{\dim \tau} (-1)^{\dim \sigma + j} \lambda(\sigma \times \tau^j)$$

is linear in the coefficients of the deformation cochain λ. So for every pair $\sigma \times \tau$ of simplices of complementary dimensions with disjoint images we get a linear equation that is valid for the coefficients of $\lambda_{f,g}$.

This is particularly useful when we assume the existence of a geometric realization but can also be used to express geometric immersability. So we gather all information we have on the deformation cochain in our main Theorem:

Theorem 5.3 (Obstruction Polytope). *If there is a geometric realization of the simplicial complex* K *in* $\mathbb{R}^m$, *then the obstruction polytope in the cochain space* $C^{m-1}(S^2_\Delta, \mathbb{R})$ *given by the following inequalities contains a point* $\lambda \in C^{m-1}(S^2_\Delta, \mathbb{Z})$ *with integer coefficients.*

1. (The symmetries of Lemma 3.7)

$$\lambda(\tau_1 \times \tau_2) = (-1)^{(\dim \tau_1 + 1)(\dim \tau_2 + 1)} \lambda_{f,g}(\tau_2 \times \tau_1)$$

for all $\tau_1 \times \tau_2 \in S^2_\Delta$.

2. (The deformation inequalities of Theorem 4.8) *For every subset* $J \subset \langle N \rangle$,
   (a)
$$\sum_{\tau_+ \times \tau_- \in \mathcal{P}_J} |\lambda(\tau_+ \times \tau_-)| \leq m - \ell^m_J,$$

   (b)
$$-1 \leq \sum_{\tau_+ \times \tau_- \in \mathcal{P}_J} \text{sgn}(\tau_+, \tau_-)\lambda(\tau_+ \times \tau_-) \leq 0,$$

   and for every $\tau_+ \times \tau_- \in \mathcal{P}_J$,
   (c)
$$-\left\lceil \frac{m - \ell^m_J}{2} \right\rceil \leq \text{sgn}(\tau_+, \tau_-)\lambda(\tau_+ \times \tau_-) \leq \left\lfloor \frac{m - \ell^m_J}{2} \right\rfloor.$$

3. (The intersection and linking inequalities of Proposition 4.3)
   (a) $\varphi_c(\sigma \times \tau) - 1 \leq \delta\lambda(\sigma \times \tau) \leq \varphi_c(\sigma \times \tau) + 1$
       for all $\sigma \times \tau \in S^2_\Delta$, $\dim \sigma + \dim \tau = m$,
   (b) $\varphi_c(\sigma \times \partial\tau) - 1 \leq \lambda(\partial\sigma \times \partial\tau) \leq \varphi_c(\sigma \times \partial\tau) + 1$
       for all $\sigma \times \tau \in S^2_\Delta$, $\dim \sigma = m - 1$ *and* $\dim \tau = 2$.

4. (The equations of Lemma 5.2) *For every pair* $\sigma \times \tau$ *of simplices in* K^2_Δ:

$$\delta\lambda(\sigma \times \tau) = -\varphi_c(\sigma \times \tau).$$

Proof. If there is a geometric realization $f : \langle N \rangle \to \mathbb{R}^m$, then there also is a geometric realization $\tilde{f}$ such that the first m vertices satisfy $\tilde{f}(i) = c(i)$ for $i \in \langle m \rangle$ and such that the set $\{\tilde{f}(0), \ldots, \tilde{f}(N), c(m+1), \ldots, c(N)\}$ is in general position. The deformation cochain $\lambda_{c,\tilde{f}}$ has the desired properties as $\varepsilon_c \equiv 1$. $\square$

Remark. The system Novik described consists of the equations 4. along with the equations 1. and the bounds from Remark 4.2.2 for pairs of simplices in K^2_Δ only.

6. Subsystems and experiments

Theorem 5.3 provides us with a system of linear equations and inequalities that has an integer solution if the complex K has a geometric realization. So we can attack non-realizability proofs by solving integer programming feasibility problems. However, the system sizes grow rapidly with the number of vertices. There are $\mathcal{O}(n^{m+1})$ variables in the system associated to a complex with n vertices and target ambient dimension m. For Brehm's triangulated Möbius strip (and all other complexes on 9 vertices) we already get 1764 variables. The integer feasibility problems—even for complexes with few vertices—are therefore much too big to be sucessfully solved with standard integer programming software. On the other hand, for a non-realizability proof it suffices to exhibit a subsystem of the obstruction system that has no solution.

In this section we therefore look at subsystems of the obstruction system, that only use those variables associated to simplices that belong to the complex K and certain sums of the other variables.

Subsystem 6.1. *If there is a geometric realization of the simplicial complex K in $\mathbb{R}^m$, then there is a cochain $\lambda \in C^{m-1}(K_\Delta^2)$ that satisfies the equations of Lemma 5.2 for every pair of simplices in K_Δ^2 and the linking inequalities (3b) of Proposition 4.3 that only use values of λ on K_Δ^2.*

The deformation inequalities of Theorem 4.8 imply the following for the variables under consideration: For every subset $J \subset \langle N \rangle$ we have

$$\sum_{\tau_+ \times \tau_- \in \mathcal{P}_J \cap K_\Delta^2} |\lambda(\tau_+ \times \tau_-)| + |y_J| \leq m - \ell_J^m, \tag{6.1}$$

$$-1 \leq \sum_{\tau_+ \times \tau_- \in \mathcal{P}_J \cap K_\Delta^2} \operatorname{sgn}(\tau_+, \tau_-)\lambda(\tau_+ \times \tau_-) + y_J \leq 0, \tag{6.2}$$

by introducing the new variable y_J for 'the rest of the sum'. We still have for every $\tau_+ \times \tau_- \in K_\Delta^2$,

$$-\left\lceil \frac{m - \ell_J^m}{2} \right\rceil \leq \operatorname{sgn}(\tau_+, \tau_-)\lambda_{f,g}(\tau_+ \times \tau_-) \leq \left\lfloor \frac{m - \ell_J^m}{2} \right\rfloor, \tag{6.3}$$

and the same bounds hold for y_J.

We can even do with less variables at the expense of more inequalities.

Subsystem 6.2. *If there is a geometric realization of the simplicial complex K in $\mathbb{R}^m$, then there is a cochain $\lambda \in C^{m-1}(K_\Delta^2)$ that satisfies the equations of Lemma 5.2 for every pair of simplices in K_Δ^2 and the linking inequalities (3b) of Proposition 4.3 that only use values of λ on K_Δ^2. The deformation inequalities of Theorem 4.8 imply inequalities for every subset S of $\mathcal{P}_J \cap K_\Delta^2$:*

$$-\left\lceil \frac{m - \ell_J^m}{2} \right\rceil \leq \sum_{\tau_+ \times \tau_- \in S} \operatorname{sgn}(\tau_+, \tau_-)\lambda_{f,g}(\tau_+ \times \tau_-) \leq \left\lfloor \frac{m - \ell_J^m}{2} \right\rfloor \tag{6.4}$$

and

$$\sum_{\tau_+ \times \tau_- \in \mathcal{P}_J \cap K_\Delta^2} |\lambda(\tau_+ \times \tau_-)| \leq m - \ell_J^m. \tag{6.5}$$

surface	file	realizable	f-vector	var.	constr.	solv.	time
$\mathbb{R}P^2$	rp2.gap	no	(6, 15, 10)	150	1365	no	0.24 sec
$\mathcal{B}$	moebius.gap	no [5]	(9, 24, 15)	510	2262	no	46.3 sec
$\mathcal{M}_0$	bipyramid.gap	yes	(5, 9, 6)	48	500	yes	0.1 sec
$\mathcal{M}_1$	csaszar.gap	yes	(7, 21, 14)	322	2583	yes	0.78 sec
$\mathcal{M}_2$	m2_10.gap	yes [7]	(10, 36, 24)	1136	5888	yes	34.83 sec
$\mathcal{M}_3$	m3_10.gap	yes [6]	(10, 42, 28)	1490	9847	yes	143 sec
$\mathcal{M}_4$	m4_11.gap	yes [3]	(11, 51, 34)	2248	15234	yes	564 min
$\mathcal{M}_5$	m5_12.gap	?	(12, 60, 40)	3180	21840	?	
$\mathcal{M}_6$	altshuler54.gap	no [4]	(12, 66, 44)	3762	33473	?	

TABLE 1. Computational results

The systems of the above corollaries are generated by a gap program `generate_obstructions.gap` that can be obtained via my homepage `http://www.math.tu-berlin.de/~timmreck`.

The resulting systems can be examined further by integer programming software. I ran several experiments using SCIP [1] to examine the resulting systems. Table 1 gives an overview on system sizes and solution times. $\mathcal{M}_g$ denotes an orientable surface of genus g. The triangulations under consideration have the minimum number of vertices and can be found in the file. $\mathcal{B}$ denotes the triangulated Möbius strip [5] by Brehm. The systems under consideration are those of Subsystem 6.2 expressing the inequalities involving absolute values without the use of new variables.

The smallest system showing the non-realizability of the Möbius strip only uses the parts 1, 2c, 3b and 4 of Theorem 5.3 and has 510 variables and 426 constraints. The systems for genus 5 and 6 using only these parts of Theorem 5.3 are solvable.

References

[1] T. Achterberg, *SCIP – a framework to integrate constraint and mixed integer programming*, ZIB report 04–19 (Berlin), 2004.

[2] A. Altshuler, J. Bokowski, and P. Schuchert, *Neighborly 2-manifolds with 12 vertices*, J. Combin. Theory Ser. A **75** (1996), no. 1, 148–162.

[3] J. Bokowski and U. Brehm, *A polyhedron of genus 4 with minimal number of vertices and maximal symmetry*, Geom. Dedicata **29** (1989), 53–64.

[4] J. Bokowski and A. Guedes de Oliveira, *On the generation of oriented matroids*, Discrete Comput. Geom. **24** (2000), no. 2–3, 197–208.

[5] U. Brehm, *A nonpolyhedral triangulated Möbius strip*, Proc. Amer. Math. Soc. **89** (1983), no. 3, 519–522.

[6] S. Hougardy, F.H. Lutz, and M. Zelke, *Surface realization with the intersection edge functional*, arXiv:math.MG/0608538, 2006

[7] F.H. Lutz, *Enumeration and random realization of triangulated surfaces*, Discrete Differential Geometry (A.I. Bobenko, P. Schröder, J.M. Sullivan, G.M. Ziegler, eds.), Oberwolfach Seminars, vol. 38, Birkhäuser, 2008, this volume, pp. 235–253; arXiv:math.CO/0506316v2.

[8] J. Matoušek, *Using the Borsuk–Ulam theorem*, Universitext, Springer-Verlag, Heidelberg, 2003.

[9] I. Novik, *A note on geometric embeddings of simplicial complexes in a Euclidean space*, Discrete Comp. Geom. **23** (2000), no. 2, 293–302.

[10] L. Schewe, *Satisfiability Problems in Discrete Geometry*. Dissertation at Technische Universität Darmstadt, 2007.

[11] A. Shapiro, *Obstructions to the imbedding of a complex in a euclidean space. I. The first obstruction*, Ann. of Math. (2) **66** (1957), no. 2, 256–269.

[12] W.T. Wu, *A theory of imbedding, immersion and isotopy of polytopes in a Euclidean space*, Science Press, Peking, 1965.

Dagmar Timmreck
Institut für Mathematik, MA 6–2
Technische Universität Berlin
Str. des 17. Juni 136
10623 Berlin
Germany
e-mail: timmreck@math.tu-berlin.de

Discrete Differential Geometry, A.I. Bobenko, P. Schröder, J.M. Sullivan and G.M. Ziegler, eds.
Oberwolfach Seminars, Vol. 38, 235–253

Enumeration and Random Realization of Triangulated Surfaces

Frank H. Lutz

Abstract. We discuss different approaches for the enumeration of triangulated surfaces. In particular, we enumerate all triangulated surfaces with 9 and 10 vertices. We also show how geometric realizations of orientable surfaces with few vertices can be obtained by choosing coordinates randomly.

Keywords. Triangulated surfaces, mixed lexicographic enumeration, random realization.

1. Introduction

The enumeration of triangulations of the 2-dimensional sphere S^2 was started at the end of the 19th century by Brückner [19]. It took him several years and nine thick manuscript books to compose a list of triangulated 2-spheres with up to 13 vertices; cf. [20]. His enumeration was complete and correct up to 10 vertices. On 11, 12, and 13 vertices his census comprised 1251, 7616, and 49451 triangulations, respectively. These numbers were off only slightly; they were later corrected by Grace [28] (for 11 vertices), by Bowen and Fisk [14] (for 12 vertices), and by Brinkmann and McKay [18] (for 13 vertices). In fact, Brinkmann and McKay enumerated the triangulations of the 2-sphere with up to 23 vertices with their program `plantri` [17]. Table 1 lists the respective numbers. Precise formulas for rooted triangulations of the 2-sphere with n vertices were determined by Tutte [54].

It follows from work of Steinitz [48, §46] that every triangulated 2-sphere can be reduced to the boundary of the tetrahedron by a sequence of *edge contractions*. In other words, the boundary of the tetrahedron is the only *irreducible triangulation* of the 2-sphere, and from this every n-vertex triangulation can be obtained by a suitable sequence of *vertex splits*. The program `plantri` implements this procedure and allows for a fast enumeration of triangulations of the 2-sphere.

Barnette and Edelson [7] showed that, in fact, every 2-manifold has only finitely many irreducible triangulations. The exact numbers have been determined for six surfaces

TABLE 1. Triangulated 2-spheres with up to 23 vertices.

n	Types
4	1
5	1
6	2
7	5
8	14
9	50
10	233
11	1249
12	7595
13	49566
14	339722
15	2406841
16	17490241
17	129664753
18	977526957
19	7475907149
20	57896349553
21	453382272049
22	3585853662949
23	28615703421545

of small genus; see Table 2. With his program surftri [49] (based on plantri),
Sulanke generated all triangulations with up to (at least) 14 vertices for these surfaces by
vertex splitting; see [49] for respective counts of triangulations.

In Section 2 we give a brief survey on the combinatorics of surfaces. Section 3 is
devoted to a discussion of different enumeration approaches. A complete enumeration of
all triangulated surfaces with up to 10 vertices, obtained by mixed lexicographic enumeration, is presented in Section 4.

For triangulations of orientable surfaces of genus $g \geq 1$ it is, in general, a difficult problem to decide realizability. For triangulations with few vertices, however, 3-
dimensional geometric realizations (with straight edges, flat triangles, and without self-
intersections) can be obtained by choosing coordinates randomly; see Section 5.

2. Triangulated surfaces and their f-vectors

For an arbitrary orientable or non-orientable surface M the Euler characteristic $\chi(M)$
equals, by Euler's equation, the alternating sum of the number of vertices $n = f_0$, the

TABLE 2. Irreducible triangulations of surfaces.

Surface	Irreducible triangulations	References
torus	21	Grünbaum; Lavrenchenko [38]
projective plane	2	Barnette [6]
Klein bottle	29	Lawrencenko and Negami [39]; Sulanke [50]
$M(2,+)$	396784	Sulanke [51, 52]
$M(3,-)$	9708	Sulanke [51, 52]
$M(4,-)$	6297982	Sulanke [51, 52]

number of edges f_1, and the number of triangles f_2, i.e.,

$$n - f_1 + f_2 = \chi(M). \qquad (2.1)$$

By double counting of incidences between edges and triangles of a triangulation, it follows that $2f_1 = 3f_2$. Thus, the number of vertices n determines f_1 and f_2, that is, a triangulated surface M of Euler characteristic $\chi(M)$ on n vertices has *face-vector* or *f-vector*

$$f = (f_0, f_1, f_2) = (n, 3n - 3\chi(M), 2n - 2\chi(M)). \qquad (2.2)$$

An *orientable surface* $M(g, +)$ of genus g has homology and Euler characteristic

$$H_*(M(g,+)) = (\mathbb{Z}, \mathbb{Z}^{2g}, \mathbb{Z}), \qquad \chi(M(g,+)) = 2 - 2g,$$

whereas a *non-orientable surface* $M(g, -)$ of genus g has

$$H_*(M(g,-)) = (\mathbb{Z}, \mathbb{Z}^{g-1} \oplus \mathbb{Z}_2, 0), \qquad \chi(M(g,-)) = 2 - g.$$

The smallest possible n for a triangulation of a 2-manifold M is given by Heawood's bound [30],

$$n \geq \left\lceil \tfrac{1}{2}(7 + \sqrt{49 - 24\chi(M)}) \right\rceil, \qquad (2.3)$$

except in the cases of the orientable surface of genus 2, the Klein bottle, and the non-orientable surface of genus 3, for each of which an extra vertex has to be added.

Corresponding minimal and combinatorially unique triangulations of the real projective plane $\mathbb{RP}^2$ with 6 vertices ($\mathbb{RP}^2_6$) and of the 2-torus with 7 vertices (the Möbius torus [44]) were already known in the 19th century; see Figure 1. However, it took until 1955 to complete the construction of series of examples of minimal triangulations for all non-orientable surfaces (Ringel [45]) and until 1980 for all orientable surfaces (Jungerman and Ringel [35]).

A complete classification of triangulated surfaces with up to 8 vertices was obtained by Datta [22] and Datta and Nilakantan [23].

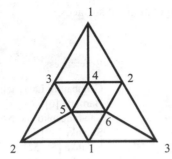 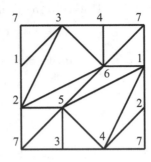

FIGURE 1. The real projective plane $\mathbb{RP}^2_6$ and the Möbius torus.

3. Enumeration of triangulated surfaces

At present, there are three essentially different enumeration schemes available to algorithmically generate triangulated surfaces with a fixed number n of vertices:

1. *Generation from irreducible triangulations*

   An edge of a triangulated surface is *contractible* if the vertices of the edge can be identified without changing the topological type of the triangulation (while remaining a simplicial complex). A triangulation of a surface is *irreducible* if it has no contractible edge.

   The triangulations of a given 2-manifold M with n vertices can therefore be obtained in two steps: First, determine all irreducible triangulations of M with up to n vertices, and, second, generate additional reducible triangulations of M from the irreducible ones with less than n vertices by *vertex-splitting*.

   Comments: Although every 2-manifold M has only finitely many irreducible triangulations [7], it is nontrivial to classify or enumerate these. In the case of S^2, however, the boundary of the tetrahedron is the only irreducible triangulation. For irreducible triangulations of other surfaces see the overview in the introduction.

2. *Strongly connected enumeration*

   The basic idea here is to start with a single triangle, which has three edges as its boundary. Then select the lexicographically smallest edge of the boundary as the pivot. This edge has to lie in a second triangle (since in a triangulated surface every edge is contained in exactly two triangles). In order to complement our pivot edge to a triangle we can choose as a third vertex either a vertex of the boundary of the current partial complex or a new vertex that has not yet been used. For every such choice we compose a new complex by adding the respective triangle to the previous partial complex. We continue with adding triangles until, eventually, we obtain a closed surface.

   Comments: At every step of the procedure the complexes that we produce are *strongly connected*, i.e., for every pair of triangles there is a connecting sequence of neighboring triangles (in other words: the dual graph of the complex is connected). By the lexicographic choice of the pivot edges, the vertex-stars are closed in lexicographic order (which helps to sort out non-manifolds at an early stage of the generation; see the discussion in the next section). Nevertheless, the generation is not very systematic: Albeit we choose the pivot edges in lexicographic order the resulting triangulations do not have to be lexicographically minimal.

Strongly connected enumeration was used by Altshuler and Steinberg [1, 4] to determine all combinatorial 3-manifolds with up to 9 vertices, by Bokowski [2, 8] to enumerate all 59 neighborly triangulations with 12 vertices of the orientable surface of genus 6, and by Lutz and Sullivan [42] to enumerate the 4787 triangulated 3-manifolds of edge degree at most 5.

3. *Lexicographic enumeration*

It is often very useful to list a collection of objects, e.g., triangulated surfaces, in *lexicographic order*: Every listed triangulated surface with n vertices is the lexicographically smallest set of triangles combinatorially equivalent to this triangulation and is lexicographically smaller than the next surface in the list.

An enumeration algorithm which yields triangulated surfaces in lexicographic order is discussed in the next section. (See [53] for an isomorphism-free approach to lexicographic enumeration.)

3'. *Mixed lexicographic enumeration*

A variant of 3; see below.

4. Lexicographic enumeration

We present in the following an algorithm for the lexicographic and the mixed lexicographic enumeration of triangulated surfaces. (Similar enumeration schemes for vertex-transitive triangulations have been described earlier by Kühnel and Lassmann [37] and Köhler and Lutz [36].)

Let $\{1, 2, \ldots, n\}$ be the ground set of n vertices. Then a *triangulation of a surface* (or a *triangulated surface*) with n vertices is

- a connected 2-dimensional simplicial complex $M \subseteq 2^{\{1,\ldots,n\}}$
- such that the link of every vertex is a circle.

In particular,

- M is *pure*, that is, every maximal face of M is 2-dimensional,
- and every edge of M is contained in exactly two triangles.

As an example, we consider the case $n = 5$. On the ground set $\{1, 2, \ldots, 5\}$ there are $\binom{5}{3} = 10$ triangles, $\binom{5}{2} = 10$ edges, and 2^{10} different sets of triangles, of which only few compose a triangulated surface with 5 vertices. Our aim will be to find those sets of triangles that indeed form a triangulated surface. One way to proceed is by *backtracking*:

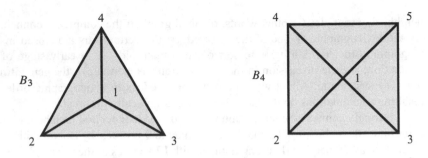

FIGURE 2. Beginning segments.

Start with some triangle and add further triangles as long as no edge is contained in more than two triangles. If this condition is violated, then backtrack. A set of triangles is *closed* if every of its edges is contained in exactly two triangles. If the link of every vertex of a closed set of triangles is a circle, then this set of triangles gives a triangulated surface: OUTPUT surface.

We are interested in enumerating triangulated surfaces *up to combinatorial equivalence*, i.e., up to relabeling the vertices. Thus we can, without loss of generality, assume that the triangle 123 should be present in the triangulation and therefore can be chosen as the starting triangle. More than that, we can assume that the triangulated surfaces that we are going to enumerate should come in lexicographic order.

In a lexicographically minimal triangulation, the collection B_d of triangles containing vertex 1 is of the form

$$123, 124, 135, 146, 157, 168, \ldots, 1\,(d-1)\,(d+1), 1\,d\,(d+1),$$

where $d = \deg(1)$ is the *degree* of vertex 1, i.e., the number of neighbors of vertex 1. Obviously, $3 \leq \deg(1) \leq n-1$. On $n = 5$ vertices, vertex 1 has possible beginning segments B_3 and B_4; see Figure 2.

In a lexicographically sorted list of (lexicographically smallest) triangulated surfaces with n vertices those with beginning segment B_k are listed before those with beginning segment B_{k+1}, etc. Thus, we start the backtracking with beginning segment B_3 and enumerate all corresponding triangulated surfaces, then restart the backtracking with beginning segment B_4, and so on. In triangulations with beginning segment B_k all vertices have degree *at least* k. Otherwise such a triangulation has a vertex of degree $j < k$. However, by relabeling the vertices, it can be achieved that the relabeled triangulation has beginning segment B_j (and thus would have appeared earlier in the lexicographically sorted list). Contradiction.

We store the triangles on the ground set as rows of a (sparse) triangle-edge-incidence matrix; see Table 3 for the triangle-edge-incidence matrix in the case of $n = 5$ vertices. The backtracking in terms of the triangle-edge-incidence matrix then can be formulated as follows: Start with the zero row vector and add to it all rows corresponding to the triangles of the beginning segment B_3; see Table 4. The resulting vector has entries:

TABLE 3. The triangle-edge-incidence matrix for $n = 5$ vertices.

	12	13	14	15	23	24	25	34	35	45
123	1	1			1					
124	1		1			1				
125	1			1			1			
134		1	1					1		
135		1		1					1	
145			1	1						1
234					1	1		1		
235					1		1		1	
245						1	1			1
345								1	1	1

0   (the corresponding edge does not appear in B_3),

1   (the corresponding edge is a boundary edge of B_3),

2   (the corresponding edge appears twice in the triangles of B_3).

We next add (the corresponding rows of) further triangles to (the sum vector of) our beginning segment. As soon as a resulting entry is larger than two, we backtrack, since in such a combination of triangles the edge corresponding to the entry is contained in at least three triangles, which is forbidden. If a resulting vector has entries 0 and 2 only, then the corresponding set of triangles is closed and thus is a candidate for a triangulated surface. In case of $n = 5$ vertices, the backtracking (in short) is as in Table 4.

It remains to check whether every candidate is indeed a triangulated surface. For this we have to verify that the link of every vertex is a circle, which is a purely combinatorial condition that can easily be tested. Moreover, we have to check for every candidate whether it is combinatorially isomorphic to a triangulation that has appeared in the backtracking before. For example, the triangulations $B_4 + 234 + 345$ and $B_4 + 235 + 245$ are isomorphic to $B_3 + 235 + 245 + 345$. In fact, we should have stopped adding triangles to $B_4 + 234$: The link of vertex 2 in this case is a closed circle of length 3, which should not occur, since the beginning segment B_4 has degree 4. Similarly, we should have backtracked at $B_4 + 235$. Finally, we neglect triangulations with less than n vertices (such as $B_3 + 234$). As result in the case $n = 5$, we obtain that there is a unique triangulation of the 2-sphere with 5 vertices.

The basic property that every edge is contained in exactly two triangles is often called the *pseudomanifold property*: Every closed set of triangles forms a 2-*dimensional pseudomanifold*.

It is rather easy to construct 2-dimensional pseudomanifolds that are not surfaces: The left pseudomanifold in Figure 3 has the middle vertex as an isolated singularity with its vertex-link consisting of two disjoint triangles. The right pseudomanifold in Figure 3 has no singularities and thus is a surface, but it is not connected.

TABLE 4. Backtracking for $n = 5$ vertices.

	12	13	14	15	23	24	25	34	35	45	
$-$:	(0	0	0	0	0	0	0	0	0	0)	
B_3 :	(2	2	2	0	1	1	0	1	0	0)	
$B_3 + 234$:	(2	2	2	0	2	2	0	2	0	0)	*Candidate!*
B_3 :	(2	2	2	0	1	1	0	1	0	0)	
$B_3 + 235$:	(2	2	2	0	2	1	1	1	1	0)	
$B_3 + 235 + 245$:	(2	2	2	0	2	2	2	1	1	1)	
$B_3 + 235 + 245 + 345$:	(2	2	2	0	2	2	2	2	2	2)	*Candidate!*
$B_3 + 235 + 245$:	(2	2	2	0	2	2	2	1	1	1)	
$B_3 + 235$:	(2	2	2	0	2	1	1	1	1	0)	
B_3 :	(2	2	2	0	1	1	0	1	0	0)	
$-$:	(0	0	0	0	0	0	0	0	0	0)	
B_4 :	(2	2	2	2	1	1	0	0	1	1)	
$B_4 + 234$:	(2	2	2	2	2	2	0	1	1	1)	
$B_4 + 234 + 345$:	(2	2	2	2	2	2	0	2	2	2)	*Candidate!*
$B_4 + 234$:	(2	2	2	2	2	2	0	1	1	1)	
B_4 :	(2	2	2	2	1	1	0	0	1	1)	
$B_4 + 235$:	(2	2	2	2	2	1	1	0	2	1)	
$B_4 + 235 + 245$:	(2	2	2	2	2	2	2	0	2	2)	*Candidate!*
$B_4 + 235$:	(2	2	2	2	2	1	1	0	2	1)	
B_4 :	(2	2	2	2	1	1	0	0	1	1)	
$-$:	(0	0	0	0	0	0	0	0	0	0)	

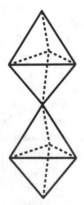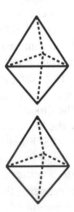

FIGURE 3. Examples of pseudomanifolds.

TABLE 5. Triangulated surfaces with up to 10 vertices.

n	Surface	Types		n	Surface	Types		n	Surface	Types
4	S^2	1		8	S^2	14		10	S^2	233
					T^2	7			T^2	2109
5	S^2	1							$M(2,+)$	865
					$\mathbb{R}P^2$	16			$M(3,+)$	20
6	S^2	2			K^2	6				
									$\mathbb{R}P^2$	1210
	$\mathbb{R}P^2$	1		9	S^2	50			K^2	4462
					T^2	112			$M(3,-)$	11784
7	S^2	5							$M(4,-)$	13657
	T^2	1			$\mathbb{R}P^2$	134			$M(5,-)$	7050
					K^2	187			$M(6,-)$	1022
	$\mathbb{R}P^2$	3			$M(3,-)$	133			$M(7,-)$	14
					$M(4,-)$	37				
					$M(5,-)$	2				

Since there are many more 2-dimensional pseudomanifolds than there are (connected) surfaces, it is necessary to sort out those pseudomanifolds that are not surfaces as early as possible in our backtracking. In a 2-dimensional pseudomanifold the link of every vertex is a 1-dimensional pseudomanifold (i.e., every vertex lies in exactly two edges). Thus, every vertex-link is a union of disjoint circles whereas the link of a vertex in a proper surface is one single circle:

- Backtrack if the link of a vertex of a current sum of triangles consists of a closed circle plus at least one extra edge.

Since at isolated singularities there are more edges in the link than we want to have, we can try to avoid this as follows:

Mixed lexicographic enumeration: Start the backtracking with the beginning segment B_{n-1} (instead of B_3), then proceed in reversed order with $B_{n-2}, \ldots$, until B_3 is processed.

In the case that the beginning segment is B_k we have the additional criterion:

- (Lexicographic enumeration:) Backtrack if the closed link of a vertex has less than k edges.
- (Mixed lexicographic enumeration:) Backtrack if the number of edges in the link of a vertex or if the degree of a vertex is larger than k.

In the mixed lexicographic approach, the sub-collections of triangulations that contain a vertex of the same maximal degree are sorted lexicographically. In particular, every triangulation in such a sub-collection begins with the same beginning segment (which is of type B_k for some k). However, the complete list of triangulations is not produced in lexicographic order anymore.

By symmetry, we can, in addition, exclude the following cases (for the lexicographic as well as for the mixed lexicographic approach):

- Do not use triangles of the form $23j$ with odd $5 \leq j \leq k$. (A resulting surface would, by relabeling, be isomorphic to a triangulation with beginning segment B_k plus triangle $23(j-1)$, which is lexicographically smaller.) If the triangle $23i$ with even $6 \leq i \leq k$ is used, then do not use the triangles $24j$ with odd $3 \leq j \leq i - 3$.

Finally, we test for every resulting (connected) surface whether it has, up to combinatorial equivalence, appeared previously in the enumeration. For this, we first compute as combinatorial invariants the f-vector, the sequence of vertex degrees, and the *Altshuler–Steinberg determinant* [3] (i.e., the determinant $\det(AA^T)$ for the vertex-triangle incidence matrix A) of an example. If these invariants are equal for two resulting surfaces, then we take one triangle of the first complex and test for all possible ways it can be mapped to the triangles of the second complex whether this map can be extended to a simplicial isomorphism of the two complexes. (Alternatively, one can use McKay's fast graph isomorphism testing program `nauty` [43] to determine whether the vertex-facet incidence graphs of the two complexes are isomorphic or not.)

Theorem 4.1. *There are exactly* 655 *triangulated surfaces with* 9 *vertices and* 42426 *triangulated surfaces with* 10 *vertices.*

The respective triangulations (in mixed lexicographic order) can be found online at [41]. Table 5 gives the detailed numbers of orientable and non-orientable surfaces that appear with up to 10 vertices. The corresponding combinatorial symmetry groups G of the examples are listed in Tables 6 and 7.

5. Random realization

It was asked by Grünbaum [29, Ch. 13.2] whether every triangulated orientable surface can be embedded geometrically in $\mathbb{R}^3$, i.e., whether it can be realized with straight edges, flat triangles, and without self intersections. By Steinitz' theorem (cf. [47, 48, 55]), every triangulated 2-sphere is realizable as the boundary complex of a convex 3-dimensional polytope.

The first explicit geometric realization (see Figure 4) of Möbius's minimal 7-vertex triangulation [44] of the 2-torus was given by Császár [21] (cf. also [26] and [40]). Bokowski and Eggert [11] showed that there are altogether 72 different "types" of realizations of the Möbius torus, and Fendrich [25] verified that triangulated tori with up to 11 vertices are realizable. Very recently, Archdeacon, Bonnington, and Ellis-Monaghan [5] have proved that all triangulated tori are realizable, which settles an old conjecture of Grünbaum [29] and Duke [24].

Bokowski and Brehm [9, 10] and Brehm [15, 16] constructed geometric realizations for several triangulated orientable surfaces of genus $g = 2, 3, 4$ with minimal numbers of vertices $n = 10, 10, 11$, respectively.

Neighborly triangulations (i.e., triangulations that have as their 1-skeleton the complete graph K_n) of higher genus were considered as candidates for counterexamples to

TABLE 6. Symmetry groups of triangulated surfaces with up to 9 vertices.

n	Surface	$|G|$	G	Types
4	S^2	24	$T^* = S_4$, 3-transitive	1
5	S^2	12	$S_3 \times \mathbb{Z}_2$	1
6	S^2	4	$\mathbb{Z}_2 \times \mathbb{Z}_2$	1
		48	$O^* = \mathbb{Z}_2 \wr S_3$, vertex-trans.	1
	$\mathbb{R}\mathbf{P}^2$	60	A_5, 2-transitive	1
7	S^2	2	$\mathbb{Z}_2$	1
		4	$\mathbb{Z}_2 \times \mathbb{Z}_2$	1
		6	S_3	2
		20	D_{10}	1
	T^2	42	$AGL(1,7)$, 2-transitive	1
	$\mathbb{R}\mathbf{P}^2$	4	$\mathbb{Z}_2 \times \mathbb{Z}_2$	1
		6	S_3	1
		24	$T^* = S_4$	1
8	S^2	1	trivial	2
		2	$\mathbb{Z}_2$	5
		4	$\mathbb{Z}_2 \times \mathbb{Z}_2$	3
		8	D_4	1
		12	$S_3 \times \mathbb{Z}_2$	1
		24	$T^* = S_4$	1
			$D_6 \times \mathbb{Z}_2$	1
	T^2	2	$\mathbb{Z}_2$	2
		3	$\mathbb{Z}_3$	1
		4	$\mathbb{Z}_2 \times \mathbb{Z}_2$	2
		6	S_3	1
		32	$[32, 43]$, vertex-trans.	1
	$\mathbb{R}\mathbf{P}^2$	1	trivial	2
		2	$\mathbb{Z}_2$	8
		4	$\mathbb{Z}_2 \times \mathbb{Z}_2$	5
		14	D_7	1
	K^2	1	trivial	1
		2	$\mathbb{Z}_2$	4
		8	D_4	1
9	S^2	1	trivial	16
		2	$\mathbb{Z}_2$	25
		4	$\mathbb{Z}_2 \times \mathbb{Z}_2$	5
		6	S_3	1
		12	$S_3 \times \mathbb{Z}_2$	2
		28	$D_7 \times \mathbb{Z}_2$	1
	T^2	1	trivial	52
		2	$\mathbb{Z}_2$	46
		3	$\mathbb{Z}_3$	1
		4	$\mathbb{Z}_2 \times \mathbb{Z}_2$	7
		6	$\mathbb{Z}_6$	1
			S_3	1
		12	$S_3 \times \mathbb{Z}_2$	2
		18	D_9, vertex-trans.	1
		108	$\mathbb{Z}_3^2 : D_6$, vertex-trans.	1
	$\mathbb{R}\mathbf{P}^2$	1	trivial	63
		2	$\mathbb{Z}_2$	52
		3	$\mathbb{Z}_3$	1
		4	$\mathbb{Z}_4$	1
			$\mathbb{Z}_2 \times \mathbb{Z}_2$	11
		6	S_3	5
		24	$T^* = S_4$	1
	K^2	1	trivial	131
		2	$\mathbb{Z}_2$	46
		4	$\mathbb{Z}_2 \times \mathbb{Z}_2$	6
		6	S_3	2
		12	$S_3 \times \mathbb{Z}_2$	2
	$M(3,-)$	1	trivial	106
		2	$\mathbb{Z}_2$	24
		3	$\mathbb{Z}_3$	1
		6	$\mathbb{Z}_6$	1
			S_3	1
	$M(4,-)$	1	trivial	19
		2	$\mathbb{Z}_2$	11
		3	$\mathbb{Z}_3$	1
		4	$\mathbb{Z}_2 \times \mathbb{Z}_2$	5
		6	S_3	1
	$M(5,-)$	6	$\mathbb{Z}_6$	1
		18	$S_3 \times \mathbb{Z}_3$, vertex-trans.	1

TABLE 7. Symmetry groups of triangulated surfaces with $n = 10$ vertices.

Surface	$|G|$	G	Types
S^2	1	trivial	137
	2	$\mathbb{Z}_2$	69
	3	$\mathbb{Z}_3$	1
	4	$\mathbb{Z}_2 \times \mathbb{Z}_2$	13
	6	S_3	6
	8	$\mathbb{Z}_2^3$	2
		D_4	2
	16	D_8	1
	24	$T^* = S_4$	1
	32	$D_8 \times \mathbb{Z}_2$	1
T^2	1	trivial	1763
	2	$\mathbb{Z}_2$	292
	3	$\mathbb{Z}_3$	10
	4	$\mathbb{Z}_4$	2
		$\mathbb{Z}_2 \times \mathbb{Z}_2$	33
	6	S_3	3
	8	$\mathbb{Z}_2^3$	2
		D_4	1
	12	$S_3 \times \mathbb{Z}_2$	1
	20	D_{10}, vertex-trans.	1
		$AGL(1,5)$	1
$M(2,+)$	1	trivial	789
	2	$\mathbb{Z}_2$	61
	3	$\mathbb{Z}_3$	7
	4	$\mathbb{Z}_4$	2
		$\mathbb{Z}_2 \times \mathbb{Z}_2$	4
	8	$\mathbb{Z}_8$	1
	16	$\langle 2,2\,|\,2 \rangle$	1
$M(3,+)$	1	trivial	8
	2	$\mathbb{Z}_2$	3
	3	$\mathbb{Z}_3$	4
	4	$\mathbb{Z}_4$	2
		$\mathbb{Z}_2 \times \mathbb{Z}_2$	1
	12	A_4	1
	21	$\mathbb{Z}_7 \times \mathbb{Z}_3$	1
$\mathbb{RP}^2$	1	trivial	923
	2	$\mathbb{Z}_2$	242
	3	$\mathbb{Z}_3$	2
	4	$\mathbb{Z}_2 \times \mathbb{Z}_2$	29
	6	S_3	10
	12	$S_3 \times \mathbb{Z}_2$	2
		A_4	1
	18	D_9	1

Surface	$|G|$	G	Types
K^2	1	trivial	4057
	2	$\mathbb{Z}_2$	367
	4	$\mathbb{Z}_4$	1
		$\mathbb{Z}_2 \times \mathbb{Z}_2$	30
	8	$\mathbb{Z}_2^3$	2
		D_4	3
	10	D_5	1
	20	D_{10}, vertex-trans.	1
$M(3,-)$	1	trivial	11308
	2	$\mathbb{Z}_2$	448
	3	$\mathbb{Z}_3$	12
	4	$\mathbb{Z}_2 \times \mathbb{Z}_2$	11
	6	S_3	5
$M(4,-)$	1	trivial	13037
	2	$\mathbb{Z}_2$	556
	3	$\mathbb{Z}_3$	6
	4	$\mathbb{Z}_4$	1
		$\mathbb{Z}_2 \times \mathbb{Z}_2$	45
	6	$\mathbb{Z}_6$	1
		S_3	5
	8	$\mathbb{Z}_2^3$	1
		D_4	4
	48	$O^* = \mathbb{Z}_2 \wr S_3$	1
$M(5,-)$	1	trivial	6792
	2	$\mathbb{Z}_2$	214
	3	$\mathbb{Z}_3$	22
	4	$\mathbb{Z}_2 \times \mathbb{Z}_2$	16
	6	$\mathbb{Z}_6$	2
		S_3	4
$M(6,-)$	1	trivial	926
	2	$\mathbb{Z}_2$	71
	3	$\mathbb{Z}_3$	18
	4	$\mathbb{Z}_2 \times \mathbb{Z}_2$	6
	12	A_4	1
$M(7,-)$	1	trivial	4
	2	$\mathbb{Z}_2$	4
	3	$\mathbb{Z}_3$	1
	5	$\mathbb{Z}_5$	1
	6	S_3	1
	9	$\mathbb{Z}_9$	1
	12	A_4	1
	60	A_5, vertex-trans.	1

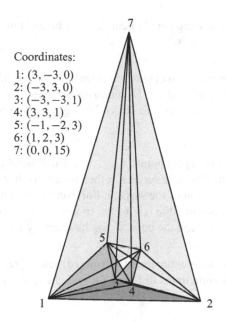

Coordinates:

1: $(3, -3, 0)$
2: $(-3, 3, 0)$
3: $(-3, -3, 1)$
4: $(3, 3, 1)$
5: $(-1, -2, 3)$
6: $(1, 2, 3)$
7: $(0, 0, 15)$

Triangles:

123 145 156 345 167 467 247
124 236 256 346 257 357 137

FIGURE 4. Császár's torus.

the Grünbaum realization problem for a while (cf. [21] and [13, p. 137]). Neighborly orientable surfaces have genus $g = (n - 3)(n - 4)/12$ and therefore $n \equiv 0, 3, 4, 7 \bmod 12$ vertices, with $g = 6$ and $n = 12$ as the first case beyond the tetrahedron and the 7-vertex torus; cf. [56].

Theorem 5.1. (Bokowski and Guedes de Oliveira [12]) *The triangulated orientable surface N_{54}^{12} of genus 6 with 12 vertices of Altshuler's list [2] is not geometrically realizable in $\mathbb{R}^3$.*

In fact, Bokowski and Guedes de Oliveira showed that there is no oriented matroid compatible with the triangulation N_{54}^{12}, from which the non-realizability follows. Recently, Schewe [46] reimplemented the approach of Bokowski and Guedes de Oliveira and was able to show that none of the 59 triangulations of the orientable surface of genus 6 with 12 vertices is realizable. Moreover:

Theorem 5.2. (Schewe [46]) *Every orientable surface of genus $g \geq 5$ has a triangulation which is not geometrically realizable in $\mathbb{R}^3$.*

From an algorithmic point of view the realizability problem for triangulated surfaces is decidable (cf. [13, p. 50]), but there is *no* algorithm known that would solve the realization problem for instances with, say, 10 vertices in reasonable time.

Surprisingly, the following simple heuristic can be used to realize tori and surfaces of genus 2 with up to 10 vertices.

Random Realization. *For a given orientable surface with n vertices pick n integer points in a cube of size k^3 (for some fixed $k \in \mathbb{N}$) uniformly at random (in general position). Test whether this (labeled) set of n points in $\mathbb{R}^3$ yields a geometric realization of the surface. If not, try again.*

Suppose, we are given n (integer) points $\vec{x}_1, \ldots, \vec{x}_n$ (in general position) in $\mathbb{R}^3$ together with a triangulation of an orientable surface. It then is easy to check whether these n points provide a geometric realization of the surface: For every pair of a triangle $i_1 i_2 i_3$ together with a combinatorially disjoint edge $i_4 i_5$ of the triangulation we have to test whether the geometric triangle $\vec{x}_{i_1} \vec{x}_{i_2} \vec{x}_{i_3}$ and the edge $\vec{x}_{i_4} \vec{x}_{i_5}$ have empty intersection.

Since the surfaces were enumerated in (mixed) lexicographic order with different triangulations in the list sometimes differing only slightly, it is promising to try the following:

Recycling of Coordinates. *As soon as a realization has been found for a triangulated orientable surface with n vertices, test whether the corresponding set of coordinates yields realizations for other triangulations with n vertices as well. Moreover, perturb the coordinates slightly and try again.*

Using random realization and recycling of coordinates we obtained realizations for 864 of the 865 triangulations of the orientable surface of genus 2 with 10 vertices, leaving one case open. This last example (`manifold_2_10_41348` in the catalog [41]) has the highest symmetry group of order 16 among the 865 examples; see Figure 5. It was realized by Jürgen Bokowski [8] by construction of an explicit geometric model.

Theorem 5.3. (Bokowski and Lutz; cf. [8]) *All 865 vertex-minimal 10-vertex triangulations of the orientable surface of genus 2 can be realized geometrically in $\mathbb{R}^3$.*

Conjecture 5.4. *Every triangulation of the orientable surface of genus 2 is realizable in $\mathbb{R}^3$.*

For the computational experiments on random realization of the triangulated orientable surfaces with 10 vertices, we chose $k = 2^{15} = 32768$ as the side length of the cube.

- For the 233 triangulations of S^2 with 10 vertices it took on average about 700 tries per example to obtain a realization (in non-convex position).
- It took on average about 418000 tries to realize one of the 2109 triangulated tori with 10 vertices.

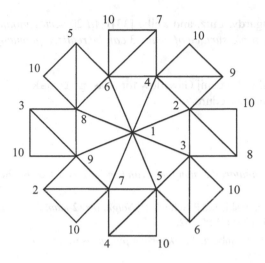

FIGURE 5. The double torus manifold_2_10_41348.

- For the 865 vertex-minimal 10-vertex triangulations of the orientable surface of genus 2 we initially set a limit of 200 million tries for every example. Random realizations were found for about one fifth of the triangulations. All other of the 864 realizations were found by recycling of coordinates. The computations were run for 3 months on ten Pentium R 2.8 GHz processors.

- It happened a few times that for a given triangulation realizations with identical coordinates were found by different processors after different numbers of tries. (Thus the length of cycles of the random generator plays a role.)

By successively rounding the coordinates of one of the randomly realized surfaces it is in most cases relatively easy to obtain realizations of the respective surface with much smaller coordinates. (For triangulations of S^2 it is an open problem whether there are *convex* realizations with small coordinates; cf. [55, Ex. 4.16].) An enumerative search for realizations with small coordinates was carried out in [31] (this time, the computations were run for eight days on ten Pentium R 2.8 GHz processors) and in [32, 34]; see these references and [41] for explicit coordinates and visualizations.

Theorem 5.5. (Hougardy, Lutz, and Zelke [31]) *All the 865 vertex-minimal 10-vertex triangulations of the orientable surface of genus 2 have realizations in the (4 × 4 × 4)-cube, but none can be realized in general position in the (3 × 3 × 3)–cube.*

A posteriori, the existence of small triangulations explains the success of the described random search for realizations.

Realizations with small coordinates were also obtained for 17 of the 20 vertex-minimal 10-vertex triangulations of the orientable surface of genus 3; see [32]. With a more advanced simulated annealing approach, realizations of surfaces can be obtained much more easily:

Theorem 5.6. (Hougardy, Lutz, and Zelke [33]) *All* 20 *vertex-minimal* 10-*vertex triangulations of the orientable surface of genus* 3 *can be realized geometrically in* $\mathbb{R}^3$.

Acknowledgments

The author is grateful to Ewgenij Gawrilow for a fast polymake [27] implementation of the random realization procedure.

References

[1] A. Altshuler, *Combinatorial* 3-*manifolds with few vertices*, J. Comb. Theory, Ser. A **16** (1974), 165–173.

[2] A. Altshuler, J. Bokowski, and P. Schuchert, *Neighborly* 2-*manifolds with* 12 *vertices*, J. Comb. Theory, Ser. A **75** (1996), 148–162.

[3] A. Altshuler and L. Steinberg, *Neighborly* 4-*polytopes with* 9 *vertices*, J. Comb. Theory, Ser. A **15** (1973), 270–287.

[4] ———, *An enumeration of combinatorial* 3-*manifolds with nine vertices*, Discrete Math. **16** (1976), 91–108.

[5] D. Archdeacon, C.P. Bonnington, and J.A. Ellis-Monaghan, *How to exhibit toroidal maps in space*, Discrete Comput. Geometry, **38** (2007), 573–594.

[6] D.[W.] Barnette, *Generating the triangulations of the projective plane*, J. Comb. Theory, Ser. B **33** (1982), 222–230.

[7] D.W. Barnette and A.L. Edelson, *All* 2-*manifolds have finitely many minimal triangulations*, Isr. J. Math. **67** (1988), 123–128.

[8] J. Bokowski, *On heuristic methods for finding realizations of surfaces*, Discrete Differential Geometry (A.I. Bobenko, P. Schröder, J.M. Sullivan, G.M. Ziegler, eds.), Oberwolfach Seminars, vol. 38, Birkhäuser, 2008, this volume, pp. 255–260.

[9] J. Bokowski and U. Brehm, *A new polyhedron of genus* 3 *with* 10 *vertices*, Intuitive Geometry, Internat. Conf. on Intuitive Geometry, Siófok, Hungary, 1985 (K. Böröczky and G. Fejes Tóth, eds.), Colloquia Mathematica Societatis János Bolyai, vol. 48, North-Holland, Amsterdam, 1987, pp. 105–116.

[10] ———, *A polyhedron of genus* 4 *with minimal number of vertices and maximal symmetry*, Geom. Dedicata **29** (1989), 53–64.

[11] J. Bokowski and A. Eggert, *Toutes les réalisations du tore de Moebius avec sept sommets/All realizations of Moebius' torus with* 7 *vertices*, Topologie Struct. **17** (1991), 59–78.

[12] J. Bokowski and A. Guedes de Oliveira, *On the generation of oriented matroids*, Discrete Comput. Geom. **24** (2000), 197–208.

[13] J. Bokowski and B. Sturmfels, *Computational Synthetic Geometry*, Lecture Notes in Mathematics, vol. 1355, Springer-Verlag, Berlin, 1989.

[14] R. Bowen and S. Fisk, *Generation of triangulations of the sphere*, Math. Comput. **21** (1967), 250–252.

[15] U. Brehm, *Polyeder mit zehn Ecken vom Geschlecht drei*, Geom. Dedicata **11** (1981), 119–124.

[16] ———, *A maximally symmetric polyhedron of genus* 3 *with* 10 *vertices*, Mathematika **34** (1987), 237–242.

[17] G. Brinkmann and B. McKay, `plantri`: *a program for generating planar triangulations and planar cubic graphs*, `cs.anu.edu.au/people/bdm/plantri/`, 1996–2001, version 4.1.

[18] ———, *Fast generation of planar graphs*, MATCH Commun. Math. Comput. Chem. **58** (2007), 323–357.

[19] M. Brückner, *Geschichtliche Bemerkungen zur Aufzählung der Vielflache*, Pr. Realgymn. Zwickau **578** (1897).

[20] ———, *Über die Anzahl $\psi(n)$ der allgemeinen Vielflache*, Atti Congresso Bologna 4 (1931), 5–11.

[21] A. Császár, *A polyhedron without diagonals*, Acta Sci. Math., Szeged **13** (1949–1950), 140–142.

[22] B. Datta, *Two dimensional weak pseudomanifolds on seven vertices*, Bol. Soc. Mat. Mex., III. Ser. **5** (1999), 419–426.

[23] B. Datta and N. Nilakantan, *Two-dimensional weak pseudomanifolds on eight vertices*, Proc. Indian Acad. Sci., Math. Sci. **112** (2002), 257–281.

[24] R.A. Duke, *Geometric embedding of complexes*, Am. Math. Mon. **77** (1970), 597–603.

[25] S. Fendrich, *Methoden zur Erzeugung und Realisierung von triangulierten kombinatorischen 2-Mannigfaltigkeiten*, Diplomarbeit, Technische Universität Darmstadt, 2003, 56 pages.

[26] M. Gardner, *Mathematical Games. On the remarkable Császár polyhedron and its applications in problem solving*, Scientific American **232** (1975), no. 5, 102–107.

[27] E. Gawrilow and M. Joswig, `polymake`, *version 2.2, 1997–2006, with contributions by T. Schröder and N. Witte*, `www.math.tu-berlin.de/polymake`.

[28] D.W. Grace, *Computer search for non-isomorphic convex polyhedra*, Report CS 15, Computer Science Department, Stanford University, 1965.

[29] B. Grünbaum, *Convex Polytopes*, Pure and Applied Mathematics, vol. 16, Interscience Publishers, London, 1967, second edition (V. Kaibel, V. Klee, and G.M. Ziegler, eds.), Graduate Texts in Mathematics, vol. 221, Springer-Verlag, New York, NY, 2003.

[30] P.J. Heawood, *Map-colour theorem*, Quart. J. Pure Appl. Math. **24** (1890), 332–338.

[31] S. Hougardy, F.H. Lutz, and M. Zelke, *Polyhedra of genus 2 with 10 vertices and minimal coordinates*, Electronic Geometry Models No. 2005.08.001 (2007), `www.eg-models.de/2005.08.001`.

[32] ———, *Polyhedra of genus 3 with 10 vertices and minimal coordinates*, Electronic Geometry Models No. 2006.02.001 (2007), `www.eg-models.de/2006.02.001`.

[33] ———, *Surface realization with the intersection edge functional*, `arXiv:math.MG/0608538`, 2006, 19 pages; Exp. Math., to appear.

[34] ———, *Polyhedral tori with minimal coordinates*, `arXiv:0709.2794`, 2007, 5 pages.

[35] M. Jungerman and G. Ringel, *Minimal triangulations on orientable surfaces*, Acta Math. **145** (1980), 121–154.

[36] E.G. Köhler and F.H. Lutz, *Triangulated Manifolds with Few Vertices: Vertex-Transitive Triangulations I*, `arXiv:math.GT/0506520`, 2005, 74 pages.

[37] W. Kühnel and G. Lassmann, *Neighborly combinatorial 3-manifolds with dihedral automorphism group*, Isr. J. Math. **52** (1985), 147–166.

[38] S.A. Lavrenchenko, *Irreducible triangulations of the torus*, J. Sov. Math. **51** (1990), 2537–2543, translation from Ukr. Geom. Sb. **30** (1987), 52–62.

[39] S.[A.] Lawrencenko and S. Negami, *Irreducible triangulations of the Klein bottle*, J. Comb. Theory, Ser. B **70** (1997), 265–291.

[40] F.H. Lutz, *Császár's torus*, Electronic Geometry Models No. 2001.02.069 (2002), www.eg-models.de/2001.02.069.

[41] _____, *The Manifold Page, 1999–2007*, www.math.tu-berlin.de/diskregeom/stellar/.

[42] F.H. Lutz and J.M. Sullivan, *Simplicial manifolds with small valence*, in preparation.

[43] B.D. McKay, *nauty, version 2.2*, cs.anu.edu.au/people/bdm/nauty/, 1994–2003.

[44] A.F. Möbius, *Mittheilungen aus Möbius' Nachlass: I. Zur Theorie der Polyëder und der Elementarverwandtschaft*, Gesammelte Werke II (F. Klein, ed.), Verlag von S. Hirzel, Leipzig, 1886, pp. 515–559.

[45] G. Ringel, *Wie man die geschlossenen nichtorientierbaren Flächen in möglichst wenig Dreiecke zerlegen kann*, Math. Ann. **130** (1955), 317–326.

[46] L. Schewe, *Satisfiability Problems in Discrete Geometry*, Dissertation, Technische Universität Darmstadt, 2007, 101 pages.

[47] E. Steinitz, *Polyeder und Raumeinteilungen*, Encyklopädie der mathematischen Wissenschaften mit Einschluss ihrer Anwendungen, Dritter Band: Geometrie, III.1.2., Heft 9 (W.Fr. Meyer and H. Mohrmann, eds.), B. G. Teubner, Leipzig, 1922, pp. 1–139.

[48] E. Steinitz and H. Rademacher, *Vorlesungen über die Theorie der Polyeder unter Einschluß der Elemente der Topologie*, Grundlehren der mathematischen Wissenschaften, vol. 41, Springer-Verlag, Berlin, 1934, reprint, 1976.

[49] T. Sulanke, *Source for surftri and lists of irreducible triangulations*, hep.physics.indiana.edu/~tsulanke/graphs/surftri/, 2005, version 0.96.

[50] _____, *Note on the irreducible triangulations of the Klein bottle*, J. Comb. Theory, Ser. B **96** (2006), 964–972.

[51] _____, *Generating irreducible triangulations of surfaces*, arXiv:math.CO/0606687, 2006, 11 pages.

[52] _____, *Irreducible triangulations of low genus surfaces*, arXiv:math.CO/0606690, 2006, 10 pages.

[53] T. Sulanke and F.H. Lutz, *Isomorphism-free lexicographic enumeration of triangulated surfaces and 3-manifolds*, arXiv:math.CO/0610022, 2007, 24 pages; Eur. J. Comb., to appear.

[54] W.T. Tutte, *A census of planar triangulations*, Can. J. Math. **14** (1962), 21–38.

[55] G.M. Ziegler, *Lectures on Polytopes*, Graduate Texts in Mathematics, vol. 152, Springer-Verlag, New York, NY, 1995, revised edition, 1998.

[56] _____, *Polyhedral surfaces of high genus*, Discrete Differential Geometry (A.I. Bobenko, P. Schröder, J.M. Sullivan, G.M. Ziegler, eds.), Oberwolfach Seminars, vol. 38, Birkhäuser, 2008, this volume, pp. 191–213.

Frank H. Lutz
Institut für Mathematik, MA 3–2
Technische Universität Berlin
Str. des 17. Juni 136
10623 Berlin
Germany
e-mail: lutz@math.tu-berlin.de

Discrete Differential Geometry, A.I. Bobenko, P. Schröder, J.M. Sullivan and G.M. Ziegler, eds.
Oberwolfach Seminars, Vol. 38, 255–260

On Heuristic Methods
for Finding Realizations of Surfaces

Jürgen Bokowski

Abstract. This article discusses heuristic methods for finding realizations of oriented matroids of rank 3 and 4. These methods can be applied for the spatial embeddability problem of 2-manifolds. They have proven successful in previous realization problems in which finally only the result was published.

Keywords. Geometric realization, heuristic methods, oriented matroids, pseudoline arrangents.

1. Introduction

Branko Grünbaum [11] asked whether each triangulated orientable (closed) surface of genus g, $g > 0$, can be embedded with flat triangles and without self-intersections in Euclidean 3-space. For genus 6 the first example of a surface that cannot be embedded was presented in [8]. A recent investigation of Lars Schewe in [15] has shown that there even exists a triangulated orientable surface of genus 5 that cannot be embedded in 3-space. In fact, his example also settles the problem for $g > 5$. (He has detected a fault in the corresponding previous claim in [8]. The symmetry assumption in that case was not applicable.) For $1 < g < 5$, Grünbaum's question is still open; very recent work of Archdeacon et al. [1] shows that every triangulated torus ($g = 1$) can be embedded.

The enumeration and classification of combinatorial manifolds by Frank Lutz [14] provides example classes of triangulated orientable 2-manifolds in great profusion. It can also be viewed in context with the above Grünbaum problem for smaller genus. At the beginning of Lutz's investigation, it was hoped that his triangulated 2-manifolds with up to 10 vertices would provide interesting candidates for finding a first non-realizable triangulated orientable 2-manifold with genus g, $1 < g < 4$, provided there is one at all.

Deciding the realizability problem of combinatorial 2-manifolds and, more generally, deciding the realizability for abstract objects, is of general interest in its own right. The problem of finding (flat spatial) realizations of triangulated orientable 2-manifolds requires not only a careful analysis of known realization methods but forces us to find enhanced heuristic methods.

We find methods for realizing combinatorial objects under the heading *computational synthetic geometry* in [9]. The survey article [6] describes various methods in connection with the theory of oriented matroids.

Random methods designed by Frank Lutz [14] and Stefan Hougardy, Frank Lutz, and Mariano Zelke [12] were very successful for triangulated orientable surfaces with up to 10 vertices. However, in general, we have to face the non-realizable case as well. An apparently difficult remaining case of genus 2 with 10 vertices was investigated by hand by the author with the rubber band techniques described in this article.

2. Abstract 2-manifolds that exist on the oriented matroid level

From [8] we know that certain triangulated orientable 2-manifolds cannot be realized with flat triangles without self-intersections, because there is no corresponding oriented matroid, that is, there is no assignment of orientations ± 1 to the quadruples of points that would be consistent on any subset of six points. We say in this case that the 2-manifold is not compatible with the oriented matroid condition; see [7].

Definition 2.1. We call a triangulated 2-manifold with n vertices compatible with the *oriented matroid condition* when there is a uniform acyclic oriented matroid of rank 4 with n elements such that the set of circuits (abstract Radon partitions) of the oriented matroid do not contain the pair of a triangle and an edge of the 2-manifold.

It would be interesting to have a first triangulated orientable 2-manifold that is, on the one hand, compatible with the oriented matroid condition and that, on the other hand, cannot be embedded with flat triangles without self-intersections.

The list of 2-manifolds with 10 vertices does not contain such an example. We had the following indication that realizations were likely for all triangular orientable surfaces with up to 10 vertices.

Proposition 2.2 (Bokowski and Lutz, 2002). *All triangulated orientable 2-manifolds with up to 10 vertices are compatible with the oriented matroid condition.*

Proof. This proposition was the outcome of a computation using a C++ program of the author, the same program used in [8].   □

The opposite result in one of these cases would have shown that the corresponding 2-manifold cannot be realized in 3-space.

Such a test or the corresponding enhanced method of Lars Schewe by using a satisfiability solver, can be recommended as a fast preprocessing for similar investigations. In the above case the outcome did not lead to a negative decision. So, it was just an indication that random techniques are a good step to pursue next.

In this context it is interesting that none of the embeddings of the complete graph with 12 vertices on a surface of genus 6 are compatible with the oriented matroid condition, due to the result of Lars Schewe [15].

A Haskell code to obtain these 59 surfaces can be found at juergen.bokowski.de. They were presented in [3] without mentioning the method of finding them. Here we describe the corresponding method in short.

Consider a star 11 with a fixed order $0, 1, \ldots, 10$ in its link. The remaining elements in the link of the i-th star $0 \leq i \leq 10$ have to be ordered. The left edge of the link i consists of element 11 and a well-defined adjacent element. The last element within the link is the other neighbor of 11. Eight elements are not ordered at the beginning. In the general case, we have ordered some subsets of the incomplete link of smallest index i already, and the left and right boundaries of the subsets are known. Let l_i be a left bound. We then have the following choices. If there is an element n_i that has not yet been used in the link i, we can add the (ordered) triangle i, l_i, n_i to star i. Or we add for some right bound r_i the triangle i, l_i, r_i, thus gluing together the left side l_i with the right side r_i in link i. We insert the new triangle in the other two links as well provided this is possible. We continue with adding triangles to star i until star i is complete and then proceed to the next incomplete star of smallest index, until we get a solution.

3. Stretchability of pseudoline arrangements

Stretchability of a pseudoline arrangement and realizability of an oriented matroid of rank 3 describe the same problem. We know (compare [4]) that this problem is hard in the general case and we have to find heuristic methods.

Even for oriented matroids of rank 4, one might first try to find realizations of all rank-3 contractions provided such a decision is easy to obtain.

We mention a method that has proven successful in the rank-3 case in the past and that uses the stretchability property in the literal sense.

We take as a model for the projective plane a half 2-sphere S_H with identified antipodal points on its boundary, a great circle C. Assume that we have, within the plane of C, line segments S_i ending on C and rotating around the midpoint; they will later ensure that the antipodal point property is kept. We use rubber bands on S_H that connect the antipodal end points of S_i. Thus, we have a model for a line in the plane for each line segment S_i. Rotating the line segment S_i corresponds to rotating the line in the plane, and using a different half great circle for the corresponding rubber band on the 2-sphere corresponds to a different parallel line in the plane.

When we use piecewise great circle segments on the 2-sphere, the rubber band on the 2-sphere might form a model for a non-stretched line in the plane, a polygon. However, the pseudoline property can still be checked. Now start with a pseudoline arrangement and use for each element a line segment together with piecewise great circle segments that form the pseudoline arrangement.

It turns out that all the mutations are decisive for keeping this pseudoline arrangement. This means that when you use a pin on the 2-sphere for each mutation such that the rubber bands obey these mutation obstructions (i.e., the rubber bands keep the pin always on the proper side) the corresponding stretched version is again a model of the given pseudoline arrangement.

You can now try to rotate the line segments and you can try to move the rubber bands (inductively one by one) such that you finally can see a stretched version of your pseudoline arrangement, provided it was stretchable.

This method of the author has been successfully applied in an unpublished investigation with Jürgen Richter-Gebert. The method often indicates why the example cannot be stretched. An additional investigation of all mutations that are decisive in this context can be appended and has often led to a final polynomial. The fact that the set of mutations is decisive for keeping the oriented matroid property whilst changing the line segments and corresponding rubber bands has been proven in [6].

An implementation of this idea was done by Klaus Pock and Jens Scharnbacher. A corresponding rank-4 version of Scharnbacher has been applied for neighborly pinched 2-spheres and their geometric realization [2], but the underlying method was not described in the literature.

4. Realizability of oriented matroids in rank 4

In architecture it is known that even sophisticated methods of computer-aided design do not replace spatial models completely. The same holds true for 3-dimensional mathematical models.

A stretched long thin rubber band (perhaps up to three meters long) can serve as a practically precise model for a line segment. When we have all edges of a triangulated orientable 2-manifold as stretched long thin rubber bands, we can check the intersection property of triangles and edges easily. The model is cheap, and the line segments can have different colors. It is essential that we can change the coordinates in a wide range easily. We find the limits up to which the 2-manifold property holds. Moreover, we see the violations when we change the coordinates more than we are allowed to do. We can form such a model in which, say for one point, not all conditions of the oriented matroid are fulfilled. From all violations we often see how we have to change the coordinates in order to find an allowable realization. This, in short, is the main idea how realizations were found in various cases by the author in the past. When we finally have coordinates, we do not have to describe how they were found. For a triangulated orientable 2-manifold, we can start with a star with the maximal number of points in its link. This cone-like structure soon gets a different shape when we require additional triangles. When the oriented matroid is given, we can use in addition cocircuit information to construct a realization inductively.

Random techniques of Frank Lutz and a rubber band model decision for the last case, similar to the first realization [5] of the regular map of Walther Dyck, has lead to the following theorem.

Theorem 4.1 (Bokowski and Lutz, 2005). *All triangulated orientable 2-manifolds with* 10 *vertices are realizable.*

We mention finally that all tori with up to 11 elements are realizable, as found by Sascha Fendrich [10]. He first generated all combinatorial types by applying vertex splits to the 21 irreducible triangulations of the torus [13]. In particular, he found 37867 triangulations of the torus with 11 vertices. For most of the examples, realizations were obtained via embeddings in the 2-skeleta of random 4-polytopes. Some remaining examples were realized with a program that tries to mimic the rubber band idea; it can be downloaded from `juergen.bokowski.de`. The program shows for the input of an abstract 2-manifold front view, top view, and side view of the 2-manifold for a first set of points. The violations in terms of Radon partitions are indicated in a fourth window. One can move points within all three orthogonal projections with the mouse until (hopefully) a violation vanishes and no additional violations appear.

References

[1] D. Archdeacon, C.P. Bonnington, and J.A. Ellis-Monaghan, *How to exhibit toroidal maps in space*, Discrete Comput. Geometry, **38** (2007), 573–594.

[2] A. Altshuler, J. Bokowski, and P. Schuchert, *Spatial polyhedra without diagonals*, Isr. J. Math. **86** (1994), 373–396.

[3] ———, *Neighborly 2-manifolds with 12 vertices*, J. Comb. Theory, Ser. A **75** (1996), 148–162.

[4] A. Björner, M. Las Vergnas, B. Sturmfels, N. White, and G.M. Ziegler, *Oriented Matroids*, 2nd ed., Encyclopedia of Mathematics and Its Applications, vol. 46, Cambridge University Press, Cambridge, 1999.

[5] J. Bokowski, *A geometric realization without self-intersections does exist for Dyck's regular map*, Discrete Comput. Geom. **4** (1989), 583–589.

[6] ———, *Effective methods in computational synthetic geometry*, Automated Deduction in Geometry, Proc. 3rd Internat. Workshop (ADG 2000), Zürich, 2000 (Berlin) (J. Richter-Gebert and D. Wang, eds.), Lecture Notes in Computer Science, vol. 2061, Springer-Verlag, 2001, pp. 175–192.

[7] ———, *Computational Oriented Matroids*, Cambridge University Press, Cambridge, 2006.

[8] J. Bokowski and A. Guedes de Oliveira, *On the generation of oriented matroids*, Discrete Comput. Geom. **24** (2000), 197–208.

[9] J. Bokowski and B. Sturmfels, *Computational Synthetic Geometry*, Lecture Notes in Mathematics, vol. 1355, Springer-Verlag, Berlin, 1989.

[10] S. Fendrich, *Methoden zur Erzeugung und Realisierung von triangulierten kombinatorischen 2-Mannigfaltigkeiten*, Diplomarbeit, Technische Universität Darmstadt, 2003, 56 pages.

[11] B. Grünbaum, *Arrangements and Spreads*, Conference Board of the Mathematical Sciences. Regional Conference Series in Mathematics, vol. 10, American Mathematical Society, Providence, RI, 1972.

[12] S. Hougardy, F.H. Lutz, and M. Zelke, *Surface realization with the intersection edge functional*, `arXiv:math.MG/0608538`, 2006, 19 pages.

[13] S.A. Lavrenchenko, *Irreducible triangulations of the torus*, J. Sov. Math. **51** (1990), 2537–2543, translation from Ukr. Geom. Sb. **30** (1987), 52–62.

[14] F.H. Lutz, *Enumeration and random realization of triangulated surfaces*, Discrete Differential Geometry (A.I. Bobenko, P. Schröder, J.M. Sullivan, G.M. Ziegler, eds.), Oberwolfach Seminars, vol. 38, Birkhäuser, 2008, this volume, pp. 235–253.

[15] L. Schewe, *Satisfiability Problems in Discrete Geometry*, Dissertation, Technische Universität Darmstadt, 2007, 101 pages.

Jürgen Bokowski
Fachbereich Mathematik
AG 7: Diskrete Optimierung
Technische Universität Darmstadt
Schloßgartenstr. 7
64289 Darmstadt
Germany
e-mail: juergen@bokowski.de

Part IV

Geometry Processing and Modeling
with Discrete Differential Geometry

Discrete Differential Geometry, A.I. Bobenko, P. Schröder, J.M. Sullivan and G.M. Ziegler, eds.
Oberwolfach Seminars, Vol. 38, 263–273
© 2008 Birkhäuser Verlag Basel/Switzerland

What Can We Measure?

Peter Schröder

Abstract. In this chapter we approach the question of "what is measurable" from an abstract point of view using ideas from geometric measure theory. As it turns out such a first-principles approach gives us quantities such as mean and Gaussian curvature integrals in the discrete setting and more generally, fully characterizes a certain class of possible measures. Consequently one can characterize all possible "sensible" measurements in the discrete setting which may form, for example, the basis for physical simulation.

Keywords. Geometric measure theory, Steiner's polynomial, Hadwiger's theorem, invariant measures.

1. Introduction

When characterizing a shape or changes in shape we must first ask what we can measure about a shape. For example, for a region in $\mathbb{R}^3$ we may ask for its volume or its surface area. If the object at hand undergoes deformation due to forces acting on it we may need to formulate the laws governing the change in shape in terms of measurable quantities and their change over time. Usually such measurable quantities for a shape are defined with the help of integral calculus and often require some amount of smoothness on the object to be well-defined. In this contribution we will take a more abstract approach to the question of measurable quantities which will allow us to define notions such as mean curvature integrals and the curvature tensor for piecewise linear meshes without having to worry about the meaning of second derivatives in settings in which they do not exist. In fact in this contribution we will give an account of a classical result due to Hugo Hadwiger [3], which shows that for a convex, compact set in $\mathbb{R}^n$ there are only $n + 1$ independent measurements if we require that the measurements be invariant under Euclidean motions (and satisfy certain "sanity" conditions). We will see how these measurements are constructed in a very straightforward and elementary manner and that they can be read off from a characteristic polynomial due to Jakob Steiner [6]. This polynomial describes the volume of a family of shapes which arise when we "grow" a given shape. As a practical tool arising from these consideration we will see that there is a well defined notion of the curvature tensor for piecewise linear meshes and we will see very simple formulæ for quantities needed in physical simulation with piecewise linear meshes. Much of the treatment here will be limited to convex bodies to keep things simple.

The treatment in this contribution draws heavily upon work by Gian-Carlo Rota and Daniel Klein, on Hadwiger's pioneering work, and on some recent work by David Cohen-Steiner and Jean-Marie Morvan.

2. Geometric measures

To begin with let us define what we mean by a *measure*. A measure is a function μ defined on a family of subsets of some set S, and it takes on real values: $\mu : L \to \mathbb{R}$. Here L denotes this family of subsets and we require of L that it is closed under finite set union and intersection as well as that it contains the empty set, $\emptyset \in L$. The measure μ must satisfy two axioms: (1) $\mu(\emptyset) = 0$; and (2) $\mu(A \cup B) = \mu(A) + \mu(B) - \mu(A \cap B)$ whenever A and B are measurable. The first axiom provides a normalization. The second axiom captures the idea that the measure of the union of two sets should be the sum of the measures minus the measure of their overlap. For example, consider the volume of the union of two sets which clearly has this property. It will also turn out that the additivity property is the key to reducing measurements for complicated sets to measurements on simple sets. We will furthermore require that all measures we consider be invariant under Euclidean motions, i.e., translations and rotations. This is so that our measurements do not depend on where we place the coordinate origin and how we orient the coordinate axes. A measure which depended on these wouldn't be very useful.

Let's see some examples. A well-known example of such a measure is the volume of bodies in $\mathbb{R}^3$. Clearly the volume of the empty body is zero and the volume satisfies the additivity axiom. The volume also does not depend on where the coordinate origin is placed and how the coordinate frame is rotated. To uniquely tie down the volume there is one final ambiguity due to the units of measurement being used, which we must remove. To this end we enforce a normalization which states that the volume of the unit, coordinate axis-aligned parallelepiped in $\mathbb{R}^n$ be one. With this we get

$$\mu_n^n(x_1, \ldots, x_n) = x_1 \cdots x_n$$

for x_1 to x_n the side lengths of a given axis-aligned parallelepiped. The superscript n denotes this as a measure on $\mathbb{R}^n$, while the subscript denotes the type of measurement being taken. Consider now a translation of such a parallelepiped. Since such a transformation does not change the side lengths, μ^n is translation invariant. One can also verify that this measure is rotation invariant [5]. Overall we say that this measure is *rigid-motion invariant*. Notice that we have only defined the meaning of μ_n^n for axis-aligned parallelepipeds, as well as finite unions and intersections of such parallelepipeds. The definition can be extended to more general bodies through a limiting process akin to how Riemann integration fills the domain with ever smaller boxes to approach the entire domain in the limit. There is nothing here that prevents us from performing the same limit process. In fact we will see later that once we add this final requirement, that the measure is continuous in the limit, the class of such measures is completely tied down. This is Hadwiger's famous theorem. But, more on that later.

Of course the next question is, are there other such invariant measures? Here is a proposal:

$$\mu_{n-1}^n(x_1, \ldots, x_n) = x_1 x_2 + x_1 x_3 + \cdots + x_1 x_n + x_2 x_3 + \cdots + x_2 x_n + \cdots.$$

For an axis-aligned parallelepiped in $\mathbb{R}^3$ we'd get

$$\mu_2^3(x_1, x_2, x_3) = x_1 x_2 + x_2 x_3 + x_3 x_1$$

which is just half the surface area of the parallelepiped with sides x_1, x_2, and x_3. Since we have the additivity property, we can certainly extend this definition to more general bodies through a limiting process and find that we get, up to normalization, the surface area.

Continuing in this fashion we are next led to consider

$$\mu_1^3(x_1, x_2, x_3) = x_1 + x_2 + x_3$$

(and similarly for μ_1^n). For a parallelepiped this function measures one quarter the sum of lengths of its bounding edges. Once again this new measure is rigid-motion invariant since the side lengths x_1, x_2 and x_3 do not change under a rigid motion. What we need to check is whether it satisfies the additivity theorem. Indeed it does, which is easily checked for the union of two axis-aligned parallelepipeds if the result is another axis-aligned parallelepiped. What is less clear is what this measure represents if we extend it to more general shapes where the notion of "sum of edge lengths" is not clear. The resulting continuous measure is sometimes referred to as the *mean width*.

From these simple examples we can see a pattern. For Euclidean n-space we can use the elementary symmetric polynomials in edge lengths to define n invariant measures

$$\mu_k^n(x_1, \ldots, x_n) = \sum_{1 \le i_1 < i_2 < \cdots < i_k \le n} x_{i_1} x_{i_2} \cdots x_{i_k}$$

for $k = 1, \ldots, n$ for parallelepipeds. To extend this definition to more general bodies as alluded to above, we'll follow ideas from geometric probability. In particular we will extend these measures to the ring of compact convex bodies, i.e., finite unions and intersections of compact convex sets in $\mathbb{R}^n$.

3. How many points, lines, planes, ... hit a body?

Consider a compact convex set, a *convex body*, in $\mathbb{R}^n$ and surround it by a box. One way to measure its volume is to count the number of points that, when randomly thrown into the box, hit the body versus those that hit empty space inside the box. To generalize this idea we consider *affine subspaces* of dimension $k < n$ in $\mathbb{R}^n$. Recall that an affine subspace of dimension k is spanned by $k + 1$ points $p_i \in \mathbb{R}^n$ (in general position), i.e., the space consists of all points q which can be written as affine combinations $q = \sum_i \alpha_i p_i$, with $\sum_i \alpha_i = 1$. Such an affine subspace is simply a linear subspace translated, i.e., it does not necessarily go through the origin. For example, for $k = 1$, $n = 3$ we will consider all lines—a line being the set of points one can generate as affine combinations of two points on the line—in three space. Let's denote the measure of all lines going through a rectangle R in $\mathbb{R}^3$ as $\lambda_1^3(R)$. The claim is that

$$\lambda_1^3(R) = c\mu_2^3(R),$$

i.e., the measure of all lines which meet the rectangle is proportional to the area of the rectangle. To see this, note that a given line (in general position) meets the rectangle either once or not at all. Conversely for a given point in the rectangle there is a whole set of lines—a sphere's worth—which "pierce" the rectangle in the given point. The measure of those lines is proportional to the area of the unit sphere. Since this is true for all points in the rectangle we see that the total measure of all such lines must be proportional to the area of the rectangle with a constant of proportionality depending on the measure of the sphere. For now such constants are irrelevant for our considerations so we will just set them to unity. Given a more complicated shape C in a plane nothing prevents us from performing a limiting process and we see that the measure of lines meeting C is

$$\lambda_1^3(C) = \mu_2^3(C),$$

i.e., it is proportional to the area of the region C. Given a union of rectangles $D = \cup_i R_i$, each living in a different plane, we get

$$\int X_D(\omega)\, d\lambda_1^3(\omega) = \sum_i \mu_2^3(R_i).$$

Here $X_D(\omega)$ counts the number of times a line ω meets the set D and the integration is performed over all lines. Going to the limit we find for any convex body E a measure proportional to its surface area

$$\int X_E(\omega)\, d\lambda_1^3(\omega) = \mu_2^3(E).$$

Using planes ($k = 2$) we can now generalize the mean width. For a straight line segment $c \in \mathbb{R}^3$ we find $\lambda_2^3(c) = \mu_1^3(c) = l(c)$, i.e., the measure of all planes that meet the straight line segment is proportional—as before we set the constant of proportionality to unity—to the length of the line segment. The argument mimics what we said above: a plane meets the line either once or not at all. For a given point on the line there is once again a whole set of planes going through that point. Considering the normals to such planes we see that this set of planes is proportional in measure to the unit sphere, without being more precise about the actual constant of proportionality. Once again this can be generalized with a limiting process giving us the measure of all planes hitting an arbitrary curve in space as proportional to its length

$$\int X_F(\omega)\, d\lambda_2^3(\omega) = \mu_1^3(F).$$

Here the integration is performed over all planes $\omega \in \mathbb{R}^3$, and X_F counts the number of times a given plane touches the curve F.

It is easy to see that this way of measuring recovers the mean width of a parallelepiped as we had defined it before,

$$\lambda_2^3(P) = \mu_1^3(P).$$

To see this consider the integration over all planes but taken in groups. With the parallelepiped having one corner at the origin—and being axis-aligned—first consider all planes whose normal (n_x, n_y, n_z) has all non-negative entries (i.e., the normal points into the positive octant). Now consider a sequence of three edges of P, connected at their end points, going from one corner to its opposing corner. For example, first traversing an edge parallel to the first coordinate axis, then an edge parallel to the second coordinate axis and finally an edge parallel to the third coordinate axis. The total length of this curve will be the sum of lengths of the three segments (x_1, x_2 and x_3, respectively). Given a plane with normal pointing into the positive octant and meeting the parallelepiped P we see it must meet our sequence of three edges in exactly one point. From this it follows that the measure of all such planes is given by the length of the sequence of edges $\mu_1^3(P) = x_1 + x_2 + x_3$ (up to a constant of proportionality). The same argument holds for the remaining seven octants giving us the desired result up to a constant. We can now see that $\mu_1^3(E)$ for some convex body E can be written as

$$\int X_E(\omega)\, d\lambda_2^3(\omega) = \mu_1^3(E),$$

i.e., the measure of all planes which meet E. With this we have generalized the notion of mean width to more general sets.

We have now seen n different Euclidean-motion invariant measures $\mu_k^n(C)$, given as the measure of all affine subspaces of dimension $n - k$ meeting $C \subset \mathbb{R}^n$ for $k = 1, \ldots, n$. These measures are called the *intrinsic volumes*. Clearly any linear combination of these measures is also rigid-motion invariant. It is natural to wonder then whether these linear combinations generate all such measures. It turns out there is one final measure missing in our basis set of measures before we arrive at Hadwiger's theorem. This measure corresponds to the elementary symmetric polynomial of order zero,

$$\mu_0(x_1, \ldots, x_n) = \begin{cases} 1 & n > 0 \\ 0 & n = 0. \end{cases}$$

This very special measure will turn out to be the Euler characteristic of a convex body which takes on the value 1 on all non-empty compact convex bodies. To show that everything works as advertised we use induction on the dimension. In dimension $n = 1$ we consider closed intervals $[a, b]$, $a < b$. Instead of working with the set directly we consider a functional on the characteristic function $f_{[a,b]}$ of the set which does the trick,

$$\chi_1(f) = \int_{\mathbb{R}} f(\omega) - f(\omega+)\, d\omega.$$

Here $f(\omega+)$ denotes the right limiting value of f at ω: $\lim_{\epsilon \to 0} f(\omega + \epsilon)$, $\epsilon > 0$. For the set $[a, b]$, $f(\omega) - f(\omega+)$ is zero for all $\omega \in \mathbb{R}$ except b since $f(b) = 1$ and $f(b+) = 0$. Now we use induction to deal with higher dimensions. In $\mathbb{R}^n$ take a straight line L and consider the affine subspaces A_ω of dimension $n - 1$ which are orthogonal to L and parameterized by ω along L. Letting f be the characteristic function of a convex body in $\mathbb{R}^n$ we get

$$\chi_n(f) = \int_{\mathbb{R}} \chi_{n-1}(f_\omega) - \chi_{n-1}(f_{\omega+})\, d\omega.$$

Here f_ω is the restriction of f to the affine space A_ω or alternatively the characteristic function of the intersection of A_ω and the convex body of interest, while $f_{\omega+}$ is defined as before as the limit of f_ω from above. With this we define $\mu_0^n(G) = \chi_n(f)$ for any finite union of convex bodies G and f the characteristic function of the set $G \in \mathbb{R}^n$.

That this definition of μ_0^n amounts to the Euler characteristic is not immediately clear, but it is easy to show, if we convince ourselves that for any non-empty convex body $C \in \mathbb{R}^n$,

$$\mu_0^n(\text{Int}(C)) = (-1)^n.$$

For $n = 1$, i.e., the case of open intervals on the real line, this statement is obviously correct. We can now apply the recursive definition to the characteristic function of the interior of C and get

$$\mu_0^n(\text{Int}(C)) = \int_\omega \chi_{n-1}(f_\omega) - \chi_{n-1}(f_{\omega+}) \, d\omega.$$

By induction, the right-hand side is zero except for the first ω at which $A_\omega \cap C$ is non-empty. There $\chi_{n-1}(f_{\omega+}) = (-1)^{n-1}$, thus proving our assertion for all n.

The Euler–Poincaré formula for a convex polyhedron in $\mathbb{R}^3$,

$$|F| - |E| + |V| = 2,$$

which relates the number of faces, edges, and vertices to one another now follows easily. Given a convex polyhedron simply write it as the non-overlapping union of the interiors of all its cells from dimension n down to dimension 0, where the interior of a vertex (0-cell) is the vertex itself. Then

$$\mu_0^n(P) = \sum_{c \in P} \mu_0^n(\text{Int}(c)) = c_0 - c_1 + c_2 - \cdots,$$

where c_i equals the number of cells of dimension i. For the case of a polyhedron in $\mathbb{R}^3$ this is exactly the Euler–Poincaré formula as given above since for $n = 3$ we have

$$1 = \mu_0^3(P) = c_0 - c_1 + c_2 - c_3 = |V| - |E| + |F| - 1.$$

4. The intrinsic volumes and Hadwiger's theorem

The above machinery can now be used to define the intrinsic volumes as functions of the Euler characteristic alone for all finite unions G of convex bodies

$$\mu_k^n(G) = \int \mu_0^n(G \cap \omega) \, d\lambda_{n-k}^n(\omega).$$

Here $\mu_0^n(G \cap \omega)$ plays the role of $X_G(\omega)$ we used earlier to count the number of times ω hits G.

There is one final ingredient missing, continuity in the limit. Suppose C_n is a sequence of convex bodies which converges to C in the limit as $n \to \infty$. Hadwiger's theorem says that if a Euclidean-motion invariant measure μ of convex bodies in $\mathbb{R}^n$ is continuous in the sense that

$$\lim_{C_n \to C} \mu(C_n) = \mu(C),$$

then μ must be a linear combination of the intrinsic volumes μ_k^n, $k = 0, \ldots, n$. In other words, the intrinsic volumes, under the additional assumption of continuity, are the only linearly independent, Euclidean-motion invariant, additive measures on finite unions (and intersections) of convex bodies in $\mathbb{R}^n$.

What does all of this have to do with the applications we have in mind? A consequence of Hadwiger's theorem assures us that if we want to take measurements of piecewise linear geometry (surface or volume meshes, for example) such measurements should be functions of the intrinsic volumes. This assumes of course that we are looking for additive measurements which are Euclidean-motion invariant and continuous in the limit. For a triangle, for example, this would be area, edge length, and Euler characteristic. Similarly for a tetrahedron with its volume, surface area, mean width, and Euler characteristic. As the name suggests all of these measurements are intrinsic, i.e., they can be computed without requiring an embedding. All that is needed is a metric to compute the intrinsic volumes. Of course in practice the metric is often induced by an embedding.

5. Steiner's formula

We return now to questions of discrete differential geometry by showing that the intrinsic volumes are intricately linked to curvature integrals and represent their generalization to the non-smooth setting. This connection is established by Steiner's formula [6].

Consider a non-empty convex body $K \in \mathbb{R}^n$ together with its parallel bodies

$$K_\epsilon = \{x \in \mathbb{R}^n : d(x, K) \leq \epsilon\}$$

where $d(x, K)$ denotes the Euclidean distance from x to the set K. In effect K_ϵ is the body K thickened by ϵ. Steiner's formula gives the volume of K_ϵ as a polynomial in ϵ,

$$V(K_\epsilon) = \sum_{j=0}^{n} V(\mathbb{B}_{n-j}) V_j(K) \epsilon^{n-j}.$$

Here the $V_j(K)$ correspond to the measures μ_k^n we have seen earlier if we let $k = n - j$. For this formula to be correct the $V_j(K)$ are normalized so that they compute the j-dimensional volume when restricted to a j-dimensional subspace of $\mathbb{R}^n$. (Recall that we ignored normalizations in the definition of the μ_k^n.) $V(\mathbb{B}_k) = \pi^{k/2}/\Gamma(1 + k/2)$ denotes the k-volume of the unit k-ball. In particular we have $V(\mathbb{B}_0) = 1$, $V(\mathbb{B}_1) = 2$, $V(\mathbb{B}_2) = \pi$, and $V(\mathbb{B}_3) = 4\pi/3$.

In the case of a polyhedron we can verify Steiner's formula "by hand". Consider a tetrahedron in $T \in \mathbb{R}^3$ and the volume of its parallel bodies T_ϵ. For $\epsilon = 0$ we have the volume of T itself ($V_3(T)$). The first-order term in ϵ, $2V_2(T)$, is controlled by area measures: above each triangle a displacement along the normal creates additional volume proportional to ϵ and the area of the triangle. The second-order term in ϵ, $\pi V_1(T)$, corresponds to edge lengths and dihedral angles. Above each edge the parallel bodies form a wedge with radius ϵ and opening angle θ, which is the exterior angle of the faces meeting at that edge. The volume of each such wedge is proportional to edge length, exterior angle, and ϵ^2. Finally the third-order term in ϵ, $4\pi/3 V_0(T)$, corresponds to the volume of the parallel bodies formed over vertices. Each vertex gives rise to additional volume spanned by the vertex and a spherical cap above it. The spherical cap corresponds to a spherical triangle formed by the three incident triangle normals. The volume of such a spherical wedge is proportional to its solid angle and ϵ^3.

If we have a convex body with a boundary which is C^2, we can give a different representation of Steiner's formula. Consider such a convex $M \in \mathbb{R}^n$ and define the offset function

$$g(p) = p + t\vec{n}(p)$$

for $0 \le t \le \epsilon$, $p \in \partial M$ and $\vec{n}(p)$ the outward normal to M at p. We can now directly compute the volume of M_ϵ as the sum of $V_n(M)$ and the volume between the surfaces ∂M and ∂M_ϵ. The latter can be written as an integral of the determinant of the Jacobian of g,

$$\int_{\partial M} \left(\int_0^\epsilon \left| \frac{\partial g(p)}{\partial p} \right| dt \right) dp.$$

Since we have a choice of coordinate frame in which to do this integration, we may assume without loss of generality that we use principal curvature coordinates on ∂M, i.e., a set of orthogonal directions in which the curvature tensor diagonalizes. In that case,

$$\left| \frac{\partial g(p)}{\partial p} \right| = |\mathbb{I} + t\mathbb{K}(p)|$$

$$= \prod_{i=1}^{n-1} (1 + \kappa_i(p)t)$$

$$= \sum_{i=0}^{n-1} \mu_i^{n-1}(\kappa_1(p), \dots, \kappa_{n-1}(p))t^i.$$

In other words, the determinant of the Jacobian is a polynomial in t whose coefficients are the elementary symmetric functions in the principal curvatures. With this substitution we can trivially integrate over the variable t and get

$$V(M_\epsilon) = V_n(M) + \sum_{i=0}^{n-1} \frac{\epsilon^{i+1}}{i+1} \int_{\partial M} \mu_i^{n-1}(\kappa_1(p), \dots, \kappa_{n-1}(p)) \, dp.$$

Comparing the two versions of Steiner's formula we see that the intrinsic volumes generalize curvature integrals. For example, for $n = 3$ and an arbitrary convex body K we get

$$V(K_\epsilon) = 1 V_3(K) + 2 V_2(K)\epsilon + \pi V_1(K)\epsilon^2 + \frac{4\pi}{3} V_0(K)\epsilon^3,$$

while for a convex body M with C^2 smooth boundary the formula reads as

$$V(M_\epsilon) = V_3(M)$$

$$+ \left(\int_{\partial M} \underbrace{\mu_0^2(\kappa_1(p), \kappa_2(p))}_{=1} \, dp \right) \epsilon$$

$$\underbrace{\phantom{+ \left(\int_{\partial M} \mu_0^2(\kappa_1(p), \kappa_2(p)) \, dp \right) \epsilon}}_{=A}$$

$$+ \left(\int_{\partial M} \underbrace{\mu_1^2(\kappa_1(p), \kappa_2(p))}_{-2II} \, dp \right) \frac{\epsilon^2}{2}$$

$$+ \left(\int_{\partial M} \underbrace{\mu_2^2(\kappa_1(p), \kappa_2(p))}_{=K} \, dp \right) \frac{\epsilon^3}{3}$$

$$\underbrace{\phantom{+ \left(\int_{\partial M} \mu_2^2(\kappa_1(p), \kappa_2(p)) \, dp \right) \frac{\epsilon^3}{3}}}_{=4\pi}$$

$$= V_3(M) + \epsilon \int_{\partial M} dp + \epsilon^2 \int_{\partial M} H \, dp + \frac{\epsilon^3}{3} \int_{\partial M} K \, dp.$$

6. What all this machinery tells us

We began this section by considering the question of what additive, continuous, rigid-motion invariant measurements there are for convex bodies in $\mathbb{R}^n$ and learned that the $n + 1$ intrinsic volumes are the only ones and any such measure must be a linear combination of these. We have also seen that the intrinsic volumes in a natural way extend the idea of curvature integrals over the boundary of a smooth body to general convex bodies without regard to a differentiable structure. These considerations become one possible basis on which to claim that integrals of Gaussian curvature on a triangle mesh become sums over excess angle at vertices and that integrals of mean curvature can be identified with sums over edges of dihedral angle weighted by edge length. These quantities are always *integrals*. Consequently *they do not make sense as pointwise quantities*. In the case of smooth geometry we can define quantities such as mean and Gaussian curvature as pointwise quantities. On a simplicial mesh they are only defined as integral quantities.

All this machinery was developed for convex bodies. If a given mesh is not convex, the additivity property allows us to compute the quantities anyway by writing the mesh as a finite union and intersection of convex bodies and then tracking the corresponding sums and differences of measures. For example, $V(K_\epsilon)$ is well defined for an individual triangle K, and we know how to identify the coefficients involving intrinsic volumes with the integrals of elementary polynomials in the principal curvatures. Gluing two triangles together we can perform a similar identification carefully teasing apart the intrinsic volumes of the union of the two triangles. In this way the convexity requirement is relaxed so long as the shape of interest can be decomposed into a finite union of convex bodies.

This machinery was used by Cohen–Steiner and Morvan to give formulæ for integrals of a discrete curvature tensor. We give these here together with some fairly straightforward intuition regarding the underlying geometry.

Let P be a polyhedron with vertex set V and edge set E and B a small region (e.g., a ball) in $\mathbb{R}^3$, then we can define integrated Gaussian and mean curvature measures as

$$\phi_P^K(B) = \sum_{v \in V \cap B} K_v \quad \text{and} \quad \phi_P^H(B) = \sum_{e \in E} l(e \cap B)\theta_e,$$

where $K_v = 2\pi - \sum_j \alpha_j$ is the excess angle sum at vertex v defined through all the incident triangle angles at v, while $l(\cdot)$ denotes the length and θ_e is the signed dihedral angle at e made between the incident triangle normals. Its sign is positive for convex edges and negative for concave edges (note that this requires an orientation on the polyhedron). In essence this is simply a restatement of the Steiner polynomial coefficients restricted to the intersection of the ball B and the polyhedron P. To talk about the second fundamental form II_p at some point p in the surface, it is convenient to first extend it to all of $\mathbb{R}^3$. This is done by setting it to zero if one of its arguments is parallel to the normal p. With this one may define

$$\overline{II}_P(B) = \sum_{e \in E} l(e \cap B)\theta_e e_n \otimes e_n, \quad e_n = e/\|e\|.$$

The dyad $(e_n \otimes e_n)(u, v) := \langle u, e_n \rangle \langle v, e_n \rangle$ projects given vectors u and v along the normalized edge.

What is the geometric interpretation of the summands? Consider a single edge and the associated dyad. The curvature along this edge is zero while the curvature orthogonal to the edge is θ. A vector aligned with the edge is mapped to θ_e while one orthogonal to the edge is mapped to zero. These are the principal curvatures *except they are reversed*. Hence $\overline{II}_P(B)$ is an integral measure of the curvature tensor *with the principal curvature values exchanged*. For example, we can assign each vertex a three-by-three matrix by summing the edge terms for each incident edge. As a tangent plane at the vertex, which we need to project the three-by-three matrix to the expected two-by-two matrix in the tangent plane, we may take a vector parallel to the area gradient at the vertex. Alternatively we could define $\overline{II}_P(B)$ for balls containing a single triangle and its three edges each. In that case the natural choice for the tangent plane is the support plane of the triangle.

Cohen–Steiner and Morvan show that this definition can be rigorously derived from considering the coefficients of the Steiner polynomial in particular in the presence of non-convexities (which requires some fancy footwork ...). They also show that if the polyhedron is a sufficiently fine sample of a smooth surface, the discrete curvature tensor integrals have linear precision with regards to continuous curvature tensor integrals. They also provide a formula for a discrete curvature tensor which does not have the principal curvatures swapped.

In practice one often finds that noise in the mesh vertex positions makes these discrete computations numerically delicate. One potential fix is to enlarge B to stabilize the computations. More in-depth analyses of numerically reliable methods to estimate the curvature tensor have been undertaken by Yang et al. [7] and Grinspun et al. [2].

Further reading

The material in this chapter only gives the rough outlines of what is a very fundamental theory in probability and geometric measure theory. In particular there are many other consequences which follow from relationships between intrinsic volumes which we have not touched upon. A rigorous derivation of the results of Hadwiger [3], but much shorter than the original can be found in [4]. A complete and rigorous account of the derivation of intrinsic volumes from first principles in geometric probability can be found in the short book by Klain and Rota [5], while the details of the discrete curvature tensor integrals can be found in [1]. Approximation results which discuss the accuracy of these measure vis-à-vis an underlying smooth surface are treated by Cohen–Steiner and Morvan in a series of tech reports available at www-sop.inria.fr/geometrica/publications/.

Acknowledgments

This work was supported in part by NSF (DMS-0220905, DMS-0138458, ACI-0219979), DOE (W-7405-ENG-48/B341492), the Center for Integrated Multiscale Modeling and Simulation, the Center for Mathematics of Information, the Humboldt Foundation, Autodesk, and Pixar.

References

[1] David Cohen-Steiner and Jean-Marie Morvan, *Restricted Delaunay Triangulations and Normal Cycle*, Proc. 19th Annual Sympos. Computational Geometry, 2003, pp. 312–321.

[2] Eitan Grinspun, Yotam Gingold, Jason Reisman, and Denis Zorin, *Computing Discrete Shape Operators on General Meshes*, Computer Graphics Forum (Proc. Eurographics) **25** (2006), no. 3, 547–556.

[3] H. Hadwiger, *Vorlesungen über Inhalt, Oberfläche und Isoperimetrie*, Grundlagen Math. Wiss., no. XCIII, Springer-Verlag, Berlin, Göttingen, Heidelberg, 1957.

[4] Daniel A. Klain, *A short proof of Hadwiger's characterization theorem*, Mathematika **42** (1995), no. 84, 329–339.

[5] Daniel A. Klain and Gian-Carlo Rota, *Introduction to geometric probability*, Cambridge University Press, 1997.

[6] Jakob Steiner, *Über parallele Flächen*, Monatsbericht Akad. Wiss. Berlin (1840), 114–118.

[7] Yong-Liang Yang, Yu-Kun Lai, Shi-Min Hu, and Helmut Pottmann, *Robust Principal Curvatures on Multiple Scales*, Proc. Sympos. Geometry Processing, 2006, pp. 223–226.

Peter Schröder
Caltech
Pasadena, CA 91125
USA
e-mail: ps@cs.caltech.edu

Discrete Differential Geometry, A.I. Bobenko, P. Schröder, J.M. Sullivan and G.M. Ziegler, eds.
Oberwolfach Seminars, Vol. 38, 275–286

Convergence of the Cotangent Formula: An Overview

Max Wardetzky

Abstract. The cotangent formula constitutes an intrinsic discretization of the Laplace–Beltrami operator on polyhedral surfaces in a finite-element sense. This note gives an overview of approximation and convergence properties of discrete Laplacians and mean curvature vectors for polyhedral surfaces located in the vicinity of a smooth surface in euclidean 3-space. In particular, we show that mean curvature vectors converge in the sense of distributions, but fail to converge in L^2.

Keywords. Cotangent formula, discrete Laplacian, Laplace–Beltrami operator, convergence, discrete mean curvature.

1. Introduction

There are various approaches toward a purely discrete theory of surfaces for which classical differential geometry, and in particular the notion of *curvature*, appears as the limit case. Examples include the theory of *spaces of bounded curvature* [1, 24], *Lipschitz–Killing curvatures* [5, 12, 13], *normal cycles* [6, 7, 30, 31], *circle patterns* and *discrete conformal structures* [2, 17, 26, 28], and geometric *finite elements* [10, 11, 15, 20, 29]. In this note we take a finite-element viewpoint, or, more precisely, a functional-analytic one, and give an overview over convergence properties of weak versions of the Laplace–Beltrami operator and the mean curvature vector for embedded polyhedral surfaces.

Convergence. Consider a sequence of polyhedral surfaces $\{M_n\}$, embedded into euclidean 3-space, which converges (in an appropriate sense) to a smooth embedded surface M. One may ask: What are the measures and conditions such that metric and geometric objects on M_n—like intrinsic distance, area, mean curvature, Gauss curvature, geodesics and the Laplace–Beltrami operator—converge to the corresponding objects on M? To date no complete answer has been given to this question in its full generality. For example, the approach of *normal cycles* [6, 7], while well-suited for treating convergence of curvatures of embedded polyhedra in the sense of measures, cannot deal with convergence of elliptic operators such as the Laplacian. The *finite-element approach*, on the other hand, while well-suited for treating convergence of elliptic operators (cf. [10, 11]) and mean curvature vectors, has its difficulties with Gauss curvature.

Despite the differences between these approaches, there is a remarkable similarity: The famous *lantern of Schwarz* [27] constitutes a quite general example of what can go wrong—pointwise convergence of surfaces without convergence of their normal fields. Indeed, while one cannot expect convergence of metric and geometric properties of embedded surfaces from pointwise convergence alone, it often suffices to additionally require *convergence of normals*. The main technical step, to show that this is so, is the construction of a bi-Lipschitz map between a smooth surface M, embedded into euclidean 3-space, and a polyhedral surface M_h nearby, such that the metric distortion induced by this map is bounded in terms of the Hausdorff distance between M and M_h, the deviation of normals, and the shape operator of M. (See Theorem 3.3 and compare [19] for a similar result.) This map then allows for *explicit error estimates* for the distortion of area and length, and—when combined with a functional-analytic viewpoint—error estimates for the Laplace–Beltrami operator and the mean curvature vector.

We treat convergence of Laplace–Beltrami operators in *operator norm*, and we discuss two distinct concepts of mean curvature: a *functional* representation (in the sense of distributions) as well as a representation as a *piecewise linear function*. We observe that one concept (the functional) converges whereas the other (the function) in general does not. This is in accordance with what has been observed in geometric measure theory [5, 6, 7]: for polyhedral surfaces approximating smooth surfaces, in general, one cannot expect pointwise convergence of curvatures, but only convergence in an integrated sense.

A brief history of the cotangent formula. The cotangent representation for the Dirichlet energy of piecewise linear functions on triangular nets seems to have first appeared in Duffin's work [9] in 1959. In 1988 Dziuk [10, 11] studied linear finite elements on polyhedral surfaces—without explicit reference to the cotangent formula. In 1993 Pinkall and Polthier [20] employed the cotangent formula for a *functional representation* of the discrete mean curvature vector, leading to explicitly computable discrete minimal surfaces [16, 21, 22, 23, 25]. Later, Desbrun et al. [8, 18] used the cotangent formula for expressing the area gradient of piecewise linear surfaces. Their approach rescales with an area factor, which effectively means dealing with *functions* (pointwise quantities) instead of *functionals* (integrated quantities). Based on intrinsic Delaunay triangulations, Bobenko and Springborn [3] recently derived an intrinsic version of the cotangent formula on polyhedral surfaces that obeys the discrete maximum principle.

2. Polyhedral surfaces

By a *polyhedral surface M_h*, we mean a metric space obtained by gluing finitely many flat euclidean triangles isometrically along their edges, such that the result is homeomorphic to a 2-dimensional manifold. Metrically, polyhedral surfaces are length spaces in the sense of Gromov [14]. Two triangles which share an edge can always be unfolded such that they become coplanar, so that no intrinsic curvature occurs across edges. All intrinsic

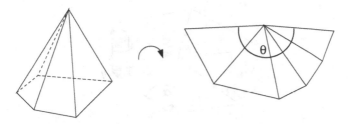

FIGURE 1. A neighborhood of an inner vertex of M_h with angle defect $2\pi - \theta$ is isometric to a metric cone with cone angle θ.

curvatures are concentrated at vertices which we treat as euclidean cone points. A *metric cone* C_θ with *cone angle* θ is the set $\{(r, \varphi)|0 \leq r; \ \varphi \in \mathbb{R}/\theta\mathbb{Z}\}/_{(0,\varphi_1)\sim(0,\varphi_2)}$ together with the (infinitesimal) metric $\mathrm{d}s^2 = \mathrm{d}r^2 + r^2 \mathrm{d}\varphi^2$, see Figure 1.

2.1. Finite elements on polyhedra

Polyhedral surfaces are piecewise linear; hence, they can be treated naturally by the finite-element method, in particular, by studying finite-dimensional subspaces of the Sobolev space $\mathcal{H}^1(M_h)$. First, we review the finite-element setting, then we discuss Sobolev spaces on polyhedral surfaces.

Given $f \in \mathcal{L}^2(M_h)$, the *Dirichlet problem* on M_h is to find $u \in \mathcal{H}_0^1(M_h)$ such that

$$\int_{M_h} g_h(\nabla_h u, \nabla_h \varphi) \, \mathrm{d} \mathrm{vol}_h = \int_{M_h} f \cdot \varphi \, \mathrm{d} \mathrm{vol}_h \quad \forall \varphi \in \mathcal{H}_0^1(M_h),$$

where g_h is the euclidean cone metric on M_h and ∇_h is the associated gradient. As in the planar case, an *abstract Galerkin scheme* is defined by restricting the space of test functions and the space of solutions to a *finite-dimensional* subspace $V_0 \subset \mathcal{H}_0^1(M_h)$. As usual, the subscript 0 denotes zero boundary conditions (we assume $\partial M_h \neq \emptyset$; the case $\partial M_h = \emptyset$ is treated similarly by setting $\mathcal{H}_0^1(M_h) = \{u \in \mathcal{H}^1(M_h) \mid \int u = 0\}$).

Definition 2.1 (Finite-element space). For vertices $p \in M_h \setminus \partial M_h$ and $q \in M_h$, let

$$\phi_p(q) := \begin{cases} 1 & \text{if } q = p \\ 0 & \text{else}, \end{cases}$$

and extend ϕ_p to all of M_h by linearly interpolating on triangles. Then $\{\phi_p\}$ is a *nodal basis* for the finite-dimensional space $S_{h,0}$.

Every $u_h \in S_{h,0}$ can be written as $u_h = \sum_q u_h^q \phi_q$ with coefficients $u_h^q \in \mathbb{R}$. Let

$$-\Delta_{pq} := \int_{M_h} g_h(\nabla_h \phi_p, \nabla_h \phi_q) \, \mathrm{d} \mathrm{vol}_h$$

$$b_p := \int_{M_h} f \cdot \phi_p \, \mathrm{d} \mathrm{vol}_h .$$

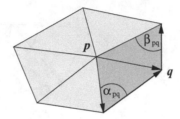

FIGURE 2. Only the angles α_{pq} and β_{pq} enter into the expression for Δ_{pq}.

Then the *discrete Dirichlet problem* amounts to a finite-dimensional linear solve: find u_h^q such that

$$-\sum_q \Delta_{pq} u_h^q = \mathsf{b}_p \ .$$

With this notation one readily verifies the *cotangent* representation of Δ_{pq}, see [20].

Lemma 2.2 (Cotangent formula). *The nonzero entries of the discrete cotan-Laplacian on a polyhedral surface are given by*

$$\Delta_{pq} = \frac{1}{2}(\cot\alpha_{pq} + \cot\beta_{pq}) \quad and \quad \Delta_{pp} = -\sum_{q_i \in link(p)} \Delta_{pq_i} \ ,$$

if p and q share an edge, and where α_{pq} and β_{pq} denote the angles opposite to the edge $\overline{pq}$ in the two triangles adjacent to $\overline{pq}$.

2.2. Sobolev spaces on polyhedra

On a smooth manifold M the definition of $\mathcal{H}^1(M)$—the space of square-integrable functions with square-integrable weak derivatives—can be based on a locally finite partition of M into smooth charts. The requirement on these charts is that the difference between the metric tensor on M and the flat metric tensor on $\mathbb{E}^n$ is uniformly bounded. (Such charts exist, for example, under the assumption of uniform curvature bounds on M.) In the polyhedral case, a difficulty arises from the fact that M_h is only of class $C^{0,1}$. A very general framework for defining $\mathcal{H}^1(M_h)$ on compact Lipschitz manifolds (of which finite polyhedra are just a special case) via local charts is provided by a consequence of Rademacher's theorem: weak differentiability is preserved under bi-Lipschitz maps. (See, e.g., Cheeger [4] and Ziemer [32].) In the following we will base our definition of $\mathcal{H}^1(M_h)$ on the assumption that there is a smooth surface M in the vicinity of M_h and a bi-Lipschitz map between M and M_h with uniformly bounded Lipschitz constant (for the existence of such a map, see Theorem 3.3), so that we can identify $\mathcal{H}^1(M_h)$ with $\mathcal{H}^1(M)$.

3. Convergence and approximation

3.1. Comparing two surfaces

If $M \subset \mathbb{E}^3$ is a compact smooth surface, and M_h is a polyhedral surface close to it, we need a map in order to compare the two surfaces. One way to define such a map is to

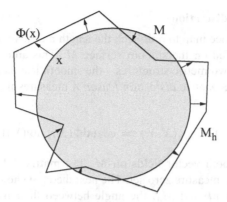

FIGURE 3. If M_h is a normal graph over M, then Φ takes a point x on M to the intersection of M_h with the normal line through x. The inverse, $\Psi = \Phi^{-1}$, thus realizes the pointwise distance from M_h to M.

map each point on M_h to its closest point on M. This is well defined for points within the *reach* of M. The reach of M is the distance of M to its *medial axis*. The medial axis of M is the set of those points in $\mathbb{E}^3$ which do not have a unique nearest point in M. The reach is related to local curvature properties of M, more precisely,

$$\text{reach}(M) \leq \inf_{x \in M} \frac{1}{|\kappa|_{\max}(x)} , \tag{3.1}$$

where $|\kappa|_{\max}(x)$ denotes the maximal absolute value of the normal curvatures at $x \in M$.

Definition 3.1 (Normal graph). A compact polyhedral surface M_h is a *normal graph* over a compact smooth surface M if it is strictly within the reach of M and the map $\Psi : M_h \to M$ which takes every point on M_h to its closest point on M is a homeomorphism.

The inverse of this map, $\Phi = \Psi^{-1} : M \to M_h$, satisfies

$$\text{dist}_{\mathbb{E}^3}(\Phi(x), M) = \|\Phi(x) - x\|_{\mathbb{E}^3} ,$$

see Figure 3. Φ is called the *shortest distance map*.

Definition 3.2 (Normal convergence). We say that a sequence of polyhedra $\{M_n\}$, with normal fields N_n, converges *normally* to a smooth surface M, with normal field N, if each M_n is a normal graph over M and the sequence of normal fields converges in $L^\infty(M)$ under the shortest distance maps,

$$\|N_n \circ \Phi_n - N\|_\infty \to 0 .$$

The sequence is said to converge *totally normally* if it converges normally and the Hausdorff distances $d_H(M_n, M)$ tend to zero as well.

3.2. Measuring metric distortion

We use the shortest distance map to pull back the length metric on the polyhedral surface M_h to a metric g_h defined on the smooth surface M. This amounts to thinking of M as being equipped with two metric structures—the smooth Riemannian metric g and the polyhedral metric g_h. The *metric distortion tensor* A measures the distortion between g and g_h. It is defined by

$$g(A(X), Y) := g_h(X, Y) := g_{\mathbb{E}^3}(d\Phi(X), d\Phi(Y)) \quad \text{a.e.}, \tag{3.2}$$

where X and Y are smooth vector fields on M. The matrix field A is symmetric and positive definite outside a measure zero set. The next theorem shows that A only depends on the distance between M and M_h, the angle between their normals, and the shape operator of M (for a proof see [15, 29]; see also Morvan and Thibert [19] for a similar result).

Theorem 3.3 (Metric distortion tensor splitting). *Let M_h be a polyhedral surface with normal field N_h. Assume M_h is a normal graph over an embedded smooth surface M with normal field N. Then the metric distortion tensor satisfies*

$$A = P \circ Q^{-1} \circ P \quad \text{a.e.}, \tag{3.3}$$

a decomposition into symmetric positive definite matrices P and Q which can pointwise be diagonalized (possibly in different orthonormal frames) by

$$P = \begin{pmatrix} 1 - d \cdot \kappa_1 & 0 \\ 0 & 1 - d \cdot \kappa_2 \end{pmatrix} \tag{3.4}$$

$$Q = \begin{pmatrix} \langle N, N_h \circ \Phi \rangle^2 & 0 \\ 0 & 1 \end{pmatrix}. \tag{3.5}$$

Here κ_1 and κ_2 denote the principal curvatures of the smooth manifold M, and $d(x)$ is the signed distance function, defined by $\Phi(x) = x + d(x) \cdot N(x)$.

3.3. Convergence of Laplace–Beltrami operators

The domain $\mathcal{H}_0^1(M)$ of the Laplace–Beltrami operator on M is the subspace of functions in $\mathcal{H}^1(M)$ with zero boundary condition; its range $\mathcal{H}^{-1}(M)$ is the space of bounded linear functionals on $\mathcal{H}_0^1(M)$. Here, $\mathcal{H}_0^1(M)$ will always be equipped with the norm

$$\|u\|_{\mathcal{H}_0^1(M)}^2 = \int_M g(\nabla u, \nabla u) \, d\text{vol}.$$

Using the shortest distance map to pull back the Laplace–Beltrami operator on M_h to the smooth surface M, we think of M as being equipped with *two* elliptic operators:

$$\Delta, \Delta_h : \mathcal{H}_0^1(M) \to \mathcal{H}^{-1}(M). \tag{3.6}$$

Let $\langle \cdot | \cdot \rangle$ denote the dual pairing between $\mathcal{H}^{-1}(M)$ and $\mathcal{H}_0^1(M)$. Then the weak definition of these operators is given by

$$\langle \Delta u | v \rangle = - \int_M g(\nabla u, \nabla v) \, d\,\mathrm{vol} \tag{3.7}$$

$$\langle \Delta_h u | v \rangle = - \int_M g(A^{-1}\nabla u, \nabla v)(\det A)^{1/2} \, d\,\mathrm{vol} . \tag{3.8}$$

Convergence of Laplace–Beltrami operators is understood in the operator norm of bounded linear operators between the spaces $\mathcal{H}_0^1(M)$ and $\mathcal{H}^{-1}(M)$.

Remark. The last formula is justified by the fact that the metric distortion tensor induces the following transformations for gradients and volume forms:

$$d\,\mathrm{vol}_h = (\det A)^{1/2} \, d\,\mathrm{vol} ,$$

$$\nabla_h = A^{-1}\nabla .$$

For a proof of the next theorem, see [15, 29].

Theorem 3.4 (Convergence of Laplacians). *Let $M_h \subset \mathbb{E}^3$ be an embedded compact polyhedral surface which is a normal graph over a smooth surface M with corresponding distortion tensor A. Define $\bar{A} := (\det A)^{1/2} A^{-1}$. Then*

$$\frac{1}{2}\|\mathrm{tr}(\bar{A} - \mathrm{Id})\|_\infty \leq \|\Delta_h - \Delta\|_{op} \leq \|\bar{A} - \mathrm{Id}\|_\infty . \tag{3.9}$$

Hence, if a sequence of polyhedral surfaces converges to M totally normally, then the corresponding Laplace–Beltrami operators converge in norm.

From here on, in order to show convergence for the discrete cotan-Laplacian, one proceeds similarly to Dziuk [10] who studies linear finite elements for *interpolating* polyhedra. For details about the extension to our case—*approximating* polyhedra—see [29].

3.4. Convergence of mean curvature

Analogously to the smooth case, we define the *mean curvature vector* as the result of applying the Laplace–Beltrami operator to the embedding function of a surface. In the polyhedral case this yields a *functional* (a distribution) rather than a function.

Definition 3.5 (Mean curvature functional). Let $\vec{I}_M : M \to \mathbb{E}^3$ and $\vec{I}_{M_h} : M_h \to \mathbb{E}^3$ denote the embeddings of M and M_h, respectively, and let $\vec{I}_h = \vec{I}_{M_h} \circ \Phi : M \to \mathbb{E}^3$. Then the mean curvature vectors are *functionals* on M defined by

$$\vec{H}_M := \Delta \vec{I}_M \in (\mathcal{H}^{-1}(M))^3 ,$$

$$\vec{H}_h := \Delta_h \vec{I}_h \in (\mathcal{H}^{-1}(M))^3 ,$$

defining one equation for each of the three components of these embeddings.

Lemma 3.6 (Connection with cotangent formula). *The mean curvature functional, when restricted to the subspace spanned by nodal basis functions, can be expressed using the cotangent formula:*

$$\langle \vec{H}_h | \phi_p \rangle = \frac{1}{2} \sum_{q \in link(p)} (\cot \alpha_{pq} + \cot \beta_{pq}) \cdot (q - p) . \tag{3.10}$$

The mean curvature functional is $\mathbb{R}^3$-valued. We need to say what we mean by the norm of such a functional. Let $\vec{F}$ be an $\mathbb{R}^n$-valued bounded linear operator on $\mathcal{H}_0^1$. We define

$$\|\vec{F}\|_{\mathcal{H}^{-1}} = \sup_{0 \neq u \in \mathcal{H}_0^1} \frac{\|\langle \vec{F} | u \rangle\|_{\mathbb{R}^n}}{\|u\|_{\mathcal{H}_0^1}} ,$$

where $\langle \cdot | \cdot \rangle$ denotes the dual pairing between $\mathcal{H}^{-1}$ and $\mathcal{H}_0^1$. The following result gives an a-priori error bound for the mean curvature functionals.

Theorem 3.7 (Convergence of mean curvature functionals). *Let M_h be a normal graph over a smooth surface M with associated shortest distance map Φ and metric distortion tensor A. Then*

$$\|\vec{H}_M - \vec{H}_h\|_{\mathcal{H}^{-1}} \leq \sqrt{|M|} \, (C_A - 1 + C_A \|\text{Id} - d\Phi\|_\infty) , \tag{3.11}$$

where $C_A = \|(\det A)^{1/2} A^{-1}\|_\infty$, $|M|$ is the total area of M, and $\|\text{Id} - d\Phi\|_\infty$ denotes the essential supremum over the pointwise operator norm of the operator $(\text{Id} - d\Phi)(x) : T_x M \to \mathbb{R}^3$. Hence, if a sequence of polyhedral surfaces converges to M totally normally, then the mean curvature functionals converge in norm.

Sketch of proof. Inequality (3.11) is a consequence of the triangle inequality applied to

$$(\Delta \vec{I}_M - \Delta_h \vec{I}_h) = (\Delta \vec{I}_M - \Delta_h \vec{I}_M) + (\Delta_h \vec{I}_M - \Delta_h \vec{I}_h) .$$

The convergence statement follows from an application of Theorem 3.3 to estimate the two terms $(C_A - 1)$ and $\|\text{Id} - d\Phi\|_\infty$, respectively. $\qquad\square$

A counterexample to the convergence of discrete mean curvature. The reason to care about functions instead of functionals (distributions) is *scaling*. The mean curvature functional scales differently from the (classical) mean curvature function: If a surface is uniformly scaled by a factor λ, then the mean curvature functional also scales with λ, whereas the mean curvature function scales with $1/\lambda$.

Definition 3.8 (Discrete mean curvature vector). The *discrete mean curvature vector* is the unique $\mathbb{R}^3$-valued piecewise linear function $\vec{H}_{dis} \in (S_{h,0})^3$, corresponding to the mean curvature functional $\vec{H}_h$, evaluated on $S_{h,0}$. It is defined by

$$(\vec{H}_{dis}, u_h)_{\mathcal{L}^2(M_h)} = \langle \vec{H}_h | u_h \rangle \quad \forall u_h \in S_{h,0} , \tag{3.12}$$

where $(\cdot, \cdot)_{\mathcal{L}^2(M_h)}$ denotes the $\mathcal{L}^2$ inner product on M_h, and $\langle \vec{H}_h | u_h \rangle$ denotes the evaluation of the mean curvature functional on u_h.

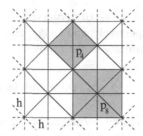

FIGURE 4. Discrete mean curvature does not converge in $\mathcal{L}^2$ for a 4–8 tessellation of a regular quad-grid, because the ratio between the areas of the stencils of p_4 and p_8 does not converge to 1.

Note that it is possible to associate a discrete function to the mean curvature functional only because the dimension of $S_{h,0}$ is finite. There is no infinite-dimensional analogue of this construction. The mean curvature function can be computed explicitly:

$$\vec{H}_{\text{dis}} = \sum_{p,q \in M_h \backslash \partial M_h} \langle \vec{H}_h | \phi_p \rangle \mathcal{M}^{pq} \phi_q \, , \tag{3.13}$$

where $\mathcal{M}^{pq}$ denotes the inverse of the *mass matrix*, $\mathcal{M}_{pq}$, which is given by

$$\mathcal{M}_{pq} = \int_{M_h} \phi_p \phi_q \, \mathrm{d} \operatorname{vol}_h \, .$$

Remark. Instead of using the full mass matrix, it is common to use a *lumped* version (such as obtained by forming a diagonal matrix with entries equal to the row sums of $\mathcal{M}$).

Counterexample. This example shows that, in general, discrete mean curvature fails to converge in $\mathcal{L}^2$ (and in fact, in any pointwise sense). Let M be a smooth cylinder of height 2π and radius 1. We construct a sequence $\{M_n\}$ of polyhedral cylinders whose vertices lie on M and which converges to M totally normally, but for which the mean curvature *functions* fail to converge. Let M be parameterized as follows:

$$x = \cos u, \quad y = \sin u, \quad z = v \, .$$

Let the vertices of M_n be given by

$$u = \frac{i\pi}{n} \qquad i = 0, \ldots, 2n - 1,$$

$$v = \begin{cases} 2j \sin \frac{\pi}{2n} & j = 0, \ldots, 2n - 1, \\ 2\pi & j = 2n. \end{cases}$$

This corresponds (up to the upper-most layer) to folding along vertical lines a regular planar quad-grid of edge length $h_n = 2 \sin(\pi/2n)$, see Figure 4. In other words, all faces of M_n are square (except for the upper-most layer). It will now depend on the *tessellation pattern* (i.e., the choice of diagonals) of this quad-grid whether there is $\mathcal{L}^2$-convergence of the mean curvature function or not. Indeed, consider the regular 4–8 tessellation scheme depicted in Figure 4. There are two kinds of vertices: those of valence 4 and those of valence 8. Call them p_4 and p_8, respectively, and let ϕ_{p_4} and ϕ_{p_8} denote the corresponding nodal basis functions. By the symmetry of the problem there exist constants $a_n, b_n \in \mathbb{R}$ such that

$$\vec{H}_{\text{dis},n} = \sum_{p_4} a_n \cdot \phi_{p_4} \cdot \partial_r + \sum_{p_8} b_n \cdot \phi_{p_8} \cdot \partial_r + \text{boundary contributions} ,$$

where ∂r is the smooth cylinder's outer normal and the contributions from the boundary include all vertices one layer away from the upper boundary. Then, as the edge lengths h_n tend to zero, it turns out that $b_n \to 0$ and $a_n \to -3$. Hence, $\vec{H}_{\text{dis},n}$ is a family of continuous functions oscillating between $a_n \approx -3$ (at the vertices of valence 4) and $b_n \approx 0$ (at the vertices of valence 8) with ever growing frequencies. Such a family, although it does converge in $\mathcal{H}^{-1}(M)$, cannot converge in $\mathcal{L}^2(M)$.

3.5. A general convergence result

We have chosen to elaborate on the concept of mean curvature here. In order to complete the picture, we conclude this overview with a summary of related results. For a more complete story, see [15, 29].

Theorem 3.9 (Geometric conditions for normal convergence). *Let $\{M_n\}$ be a sequence of compact polyhedral surfaces which are normal graphs over a compact smooth surface M, converging to M in Hausdorff distance. Then the following conditions are* equivalent*:*

1. *convergence of normals,*
2. *convergence of metric tensors ($A_n \to \text{Id}$),*
3. *convergence of area measure,*
4. *convergence of Laplace–Beltrami operators in norm.*

The fact that totally normal convergence implies convergence of metric tensors has several other consequences: convergence of shortest geodesics on M_n to geodesics on M, convergence of solutions to the Dirichlet problem, convergence of Hodge decompositions (for appropriate discrete notions of gradient, curl, and divergence), and convergence of the spectrum of the Laplacian. For details we refer to [29].

Acknowledgments

This work grew out of many stimulating discussions with Michael Herrmann, Klaus Hildebrandt, and Konrad Polthier. I wish to thank Jean-Marie Morvan for bringing to my attention the relation between the gradient of the shortest distance map and convergence of normals. This work was supported by the DFG Research Center MATHEON "Mathematics for key technologies" in Berlin.

References

[1] A.D. Aleksandrov and V.A. Zalgaller, *Intrinsic geometry of surfaces*, Translation of Mathematical Monographs, vol. 15, AMS, 1967.

[2] Alexander I. Bobenko, Tim Hoffmann, and Boris A. Springborn, *Minimal surfaces from circle patterns: geometry from combinatorics*, Annals of Mathematics **164** (2006), no. 1, 231–264.

[3] Alexander I. Bobenko and Boris A. Springborn, *A discrete Laplace–Beltrami operator for simplicial surfaces*, (2005), arXiv:math.DG/0503219, to appear in Discrete Comput. Geom.

[4] Jeff Cheeger, *Differentiability of Lipschitz functions on metric spaces*, Geom. Funct. Anal. (GAFA) **9** (1999), 428–517.

[5] Jeff Cheeger, Werner Müller, and Robert Schrader, *Curvature of piecewise flat metrics*, Comm. Math. Phys. **92** (1984), 405–454.

[6] David Cohen-Steiner and Jean-Marie Morvan, *Restricted Delaunay triangulations and normal cycle*, Sympos. Comput. Geom. (2003), 312–321.

[7] ———, *Second fundamental measure of geometric sets and local approximation of curvatures*, J. Differential Geom. **73** (2006), no. 3, 363–394.

[8] Mathieu Desbrun, Mark Meyer, Peter Schröder, and Alan H. Barr, *Implicit fairing of irregular meshes using diffusion and curvature flow*, Proceedings of ACM SIGGRAPH, 1999, pp. 317–324.

[9] R.J. Duffin, *Distributed and Lumped Networks*, Journal of Mathematics and Mechanics **8** (1959), 793–825.

[10] Gerhard Dziuk, *Finite elements for the Beltrami operator on arbitrary surfaces*, Partial Differential Equations and Calculus of Variations, Lec. Notes Math., vol. 1357, Springer, 1988, pp. 142–155.

[11] ———, *An algorithm for evolutionary surfaces*, Num. Math. **58** (1991), 603–611.

[12] Herbert Federer, *Curvature measures*, Trans. Amer. Math. **93** (1959), 418–491.

[13] Joseph H.G. Fu, *Convergence of curvatures in secant approximations*, J. Differential Geometry **37** (1993), 177–190.

[14] Mikhael Gromov, *Structures métriques pour les variétés riemanniennes*, Textes Mathématiques, Cedic/Fernand Nathan, 1981.

[15] Klaus Hildebrandt, Konrad Polthier, and Max Wardetzky, *On the convergence of metric and geometric properties of polyhedral surfaces*, Geometricae Dedicata **123** (2006), 89–112.

[16] Hermann Karcher and Konrad Polthier, *Construction of triply periodic minimal surfaces*, Phil. Trans. Royal Soc. Lond. **354** (1996), 2077–2104.

[17] Christian Mercat, *Discrete Riemann surfaces and the Ising model*, Communications in Mathematical Physics **218** (2001), no. 1, 177–216.

[18] Mark Meyer, Mathieu Desbrun, Peter Schröder, and Alan H. Barr, *Discrete differential-geometry operators for triangulated 2-manifolds*, Visualization and Mathematics III (H.-C. Hege and K. Polthier, eds.), Springer, Berlin, 2003, pp. 35–57.

[19] Jean-Marie Morvan and Boris Thibert, *Approximation of the normal vector field and the area of a smooth surface*, Discrete and Computational Geometry **32** (2004), no. 3, 383–400.

[20] Ulrich Pinkall and Konrad Polthier, *Computing discrete minimal surfaces and their conjugates*, Experim. Math. **2** (1993), 15–36.

[21] Konrad Polthier, *Unstable periodic discrete minimal surfaces*, Geometric Analysis and Nonlinear Partial Differential Equations (S. Hildebrandt and H. Karcher, eds.), Springer, 2002, pp. 127–143.

[22] ———, *Computational aspects of discrete minimal surfaces*, Global Theory of Minimal Surfaces (David Hoffman, ed.), CMI/AMS, 2005.

[23] Konrad Polthier and Wayne Rossman, *Index of discrete constant mean curvature surfaces*, J. Reine Angew. Math. **549** (2002), 47–77.

[24] Yuriĭ Grigor'evich Reshetnyak, *Geometry IV. Non-regular Riemannian geometry*, Encyclopaedia of Mathematical Sciences, vol. 70, Springer-Verlag, 1993, pp. 3–164.

[25] Wayne Rossman, *Infinite periodic discrete minimal surfaces without self-intersections*, Balkan J. Geom. Appl. **10** (2005), no. 2, 106–128.

[26] Oded Schramm, *Circle patterns with the combinatorics of the square grid*, Duke Math. J. **86** (1997), 347–389.

[27] Hermann Amandus Schwarz, *Sur une définition erronée de l'aire d'une surface courbe*, Gesammelte Mathematische Abhandlungen, vol. 2, Springer-Verlag, 1890, pp. 309–311.

[28] William Peter Thurston, *The geometry and topology of three-manifolds*, www.msri.org/publications/books/gt3m.

[29] Max Wardetzky, *Discrete Differential Operators on Polyhedral Surfaces—Convergence and Approximation*, Ph.D. Thesis, Freie Universität Berlin, 2006.

[30] P. Wintgen, *Normal cycle and integral curvature for polyhedra in Riemannian manifolds*, Differential Geometry (Soós and Szenthe, eds.), 1982, pp. 805–816.

[31] M. Zähle, *Integral and current representations of Federer's curvature measures*, Arch. Math. **46** (1986), 557–567.

[32] William P. Ziemer, *Weakly differentiable functions: Sobolev spaces and functions of bounded variation*, GTM, Springer, New York, 1989.

Max Wardetzky
Freie Universität Berlin
Arnimalle 3
14195 Berlin
Germany
e-mail: wardetzky@mi.fu-berlin.de

Discrete Differential Geometry, A.I. Bobenko, P. Schröder, J.M. Sullivan and G.M. Ziegler, eds.
Oberwolfach Seminars, Vol. 38, 287–323

Discrete Differential Forms
for Computational Modeling

Mathieu Desbrun, Eva Kanso and Yiying Tong

Abstract. This chapter introduces the background needed to develop a geometry-based, principled approach to computational modeling. We show that the use of discrete differential forms often resolves the apparent mismatch between differential and discrete modeling, for applications varying from graphics to physical simulations.

Keywords. Discrete differential forms, exterior calculus, Hodge decomposition.

1. Motivation

The emergence of computers as an essential tool in scientific research has shaken the very foundations of differential modeling. Indeed, the deeply-rooted abstraction of smoothness, or *differentiability*, seems to inherently clash with a computer's ability of storing only finite sets of numbers. While there has been a series of computational techniques that proposed discretizations of differential equations, the geometric structures they are simulating are often lost in the process.

1.1. The role of geometry in science

Geometry is the study of space and of the properties of shapes in space. Dating back to Euclid, models of our surroundings have been formulated using simple, geometric descriptions, formalizing apparent *symmetries* and experimental *invariants*. Consequently, geometry is at the foundation of many current physical theories: general relativity, electromagnetism (E&M), gauge theory as well as solid and fluid mechanics all have strong underlying geometrical structures. Einstein's theory, for instance, states that gravitational field strength is directly proportional to the *curvature of space-time*. In other words, the physics of relativity is *directly modeled* by the shape of our 4-dimensional world, just as the behavior of soap bubbles is modeled by their shapes. Differential geometry is thus, de facto, the mother tongue of numerous physical and mathematical theories.

Unfortunately, the inherent geometric nature of such theories is often obstructed by their formulation in vectorial or tensorial notations: the traditional use of a coordinate system, in which the defining equations are expressed, often obscures the underlying structures by an overwhelming usage of indices. Moreover, such complex expressions entangle the topological and geometrical content of the model.

1.2. Geometry-based exterior calculus

The geometric nature of these models is best expressed and elucidated through the use of the *exterior calculus* of *differential forms*, first introduced by Cartan [7]. This geometry-based calculus was further developed and refined over the twentieth century to become the foundation of modern differential geometry. The calculus of exterior forms allows one to express differential and integral equations on smooth and curved spaces in a consistent manner, while revealing the geometrical invariants at play. For example, the classical operations of gradient, divergence, and curl as well as the theorems of Green, Gauss and Stokes can all be expressed concisely in terms of differential forms and an operator on these forms called the exterior derivative—hinting at the generality of this approach.

Compared to classical tensorial calculus, this exterior calculus has several advantages. First, it is often difficult to recognize the coordinate-independent nature of quantities written in tensorial notation: local and global invariants are hard to notice by just staring at the indices. On the other hand, invariants are easily discovered when expressed as differential forms by invoking either Stokes's theorem, the Poincaré lemma, or by applying exterior differentiation. Note also that the exterior derivative of differential forms—the antisymmetric part of derivatives—is one of the most important parts of differentiation, since it is invariant under coordinate system change. In fact, Sharpe states in [37] that every differential equation may be expressed in term of the exterior derivative of differential forms. As a consequence, several recent initiatives have been aimed at formulating physical laws in terms of differential forms. For recent work along these lines, the reader is invited to refer to [5, 1, 27, 13, 32, 6, 16] for books offering a theoretical treatment of various physical theories using differential forms.

1.3. Differential vs. discrete modeling

We have seen that a large amount of our scientific knowledge relies on a deeply-rooted differential (i.e., smooth) comprehension of the world. This abstraction of differentiability allows researchers to model complex physical systems via concise equations. With the sudden advent of the digital age, it was therefore only natural to resort to computations based on such differential equations.

However, since digital computers can only manipulate finite sets of numbers, their capabilities seem to clash with the basic foundations of differential modeling. In order to overcome this hurdle, a first set of computational techniques (e.g., finite difference or particle methods) focused on satisfying the continuous equations at a discrete set of spatial and temporal samples. Unfortunately, focusing on accurately discretizing the local laws often fails to respect important global structures and invariants. Later methods such as Finite Elements (FEM), drawing from developments in the calculus of variations, remedied this inadequacy to some extent by satisfying local conservation laws on average and preserving some important invariants. Coupled with a finer ability to deal with arbitrary boundaries, FEM became the de facto computational tool for engineers. Even with significant advances in error control, convergence, and stability of these finite approximations, the underlying structures of the simulated continuous systems are often destroyed: a moving rigid body may gain or loose momentum; or a cavity may exhibit fictitious eigenmodes in an electromagnetism (E&M) simulation. Such examples illustrate some of the loss of fidelity that can follow from a standard discretization process, failing to preserve some fundamental geometric and topological structures of the underlying continuous models.

The cultural gap between theoretical and applied science communities may be partially responsible for the current lack of proper discrete, computational modeling that could mirror and leverage the rich developments of its differential counterpart. In particular, it is striking that the calculus of differential forms has not yet had an impact on the mainstream computational fields, despite excellent initial results in E&M [4] or Lagrangian mechanics [29]. It should also be noticed that some basic tools necessary for the definition of a discrete calculus already exist, probably initiated by Poincaré when he defined his cell decomposition of smooth manifolds. The study of the structure of ordered sets or simplices now belongs to the well-studied branch of mathematics known as *Combinatorial Differential Topology and Geometry*, which is still an active area of research (see, e.g., [15] and [2] and references therein).

1.4. Calculus ex geometrica

Given the overwhelming geometric nature of the most fundamental and successful calculus of these last few centuries, it seems relevant to *approach computations from a geometric standpoint*.

One of the key insights that percolated down from the theory of differential forms is rather simple and intuitive: one needs to recognize that different physical quantities have different properties, and must be treated accordingly. Fluid mechanics or electromagnetism, for instance, make heavy use of line integrals, as well as surface and volume integrals; even physical measurements are performed as specific local integrations or averages (think of flux for magnetic field, or current for electricity, or pressure for atoms' collisions). Point-wise evaluations or approximations for such quantities are not the appropriate discrete analogs, since the defining geometric properties of their physical meaning cannot be enforced naturally. Instead, *one should store and manipulate those quantities at their geometrically-meaningful location*: in other words, we should consider values on vertices, edges, faces, and tetrahedra as proper discrete versions of, respectively, pointwise functions, line integrals, surface integrals, and volume integrals: only then will we be able to manipulate those values without violating the symmetries that the differential modeling tried to exploit for predictive purposes.

1.5. Similar endeavors

The need for improved numerics have recently sprung a (still limited) number of interesting related developments in various fields. Although we will not try to be exhaustive, we wish to point the reader to a few of the most successful investigations with the same "flavor" as our discrete geometry-based calculus, albeit their approaches are rarely similar to ours. First, the field of *Mimetic Discretizations of Continuum Mechanics*, led by Shashkov, Steinberg, and Hyman [24], started on the premise that spurious solutions obtained from finite element or finite difference methods often originate from inconsistent discretizations of the operators div, curl, and grad, and that addressing this inconsistency pays off numerically. Similarly, *Computational Electromagnetism* has also identified the issue of field discretization as the main reason for spurious modes in numerical results. An excellent treatment of the discretization of the Maxwell's equations resulted [4], with a clear relationship to the differential case. Finally, recent developments in *Discrete Lagrangian Mechanics* have demonstrated the efficacy of a proper discretization of the Lagrangian of a dynamical system, rather than the discretization of its derived Euler–Lagrange equations: with a discrete Lagrangian, one can ensure that the integration scheme satisfies an exact discrete least-action principle, preserving all the momenta directly for arbitrary orders of accuracy [29]. Respecting the defining geometric properties of both the fields and the governing equations is a common link between all these recent approaches.

1.6. Advantages of discrete differential modeling

The reader will have most probably understood our bias by now: we believe that the systematic construction, inspired by Exterior Calculus, of *differential, yet readily discretizable computational foundations* is a crucial ingredient for numerical fidelity. Because many of the standard tools used in differential geometry have discrete combinatorial analogs, the *discrete versions of forms or manifolds* will be formally identical to (and should partake of the same properties as) the continuum models. Additionally, such an approach should clearly maintain the separation of the topological (metric-independent) and geometrical (metric-dependent) components of the quantities involved, keeping the geometric picture (i.e., intrinsic structure) intact.

A *discrete differential modeling approach to computations* will also be often much simpler to define and develop than its continuous counterpart. For example, the discrete notion of a differential form will be implemented simply as values on mesh elements. Likewise, the discrete notion of orientation will be more straightforward than its continuous counterpart: while the differential definition of orientation uses the notion of equivalence class of atlases determined by the sign of the Jacobian, the orientation of a mesh edge will be one of two directions; a triangle will be oriented clockwise or counterclockwise; a volume will have a direction as a right-handed helix or a left-handed one; no notion of atlas (a collection of consistent coordinate charts on a manifold) will be required.

1.7. Goal of this chapter

Given these premises, this contribution was written with several purposes in mind. First, we wish to demonstrate that the foundations on which powerful methods of computations can be built are quite approachable—and are not as abstract as the reader may fear: the ideas involved are very intuitive as a side effect of the simplicity of the underlying geometric principles.

Second, we wish to help bridging the gap between applied fields and theoretical fields: we have tried to render the theoretical bases of our exposition accessible to

 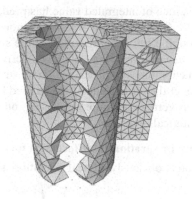

FIGURE 1. Typical 2D and 3D meshes: although the David head appears smooth, its surface is made of a triangle mesh; tetrahedral meshes (such as this mechanical part, with a cutaway view) are some typical examples of irregular meshes on which computations are performed. David's head mesh is courtesy of Marc Levoy, Stanford.

computer scientists, and the concrete implementation insights understandable by non-specialists. For this very reason, the reader should not consider this introductory exposition as a definite source of knowledge: it should instead be considered as a portal to better, more focused work on related subjects. We only hope that we will ease our readers into foundational concepts that can be undoubtedly and fruitfully applied to all sorts of computations—be it for graphics or simulation.

With these goals in mind, we will describe the background needed to develop a principled, geometry-based approach to computational modeling that gets around the apparent mismatch between differential and discrete modeling.

2. Relevance of forms for integration

The evaluation of differential quantities on a discrete space (mesh) is a nontrivial problem. For instance, consider a piecewise-linear 2-dimensional surface embedded in a three-dimensional Euclidean space, i.e., a triangle mesh. Celebrated quantities such as the Gaussian and mean curvatures are delicate to define on it. More precisely, the Gaussian curvature can be easily proven to be zero everywhere *except* on vertices, where it is a Dirac delta function. Likewise, the mean curvature can only be defined in the distributional sense, as a Dirac delta function on edges. However, through local *integrations*, one can easily manipulate these quantities numerically: if a careful choice of non-overlapping regions is made, the delta functions can be properly integrated, rendering the computations relatively simple as shown, for example, in [31, 22]. Note that the process of integration to suppress discontinuity is, in spirit, equivalent to the idea of weak form used in the finite-element method.

This idea of integrated value has predated in some cases the equivalent differential statements: for instance, it was long known that the genus of a surface can be calculated through a cell decomposition of the surface via the Euler characteristic. The actual Gauss–Bonnet theorem was, however, derived later on. Now, if one tries to discretize the Gaussian curvature of a piecewise-linear surface in an arbitrary way, it is not likely that its integral over the surface equals the desired Euler characteristic, while its discrete version, defined on vertices (or, more precisely, on the dual of each vertex), naturally preserves this topological invariant.

2.1. From integration to differential forms

Integration is obviously a linear operation, since for any disjoint sets A and B,

$$\int_{A \cup B} = \int_A + \int_B.$$

Moreover, the integration of a smooth function over a subset of measure zero is always zero; for example, an area integral of (a lower-dimensional object such as) a curve or a point is equal to zero. Finally, integration is *objective* (i.e., relevant) only if its evaluation is invariant under change of coordinate systems. These three properties combined directly imply that the integrand (i.e., the whole expression after the integral sign) has to be *antisymmetric*. That is, the basic building blocks of any type of integration are *differential forms*. Chances are, the reader is already very well acquainted with forms, maybe without even knowing it.

2.1.1. An intuitive definition.
A differential form (also denoted as exterior[1] differential form) is, informally, an integrand, i.e., a quantity that can be integrated. It is the dx in $\int dx$ and the $dx\, dy$ in $\iint dx\, dy$. More precisely, consider a smooth function $F(x)$ over an interval in $\mathbb{R}$. Now, define $f(x)$ to be its derivative, that is,

$$f(x) = \frac{dF}{dx},$$

Rewriting this last equation (with slight abuse of notation for simplicity) yields $dF = f(x)dx$, which leads to:

$$\int_a^b dF = \int_a^b f(x)dx = F(b) - F(a). \tag{2.1}$$

This last equation is known as the Newton–Leibnitz formula, or the first fundamental theorem of calculus. The integrand $f(x)dx$ is called a 1-*form*, because it can only be integrated over any 1-dimensional (1D) real interval. Similarly, for a function $G(x, y, z)$, we have:

$$dG = \frac{\partial G}{\partial x}dx + \frac{\partial G}{\partial y}dy + \frac{\partial G}{\partial z}dz,$$

which can be integrated over any 1D curve in $\mathbb{R}^3$, and is also a 1-form. More generally, *a k-form can be described as an entity ready (or designed, if you prefer) to be integrated on a kD (sub)region*. Note that forms are valued zero on (sub)regions that are of higher or

---

[1] The word "exterior" is used as the exterior algebra is basically built out of an *outer* product.

lower order dimension than the original space; for example, 4-forms are zero on $\mathbb{R}^3$. These differential forms are extensively used in mathematics, physics and engineering, as we already hinted at the fact in Section 1.4 that most of our measurements of the world are of an integral nature: even digital pictures are made out of local area integrals of the incident light over each of the sensors of a camera to provide a set of values at each pixel on the final image (see inset). The importance of this notion of forms in science is also evidenced by the fact that operations like gradient, divergence, and curl can all be expressed in terms of forms only, as well as fundamental theorems like Green's or Stokes's.

2.1.2. A formal definition. For concreteness, consider n-dimensional Euclidean space $\mathbb{R}^n$, $n \in \mathbb{N}$ and let $\mathcal{M}$ be an open region $\mathcal{M} \subset \mathbb{R}^n$; $\mathcal{M}$ is also called an *n-manifold*. The vector space $T_x\mathcal{M}$ consists of all the (tangent) vectors at a point $x \in \mathcal{M}$ and can be identified with $\mathbb{R}^n$ itself. A k-form ω^k is a rank-k, antisymmetric tensor field over $\mathcal{M}$. That is, at each point $x \in \mathcal{M}$, it is a multi-linear map that takes k tangent vectors as input and returns a real number:

$$\omega^k : T_x M \times \cdots \times T_x M \longrightarrow \mathbb{R}$$

which *changes sign for odd permutations of the variables* (hence the term antisymmetric). Any k-form naturally induces a k-form on a submanifold, through restriction of the linear map to the domain that is the product of tangent spaces of the submanifold.

Comments on the notion of pseudoforms. There is a closely related concept named pseudoform. Pseudoforms change sign when we change the orientation of coordinate systems, just like pseudovectors. As a result, the integration of a pseudoform does not change sign when the orientation of the manifold is changed. Unlike k-forms, a k-pseudoform induces a k-pseudoform on a submanifold *only* if a transverse direction is given. For example, fluid flux is sometimes called a 2-pseudoform: indeed, given a transverse direction, we know how much flux is going through a piece of surface; it does not depend on the orientation of the surface itself. Vorticity is, however, a true 2-form: given an orientation of the surface, the integration gives us the circulation around that surface boundary induced by the surface orientation. It does *not* depend on the transverse direction of the surface. But if we have an orientation of the ambient space, we can always associate transverse direction with internal orientation of the submanifold. Thus, in our case, we may treat pseudoforms simply as forms because we can consistently choose a representative from the equivalence class.

2.2. The differential structure

Differential forms are the building blocks of a whole calculus. To manipulate these basic blocks, Exterior Calculus defines seven operators:

- d: the exterior derivative, that extends the notion of the differential of a function to differential forms;
- $\star$: the Hodge star, that transforms k-forms into $(n - k)$-forms;
- $\wedge$: the wedge product, that extends the notion of exterior product to forms;

- $\sharp$ and $\flat$: the sharp and flat operators, that, given a metric, transform a 1-form into a vector and vice-versa;
- i_X: the interior product with respect to a vector field X (also called contraction operator), a concept dual to the exterior product;
- $\mathcal{L}_X$: the Lie derivative with respect to a vector field X, that extends the notion of directional derivative.

In this chapter, we will restrict our discussions to the first three operators, to provide the most basic tools necessary in computational modeling.

2.3. A taste of exterior calculus in $\mathbb{R}^3$

To give the reader a taste of the relative simplicity of the exterior calculus, we provide a list of equivalences (in the continuous world!) between traditional operations and their exterior calculus counterpart in the special case of $\mathbb{R}^3$. We will suppose that we have the usual euclidean metric. Then, forms are actually quite simple to conceive:

$$0\text{-form} \longleftrightarrow \text{scalar field}$$
$$1\text{-form} \longleftrightarrow \text{vector field}$$
$$2\text{-form} \longleftrightarrow \text{vector field}$$
$$3\text{-form} \longleftrightarrow \text{scalar field.}$$

To be clear, we will add a superscript on the forms to indicate their rank. Then applying forms to vector fields amounts to:

$$1\text{-form: } u^1(v) \longleftrightarrow u \cdot v$$
$$2\text{-form: } u^2(v, w) \longleftrightarrow u \cdot (v \times w)$$
$$3\text{-form: } f^3(u, v, w) \longleftrightarrow f u \cdot (v \times w).$$

Furthermore, the usual operations like gradient, curl, divergence and cross product can all be expressed in terms of the basic exterior calculus operators. For example:

$$d^0 f = \nabla f, \; d^1 u = \nabla \times u, \; d^2 u = \nabla \cdot u$$
$$\star^0 f = f, \; \star^1 u = u, \; \star^2 u = u, \; \star^3 f = f$$
$$\star^0 d^2 \star^1 u^1 = \nabla \cdot u, \; \star^1 d^1 \star^2 u^2 = \nabla \times u, \; \star^2 d^0 \star^3 f = \nabla f$$
$$f^0 \wedge u = f u, \; u^1 \wedge v^1 = u \times v, \; u^1 \wedge v^2 = u^2 \wedge v^1 = u \cdot v$$
$$i_v u^1 = u \cdot v, \; i_v u^2 = u \times v, \; i_v f^3 = f v.$$

Now that we have established the relevance of differential forms even in the most basic vector operations, time has come to turn our attention to make this concept of forms readily usable for computational purposes.

3. Discrete differential forms

Finding a discrete counterpart to the notion of differential forms is a delicate matter. If one was to represent differential forms using their coordinate values and approximate the exterior derivative using finite differences, basic theorems such as Stokes's theorem would not hold numerically. The main objective of this section is therefore to present a proper discretization of forms on what are known as simplicial complexes. We will show how this discrete geometric structure, well suited for computational purposes, is designed to preserve all the fundamental differential properties. For simplicity, we restrict the discussion to forms on 2D surfaces or 3D regions embedded in $\mathbb{R}^3$, but the construction is applicable to general manifolds in arbitrary spaces. In fact, the only necessary assumption is that the embedding space must be a vector space, a natural condition in practice.

3.1. Simplicial complexes and discrete manifolds

For the interested reader, the notions we introduce in this section are defined formally in much more details (for the general case of k-dimensional spaces) in references such as [33] or [21].

0-simplex (vertices)	1-simplex (edges)	2-simplex (triangles)	3-simplex (tets)

FIGURE 2. A 1-simplex is a line segment, the convex hull of two points. A 2-simplex is a triangle, i.e., the convex hull of three distinct points. A 3-simplex is a tetrahedron, as it is the convex hull of four points.

3.1.1. Notion of simplex.
A k-*simplex* is the generic term to describe the simplest mesh element of dimension k—hence the name. By way of motivation, consider a three dimensional mesh in space. This mesh is made of a series of adjacent tetrahedra (denoted *tets* for simplicity throughout). The vertices of the tets are called 0-simplices. Similarly, the line segments or edges form 1-simplices, the triangles or faces form 2-simplices, and the tets form 3-simplices. Note that we can define these simplices in a top-down manner too: faces (2-simplices) can be thought of as boundaries of tets (3-simplices), edges (1-simplices) as boundaries of faces, and vertices (0-simplices) as boundaries of edges.

The definition of a simplex can be made more abstract as a series of k-tuples (referring to the vertices they are built upon). However, for the type of applications that we are targeting in this chapter, we will often not make any distinction between an abstract simplex and its topological realization (connectivity) or geometrical realization (positions in space).

Formally, a k-simplex σ_k is the non-degenerate convex hull of $k+1$ geometrically distinct points $v_0, \ldots, v_k \in \mathbb{R}^n$ with $n \geq k$. In other words, it is the intersection of all convex sets containing $(v_0, \ldots, v_k)$; namely:

$$\sigma_k = \{x \in \mathbb{R}^n \mid x = \sum_{i=0}^{k} \alpha^i \, v_i \text{ with } \alpha^i \geq 0 \text{ and } \sum_{i=0}^{k} \alpha^i = 1\}.$$

The entities $v_0, \ldots, v_k$ are called the *vertices* and k is called the dimension of the k-simplex, which we will denote as:

$$\sigma_k = \{v_0 v_1 \cdots v_k\}.$$

3.1.2. Orientation of a simplex. Note that all orderings of the $k + 1$ vertices of a k-simplex can be divided into two equivalent classes, i.e., two orderings differing by an even permutation. Such a class of orderings is called an *orientation*. In the present work, we always assume that local orientations are *given* for each simplex; that is, each element of the mesh has been given a particular orientation. For example, an edge $\sigma_1 = \{v_0 v_1\}$ in Figure 2 has an arrow indicating its default orientation. If the opposite orientation is needed, we will denote it as $\{v_1 v_0\}$, or, equivalently, by $-\{v_0 v_1\}$. For more details and examples, the reader is referred to [33, 23].

3.1.3. Boundary of a simplex. Any $(k - 1)$-simplex spanned by a subset of $\{v_0, \ldots v_k\}$ is called a $(k - 1)$-*face* of σ_k. That is, a $(k - 1)$-*face* is simply a $(k - 1)$-simplex whose k vertices are all from the $k + 1$ vertices of the k-simplex. The union of the $(k - 1)$-faces is what is called the *boundary* of the k-simplex. One should be careful here: because of the default orientation of the simplices, the formal *signed* sum of the $(k - 1)$-faces defines the boundary of the k-simplex. Therefore, the boundary operator takes a k-simplex and gives the sum of all its $(k - 1)$-faces with 1 or -1 as coefficients depending on whether their respective orientations match or not, see Figure 4.

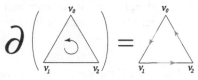

FIGURE 3. The boundary operator ∂ applied to a triangle (a 2-simplex) is equal to the signed sum of the edges (i.e., the 1-faces of the 2-simplex).

To remove possible mistakes in orientation, we can define the *boundary operator* as follows:

$$\partial\{v_0 v_1 \cdots v_k\} = \sum_{j=0}^{k} (-1)^j \{v_0, \ldots, \widehat{v}_j, \ldots, v_k\}, \tag{3.1}$$

where $\widehat{v}_j$ indicates that v_j is missing from the sequence, see Figure 3. Clearly, each k-simplex has $k + 1$ facets or $(k - 1)$-faces. For this statement to be valid even for $k = 0$, the empty set $\emptyset$ is usually defined as a (-1)-simplex face of every 0-simplex. The reader is invited to verify this definition on the triangle $\{v_0, v_1, v_2\}$ in Figure 3:

$$\partial\{v_0, v_1, v_2\} = \{v_1, v_2\} - \{v_0, v_2\} + \{v_0, v_1\}.$$

3.1.4. Simplicial complex. A simplicial complex is a collection $\mathcal{K}$ of simplices, which satisfies the following two simple conditions:

- every face of each simplex in $\mathcal{K}$ is in $\mathcal{K}$;
- the intersection of any two simplices in $\mathcal{K}$ is either empty, or an entire common face.

Computer graphics makes heavy use of what is called *realizations* of simplicial complexes. Loosely speaking, a realization of a simplicial complex is an embedding of this complex into the underlying space $\mathbb{R}^n$. Triangle meshes in 2D and tet meshes in 3D

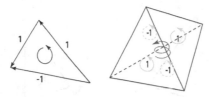

FIGURE 4. Boundary operator applied to a triangle (left), and a tetrahedron (right). Orientations of the simplices are indicated with arrows.

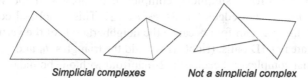

Simplicial complexes   *Not a simplicial complex*

are examples of such simplicial complexes (see Figure 1). Notice that polygonal meshes
can be easily triangulated, thus can be easily turned into simplicial complexes. One can
also use the notion of *cell complex* by allowing the elements of $\mathcal{K}$ to be non-simplicial;
we will restrict our explanations to the case of simplicial complexes for simplicity.

3.1.5. Discrete manifolds. An n-dimensional discrete manifold $\mathcal{M}$ is an n-dimensional
simplicial complex that satisfies the following condition: for each simplex, the union of all
the incident n-simplices forms an n-dimensional ball (*i.e.*, a disk in 2D, a ball in 3D, etc),
or half a ball if the simplex is on the boundary. As a consequence, each $(n-1)$-simplex
has exactly two adjacent n-simplices—or only one if it is on a boundary.

Basically, the notion of discrete manifold corresponds to the usual computer graphics acceptation of "manifold mesh". For example, in 2D, discrete manifolds cannot have
isolated edges (also called sticks or hanging edges) or isolated vertices, and each of their
edges is adjacent to two triangles (except for the boundary; in that case, the edge is adjacent to only one triangle). A surface mesh in 3D cannot have a "fin", i.e., an edge with
more than two adjacent triangles. To put it differently, infinitesimally-small, imaginary
inhabitants of an n-dimensional discrete manifold would consider themselves living in
$\mathbb{R}^n$ as any small neighborhood of this manifold is isomorphic to $\mathbb{R}^n$.

3.2. Notion of chains

We have already encountered the notion of chain, without mentioning it. Recall that the
boundary operator takes each k-simplex and gives the *signed* sum of all its $(k-1)$-
faces. We say that the boundary of a k-simplex produces a $(k-1)$-*chain*. The following
definition is more precise and general.

3.2.1. Definition. A k-*chain* of an oriented simplicial complex $\mathcal{K}$ is a set of values, one
for *each k-simplex* of $\mathcal{K}$. That is, a k-chain c can then be thought of as a linear combination
of all the k-simplices in $\mathcal{K}$:

$$c = \sum_{\sigma \in \mathcal{K}} c(\sigma) \cdot \sigma, \qquad (3.2)$$

where $c(\sigma) \in \mathbb{R}$. We will denote the group of all k-chains as $\mathcal{C}_k$.

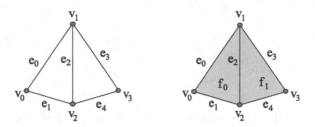

FIGURE 5. (left) A simplicial complex that consists of the vertices $\{v_0, v_1, v_2, v_3\}$ and edges $\{e_0, e_1, e_2, e_3, e_4\}$. This simplicial complex is not a discrete manifold because the neighborhoods of the vertices v_1 and v_2 are not 1D balls. (right) If we add the triangles f_0 and f_1 to the simplicial complex, it becomes a 2-manifold with one boundary.

3.2.2. Implementation of chains. Let the set of all k-simplices in $\mathcal{K}$ be denoted by $\mathcal{K}^k$, and let its cardinality be denoted as $|\mathcal{K}^k|$. A k-chain can simply be stored as a *vector* (or array) of dimension $|\mathcal{K}^k|$, *i.e.*, one number for each k-simplex $\sigma_k \in \mathcal{K}^k$.

3.2.3. Boundary operator on chains. We mentioned that the boundary operator ∂ was returning a particular type of chain, namely, a chain with coefficients equal to either 0, 1, or -1. Therefore, it should not be surprising that we can extend the notion of boundary to act also on k-chains, simply by linearity:

$$\partial \sum_k c_k \sigma_k = \sum_k c_k \partial \sigma_k.$$

That is, from one set of values assigned to all simplices of a complex, one can deduce

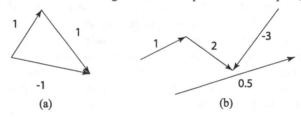

FIGURE 6. (a) An example of 1-chain being the boundary of a face (2-simplex); (b) a second example of 1-chain with 4 nonzero coefficients.

another set of values derived by weighting the boundaries of each simplex by the original value stored on it. This operation is very natural, and can thus be implemented easily as explained next.

3.2.4. Implementation of the boundary operator. Since the boundary operator is a linear mapping from the space of k-simplices to the space of $(k-1)$-simplices, it can simply be represented by a *matrix* of dimension $|\mathcal{K}^{k-1}| \times |\mathcal{K}^k|$. The reader can convince herself that this matrix is sparse, as only immediate neighbors are involved in the boundary operator. Similarly, this matrix contains only the values 0, 1, and -1. Notice that in 3D, there are three nontrivial boundary operators ∂_k (∂_1 is the boundary operator on edges, ∂_2 on triangles, ∂_3 on tets). However, the operator needed for a particular operation is obvious from the type of the argument: if the boundary of a tet is needed, the operator ∂_3 is the only one that makes sense to apply; in other words, the boundary of a k-simplex σ_k is found by invoking $\partial_k \sigma_k$. Thanks to this context-dependence, we can simplify the notation and remove the subscript when there is no ambiguity.

3.3. Notion of cochains

A k-cochain ω is the *dual* of a k-chain, that is to say, ω is a linear mapping that takes k-chains to $\mathbb{R}$. One writes:

$$\omega : C_k \to \mathbb{R},$$
$$c \mapsto \omega(c),$$

which reads as: a k-cochain ω operates on a k-chain c to give a scalar in $\mathbb{R}$. Since a chain is a linear combination of simplices, a cochain returns a linear combination of the values of that cochain on each simplex involved.

Clearly, a cochain also corresponds to one value per simplex (since all the k-simplices form a basis for the vector space C_k, and we only need to know the mapping of vectors in this basis to determine a linear mapping), and hence the notion of duality of chains and cochains is appropriate. But contrary to a chain, a k-cochain is *evaluated* on each simplex of the dimension k. In other words, a k-cochain can be thought of as a *field* that can be evaluated on each k-simplex of an oriented simplicial complex $\mathcal{K}$.

3.3.1. Implementation of cochains. The numerical representation of cochains follows from that of chains by duality. Recall that a k-chain can be represented as a vector c_k of length equal to the number of k-simplices in $\mathcal{M}$. Similarly, one may represent ω by a vector ω^k of the same size as c_k.

Now, remember that ω operates on c to give a scalar in $\mathbb{R}$. The linear operation $\omega(c)$ translates into an inner product $\omega^k \cdot c_k$. More specifically, one may continue to think of c_k as a *column vector* so that the $\mathbb{R}$-valued linear mapping ω can be represented by a *row vector* $(\omega^k)^t$, and $\omega(c)$ becomes simply the matrix multiplication of the row vector $(\omega^k)^t$ with the column vector c_k. The evaluation of a cochain is therefore trivial to implement.

3.4. Discrete forms as cochains

The attentive reader will have noticed by now: *k-cochains are discrete analogs to differential forms*. Indeed, a continuous k-form was defined as a linear mapping from k-dimensional sets to $\mathbb{R}$, as we can only integrate a k-form on a k-(sub)manifold. Note now that a kD set, when one has only a mesh to work with, is simply a *chain*. And a linear mapping from a chain to a real number is what we called a cochain: *a cochain is therefore a natural discrete counterpart of a form.*

For instance, a 0-form can be evaluated at each point, a 1-form can be evaluated on each curve, a 2-form can be evaluated on each surface, etc. Now if we *restrict* integration to take place only on the k-submanifold which is the sum of the k-simplices in the triangulation, we get a k-cochain; thus k-cochains are a discretization of k-forms. One can further map a continuous k-form to a k-cochain. To do this, first integrate the k-form on each k-simplex and assign the resulting value to that simplex to obtain a k-cochain on the k-simplicial complex. This k-cochain is a discrete representation of the original k-form.

3.4.1. Evaluation of a form on a chain. We can now naturally extend the notion of evaluation of a differential form ω on an arbitrary chain simply by linearity:

$$\int_{\sum_i c_i \sigma_i} \omega = \sum_i c_i \int_{\sigma_i} \omega. \tag{3.3}$$

As mentioned above, the integration of ω on each k-simplex σ_k provides a discretization of ω or, in other words, a mapping from the k-form ω to a k-cochain represented by:

$$\omega[i] = \int_{\sigma_i} \omega.$$

However convenient this chain/cochain standpoint is, in practical applications, one often needs a point-wise value for a k-form or to evaluate the integration on a particular k-submanifold. How do we get these values from a k-cochain? We will cover this issue of *form interpolation* in Section 6.

4. Operations on chains and cochains

4.1. Discrete exterior derivative

In the present discrete setting where the discrete differential forms are defined as cochains, defining a discrete exterior derivative can be done very elegantly: Stokes's theorem, mentioned early on in Section 2, can be used to *define* the exterior derivative d. Traditionally, this theorem states a vector identity equivalent to the well-known curl, divergence, Green's, and Ostrogradsky's theorems. Written in terms of forms, the identity becomes quite simple: it states that d applied to an arbitrary form ω is evaluated on an arbitrary simplex σ as follows:

$$\int_\sigma d\omega = \int_{\partial\sigma} \omega. \tag{4.1}$$

You surely recognize the usual property that an integral over a k-dimensional set is turned into a boundary integral (i.e., over a set of dimension $k-1$). With this simple equation relating the evaluation of $d\omega$ on a simplex σ to the evaluation of ω on the boundary of this simplex, the exterior derivative is *readily defined*: each time you encounter an exterior derivative of a form, replace any evaluation over a simplex σ by a direct evaluation of the form itself over the boundary of σ. Obviously, Stokes's theorem will be enforced by construction!

4.1.1. Coboundary operator. The operator d is called the *adjoint* of the boundary operator ∂: if we denote the integral sign as a pairing, i.e., with the convention that $\int_\sigma \omega = [\omega, \sigma]$, then applying d on the left-hand side of this operator is equivalent to applying ∂ on the right-hand side: $[d\omega, \sigma] = [\omega, \partial\sigma]$. For this very reason, d is sometimes called the coboundary operator.

Finally, by linearity of integration, we can write a more general form of Stokes's theorem, now extended to arbitrary chains, as follows:

$$\int_{\sum_i c_i \sigma_i} d\omega = \int_{\partial\left(\sum_i c_i \sigma_i\right)} \omega = \int_{\sum_i c_i \partial\sigma_i} \omega = \sum_i c_i \int_{\partial\sigma_i} \omega.$$

Consider the example shown in Figure 7. The discrete exterior derivative of the 1-form, defined as numbers on edges, is a 2-form represented by numbers on oriented faces. The orientation of the 1-forms may be opposite to that induced on the edges by the orientation of the faces. In this case, the values on the edges change sign. For instance, the 2-form associated with the d of the 1-forms surrounding the oriented shaded triangle takes the value $\omega = 2 - 1 - 0.75 = 0.25$.

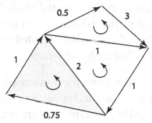

FIGURE 7. Given a 1-form as numbers on oriented edges, its discrete exterior derivative is a 2-form. In particular, this 2-form is valued 0.25 on the oriented shaded triangle.

4.1.2. Implementation of exterior derivative. Since we use vectors of dimension $|\mathcal{K}^k|$ to represent a k-cochain, the operator d can be represented by a matrix of dimension $|\mathcal{K}^{k+1}| \times |\mathcal{K}^k|$. Furthermore, this matrix has a trivial expression. Indeed, using the matrix notation introduced earlier, we have:

$$\int_{\partial c} \omega = \omega^t(\partial c) = (\omega^t \partial)c = (\partial^t \omega)^t c = \int_c d\omega.$$

Thus, the matrix d is simply equal to ∂^t. This should not come as a surprise, since we previously discussed that d is simply the adjoint of ∂. Note that care should be used when boundaries are present. However, and without digging too much into the details, it turns out that even for discrete manifolds *with* boundaries, the previous statement is valid. Implementing the exterior derivative while preserving Stokes's theorem is therefore a trivial matter in practice. Notice that just like for the boundary operator, there is actually more than one matrix for the exterior derivative operator: there is one per simplex dimension. But again, the context is sufficient to actually know which matrix is needed. A brute force approach that gets rid of these multiple matrices is to use a notion of super-chain, i.e., a vector storing *all* simplices, ordered from dimension 0 to the dimension of the space: in this case, the exterior derivative can be defined as a single, large sparse matrix that contains these previous matrices as blocks along the diagonal. We will not use this approach, as it makes the exposition less intuitive, in general.

4.2. Exact/closed forms and the Poincaré lemma

A k-form ω is called exact if there is a $(k-1)$-form α such that $\omega = d\alpha$, and it is called closed if $d\omega = 0$.

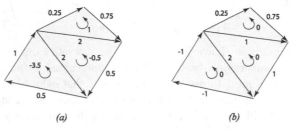

(a)   (b)

FIGURE 8. (a) The 2-form on the oriented shaded triangles defined by the exterior derivative d of the 1-form on the oriented edges is called an *exact* 2-*form*; (b) The 1-form on the oriented edges whose derivative d is identically zero is called a *closed* 1-*form*.

It is worth noting here that every exact form is closed, as will be seen in Section 4.3. Moreover, it is well known in the continuous setting that a closed form on a smooth contractible (sub)-manifold is locally exact (to be more accurate: exact over any disc-like region). This result is called the *Poincaré lemma*. The discrete analogue to this lemma can be stated as follows: given a closed k-cochain ω on a star-shaped complex, that is to say, $d\omega = 0$, there exits a $(k-1)$-cochain α such that $\omega = d\alpha$. For a formal statement and proof of this discrete version, see [8].

4.3. Introducing the deRham complex

The boundary of a boundary is the empty set. That is, the boundary operator applied twice to a k-simplex is zero. Indeed, it is easy to verify that $\partial\,\partial\sigma_k = 0$, since each $(k-2)$-simplex will appear exactly twice in this chain with different signs and, hence, cancel out (try it!). From the linearity of ∂, one can readily conclude that the property $\partial\,\partial = 0$ is true for all k-chains since the k-simplices form a basis. Similarly, one has that the discrete exterior derivative satisfies $d\,d = \partial^t\partial^t = (\partial\,\partial)^t = 0$, analogously to the exterior derivative of differential forms (notice that this last equality corresponds to the equality of mixed partial derivatives, which in turn is responsible for identities like $\nabla \times \nabla = 0$ and $\nabla \cdot \nabla\times = 0$ in $\mathbb{R}^3$).

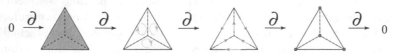

FIGURE 9. The chain complex of a tetrahedron with the boundary operator: from the tet, to its triangles, to their edges, and to their vertices.

4.3.1. Chain complex. In general, a *chain complex* is a sequence of linear spaces, connected with a linear operator D that satisfies the property $D\,D = 0$. Hence, the boundary operator ∂ (resp., the coboundary operator d) makes the spaces of chains (resp., cochains) into a chain complex, as shown in Figures 9 and 13.

When the spaces involved are the spaces of *differential* forms, and the operator is the exterior derivative d, this chain complex is called the *deRham complex*. By analogy, the chain complex for the spaces of *discrete* forms and for the coboundary operator is called the discrete deRham complex (or sometimes, the cochain complex).

4.3.2. Examples. Consider the 2D simplicial complex in Figure 10 (left) and choose the oriented basis of the i-dimensional simplices ($i = 0$ for vertices, $i = 1$ for edges and $i = 2$ for the face) as suggested by the ordering in the figure.

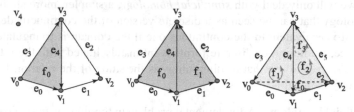

FIGURE 10. Three examples of simplicial complexes. The first one is no manifold. The two others are.

One gets $\partial(f_0) = e_0 - e_4 - e_3$; this can be identified with the vector $(1, 0, 0, -1, -1)$, representing the coefficient in front of each simplex. By repeating similar calculations for all simplices, one can readily conclude that the boundary operator ∂ is given by:

$$\partial_2 = \begin{pmatrix} 1 \\ 0 \\ 0 \\ -1 \\ -1 \end{pmatrix}, \quad \partial_1 = \begin{pmatrix} -1 & 0 & 0 & -1 & 0 \\ 1 & -1 & 0 & 0 & 1 \\ 0 & 1 & 1 & 0 & 0 \\ 0 & 0 & -1 & 1 & -1 \\ 0 & 0 & 0 & 0 & 0 \end{pmatrix},$$

That is, the chain complex under the boundary operator ∂ can be written as:

$$0 \longrightarrow C_2 \xrightarrow{\partial_2} C_1 \xrightarrow{\partial_1} C_0 \longrightarrow 0$$

where C_i, $i = 0, 1, 2$, denote the spaces of i-chains.

Consider now the domain to be the mesh shown in Figure 10 (middle). The exterior derivative operator, or the coboundary operator, can be expressed as:

$$d^0 = \begin{pmatrix} -1 & 1 & 0 & 0 \\ 0 & -1 & 1 & 0 \\ 0 & 0 & -1 & 1 \\ 1 & 0 & 0 & -1 \\ 0 & -1 & 0 & 1 \end{pmatrix}, \quad d^1 = \begin{pmatrix} 1 & 0 & 0 & 1 & 1 \\ 0 & 1 & 1 & 0 & -1 \end{pmatrix}.$$

It is worth noting that, since d is adjoint to ∂ by definition, the coboundary operator d induces a cochain complex:

$$0 \longleftarrow C^2 \xleftarrow{d^1} C^1 \xleftarrow{d^0} C^0 \longleftarrow 0$$

where C^i, $i = 0, 1, 2$, denote the spaces of i-cochains.

Finally, suppose the domain is the tetrahedron in Figure 10 (right), then the exterior derivative operators are:

$$d^0 = \begin{pmatrix} -1 & 1 & 0 & 0 \\ 0 & -1 & 1 & 0 \\ -1 & 0 & 1 & 0 \\ 1 & 0 & 0 & -1 \\ 0 & 1 & 0 & -1 \\ 0 & 0 & -1 & 1 \end{pmatrix}, \quad d^1 = \begin{pmatrix} 1 & 1 & -1 & 0 & 0 & 0 \\ 1 & 0 & 0 & 1 & -1 & 0 \\ 0 & 1 & 0 & 0 & 1 & 1 \\ 0 & 0 & 1 & 1 & 0 & 1 \end{pmatrix}, \quad d^2 = \begin{pmatrix} -1 & 1 & 1 & -1 \end{pmatrix}.$$

4.4. Notion of homology and cohomology

Homology is a concept dating back to Poincaré that focuses on studying the topological properties of a space. Loosely speaking, homology does so by counting the number of holes. In our case, since we assume that our space is a simplicial complex (i.e., triangulated), we will only deal with *simplicial homology*, a simpler, more straightforward type of homology that can be seen as a discrete version of the continuous definition (in other words, it is equivalent to the continuous one if the domain is triangulated). As we are about to see, the notion of discrete forms is intimately linked with these topological notions. In fact, we will see that (co)homology is the study of the relationship between *closed* and *exact* (co)chains.

4.4.1. Simplicial homology.
A fundamental problem in topology is that of determining, for two spaces, whether they are topologically equivalent. That is, we wish to know if one space can be morphed into the other without having to puncture it. For instance, a sphere-shaped tet mesh is not topologically equivalent to a torus-shaped tet mesh as one cannot alter the sphere-shaped mesh (i.e., deform, refine, or coarsen it locally) to make it look like a torus.

The key idea of homology is to define *invariants* (i.e., quantities that cannot change by continuous deformation) that characterize topological spaces. The simplest invariant is the number of connected components that a simplicial complex has: obviously, two simplicial complexes with different numbers of pieces cannot be continuously deformed into each other! Roughly speaking, homology groups are an extension of this idea to define more subtle invariants than the number of connected components. In general, one can say that homology is a way to *define* the notion of holes/voids/tunnels/components of an object in any dimension.

Cycles and their equivalence classes. Generalizing the previous example to other invariants is elegantly done using the notion of *cycles*. A cycle is simply a *closed k-chain*; that is, a linear combination of k-simplices so that the boundary of this chain (see Section 3.2) is the empty set. Any set of vertices is a closed chain; any set of 1D loops are too; etc. Equivalently, a k-cycle is any k-chain that belongs to Ker ∂_k, by definition.

On this set of all k-cycles, one can define *equivalence classes*. We will say that a k-cycle is *homologous* to another k-cycle (i.e., in the same equivalence class as the other) when these two chains differ by a boundary of a $(k + 1)$-chain (i.e., by an *exact chain*). Notice that this exact chain is, by definition (see Section 4.2), in the image of ∂_{k+1}, i.e., Im ∂_{k+1}. To get a better understanding of this notion of equivalence class, the reader is invited to look at Figure 11: the 1-chains L_1 and L_3 are part of the same equivalence class as their difference is indeed the boundary of a well-defined 2D chain— a rubber-band shape in this case. Notice that as a consequence, L_1 can be deformed into

L_3 without having to tear the loop apart. However, L_2 is not of the class, and thus cannot be deformed into L_3; there's no 2-chain that corresponds to their difference.

4.4.2. Homology groups.
Let us now use these definitions in the simple case of the 0th homology group $\mathcal{H}_0$.

Homology group $\mathcal{H}_0$. The boundary of any vertex is $\emptyset$. Thus, any linear combination of vertices is a 0-cycle by definition. Now if two vertices v_0 and v_1 are connected by an edge, $v_1 - v_0$ (i.e., the difference of two cycles) is the boundary of this edge. Thus, by our previous definition, two vertices linked by an edge are *homologous* as their difference *is* the boundary of this edge. By the same reasoning, any two vertices taken from the *same* connected component are, also, homologous, since there exists a chain of edges in between. Consequently, we can pick only one vertex per connected component to form a basis of this homology group. Its dimension, β_0, is therefore simply the number of connected components. The basis elements of that group are called *generators*, since they generate the whole homology group.

Homology group $\mathcal{H}_1$. Let us proceed similarly for the 1st homology class: we now have to consider 1-cycles (linear combinations of 1D loops). Again, one can easily conceive that there are different *types* of such cycles, and it is therefore possible to separate all possible cycles into different equivalence classes. For instance, the loop L_1 in Figure 11 is topologically distinct from the curve L_2: one is around a hole and the other is not, so the difference between the two is *not* the boundary of a 2-chain. Conversely, L_1 *is* in the same class as curve L_3 since they differ by one connected area. Thus, in this figure, the 1st homology group is a 1-dimensional group, and L_1 (or L_3, equivalently) is its unique generator. The reader is invited to apply this simple idea on a triangulated torus, to find two loops as generators of $\mathcal{H}_1$.

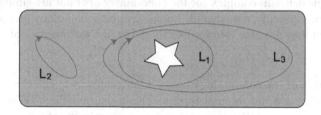

FIGURE 11. Example of homology classes: the cycles L_1 and L_2 are topologically distinct as one encloses a hole while the other does not; L_1 and L_3 are however in the same equivalence class.

Formal definition of homology groups. We are now ready to generalize this construction to all homology groups. Remember that we have a series of k-chain spaces:

$$C_n \xrightarrow{\partial_n} C_{n-1} \cdots \xrightarrow{\partial_2} C_1 \xrightarrow{\partial_1} C_0$$

with the property that $\partial\,\partial$ is the empty set. This directly implies that the image of $\mathcal{C}_j$ is always in the kernel of ∂_{j+1}—such a series is called a *chain complex*. Now, the homology groups $\{\mathcal{H}_k\}_{k=0,\ldots,n}$ of a chain complex based on ∂ are defined as the following quotient spaces:

$$\mathcal{H}_k = \text{Ker } \partial_k / \text{Im } \partial_{k+1}.$$

The reader is invited to check that this definition is *exactly* what we did for the 0th and 1st homology groups—and it is now valid for any order: indeed, we use the fact that closed chains (belonging to Ker ∂) are homologous if and only if their difference is in Im ∂, and this is exactly what this quotient vector space is.

Example. Consider the example in Figure 10(a). Geometrically, $\mathcal{H}_0$ is nontrivial because the simplicial complex σ is disconnected (it is easy to see that $\{v_0, v_4\}$ forms a basis for $\mathcal{H}_0$), while $\mathcal{H}_1$ is nontrivial since the cycle $(e_1 - e_2 + e_4)$ is not the boundary of any 2-chain of σ ($\{(e_1 - e_2 + e_4)\}$ is indeed a basis for this 1D space $\mathcal{H}_1$).

Link to Betti numbers. The dimension of the k-th cohomology group is called k-th Betti number; $\beta_k = \dim\mathcal{H}_k$. For a 3D simplicial complex embedded in $\mathbb{R}^3$, these numbers have very straightforward meanings. β_0 is the number of connected components, β_1 is the number of tunnels, β_2 is the number of voids, while β_3 is the number of 4D holes, which is 0 in the Euclidean (flat 3D) case. Finally, note that $\sum_{k=0,\ldots,n}(-1)^k \beta_k$, where β_k is the k-th Betti number, gives us the well-known Euler characteristic.

4.4.3. Cohomology groups. The definition of homology groups is much more general than what we just reviewed. In fact, the reader can take the formal definition in the previous section, replace all occurrences of chain by cochain, of ∂ by d, and reverse the direction of the operator between spaces (see Section 4.3.2): this will also define equivalence classes. Because cochains are duals of chains, and d is the adjoint of ∂, these equivalence classes define what is actually denoted as *cohomology* groups: the cohomology groups of the deRham complex for the coboundary operator are simply the quotient spaces Ker $d/\text{Im } d$. Finally, note that the homology and cohomology groups are not only dual notions, but they are also isomorphic; therefore, the cardinalities of their bases are equal.

4.4.4. Calculation of the cohomology basis. One usual way to calculate a cohomology basis is to calculate a Smith normal form to obtain the homology basis first (possibly using progressive meshes [19]), with a worst-case complexity of $O(n^3)$, and then find the corresponding cohomology basis derived from this homology basis. We provide an alternative method here with worst-case complexity also equal to $O(n^3)$. The advantage of our method is that it directly calculates the cohomology basis.

Our algorithm is a modified version of an algorithm in [11], although it was not used for the same purpose[2]. We will use row#(.) to refer to the row number of the *last nonzero coefficient* in a particular column.

The procedure is as follows:

1. Transform d^k (size $|\mathcal{K}^{k+1}| \times |\mathcal{K}^k|$) in the following manner:

---

[2]Thanks to David Cohen-Steiner for pointing us to the similarities.

```
// For each column of d^k
for (i = 0; i < |σ^k|; i++)
    // Reduce column i
    repeat
        p ← row#(d^k[i])
        find j < i such that p==row#(d^k[j])
        make d^k[i][p] zero by adding to d^k[i] a multiple of d^k[j]
    until j not_found or column i is all zeros
```

At the end of this procedure, we get $D^k = d^k N^k$, whose nonzero column vectors are linearly independent of each other and with different row#(.), and N^k is a nonsingular upper triangular matrix.

2. Construct $K^k = \{N_i^k \mid D_i^k = 0\}$ (where N_i^k and D_i^k are column vectors of matrices N^k and D^k, respectively).
   K^k is a basis for kernel of d^k.
3. Construct $I^k = \{N_i^k \mid \exists j \text{ such that } i = \text{row\#}(D_j^{k-1})\}$
4. Construct $P^k = K^k - I^k$
   P^k is a basis for the cohomology.

Short proof of correctness. First, notice that the N_i^k's are all linearly independent because N^k is nonsingular. For any nonzero linear combination of vectors in P^k, row#(.) of it (say i) equals the max of row#(.) of vectors with nonzero coefficients. But i is not row#(.) of any D_i^{k-1} (and thus any linear combination of them) by definition of P^k. Therefore, we know that the linear combination is not in the image space of d^{k-1} (since the range of d^{k-1} is the same as D^{k-1}, by construction). Thus, P^k spans a subspace of $\text{Ker}(d^k)/\text{Im}(d^{k-1})$ of dimension $\text{Card}(P^k)$.

One can also prove that I^k is a subset of K^k. Pick an N_i^k with $i = \text{row\#}(D_j^{k-1})$. We have: $d^k D_j^{k-1} = 0$ (since $d^k \circ d^{k-1} = 0$). Now row#$(\tau \equiv (N^k)^{-1} d^{k-1}{}_j) = i$ (the inverse of an upper triangular matrix is also an upper triangular matrix). So, consequently, $0 = d^k d^{k-1}{}_j = D^k (N^k)^{-1} d^{k-1}{}_j = D^k \tau$ means that $D_i^k = 0$ because the columns of D^k are linearly independent or 0. Therefore, $\text{Card}(P^k) = \text{Card}(K^k) - \text{Card}(I^k) = \text{Dim}(\text{Ker}(d^k)) - \text{Dim}(\text{Im}(d^{k-1}))$, and we conclude that, P^k spans $\text{Ker}(d^k)/\text{Im}(d^{k-1})$ as expected. □

4.4.5. Example. Consider the 2D simplicial complex in Figure 10 (left) again. We will show an example of running the same procedure described above to compute a homology basis. The only difference to the previous algorithm is that we use ∂ instead of d, since we compute the homology basis instead of the cohomology basis.

1. Compute the $D^K = \partial_k N^k$'s and N^k's: D^2 is trivial, as it is the same as ∂_2.

$$D^1 = \begin{pmatrix} -1 & 0 & 0 & 0 & 0 \\ 1 & -1 & 0 & 0 & 0 \\ 0 & 1 & 1 & 0 & 0 \\ 0 & 0 & -1 & 0 & 0 \\ 0 & 0 & 0 & 0 & 0 \end{pmatrix}, \quad N^1 = \begin{pmatrix} 1 & 0 & 0 & -1 & 0 \\ 0 & 1 & 0 & -1 & 1 \\ 0 & 0 & 1 & 1 & -1 \\ 0 & 0 & 0 & 1 & 0 \\ 0 & 0 & 0 & 0 & 1 \end{pmatrix}.$$

2. Construct the K's:

$$K^0 = \left\{ \begin{pmatrix} 1 \\ 0 \\ 0 \\ 0 \\ 0 \end{pmatrix}, \begin{pmatrix} 0 \\ 1 \\ 0 \\ 0 \\ 0 \end{pmatrix}, \begin{pmatrix} 0 \\ 0 \\ 1 \\ 0 \\ 0 \end{pmatrix}, \begin{pmatrix} 0 \\ 0 \\ 0 \\ 1 \\ 0 \end{pmatrix}, \begin{pmatrix} 0 \\ 0 \\ 0 \\ 0 \\ 1 \end{pmatrix} \right\}$$

$$= \{v_0, v_1, v_2, v_3, v_4\},$$

(N^0 is the identity)

$$K^1 = \left\{ \begin{pmatrix} -1 \\ -1 \\ 1 \\ 1 \\ 0 \end{pmatrix}, \begin{pmatrix} 0 \\ 1 \\ -1 \\ 0 \\ 1 \end{pmatrix} \right\} = \{(-e_0 - e_1 + e_2 + e_3), (e_1 - e_2 + e_4)\}.$$

3. Construct the I's:

$$I^0 = \{v_1 \ (1 = \text{row\#}(D_0^1)),$$
$$v_2 \ (2 = \text{row\#}(D_1^1)),$$
$$v_3 \ (3 = \text{row\#}(D_2^1))\},$$
$$I^1 = \{(e_1 - e_2 + e_4) \ (4 = \text{row\#}(D_0^2))\}.$$

4. Consequently, the homology basis is:

$$P^0 = \{v_0, v_1, v_2, v_3, v_4\} - \{v_1, v_2, v_3\} = \{v_0, v_4\},$$
$$P^1 = \{(-e_0 - e_1 + e_2 + e_3)\}.$$

This result confirms the basis we gave in the example of Section 4.4.2. (Note that $-(-e_0 - e_1 + e_2 + e_3) - (e_1 - e_2 + e_4) = e_0 - e_4 - e_3 = \partial f_0$, thus $(-e_0 - e_1 + e_2 + e_3)$ spans the same homology space as $(e_1 - e_2 + e_4)$.)

4.5. Dual mesh and its exterior derivative

Let us introduce the notion of a *dual mesh* of triangulated manifolds, as we will see that it is one of the key components of our discrete calculus. The main idea is to associate to each *primal k-simplex* a dual $(n - k)$-cell. For example, consider the tetrahedral mesh in Figure 13 below, we associate a dual 3-cell to each primal vertex (0-simplex), a dual polygon (2-cell) to each primal edge (1-simplex), a dual edge (1-cell) to each primal face (2-simplex), and a dual vertex (0-cell) to the primal tet (3-simplex). By construction, the number of dual $(n - k)$-cells is equal to that of primal k-simplices. The collection of dual cells is called a *cell complex*, which need not be a simplicial complex in general.

Yet, this dual complex inherits several properties and operations from the primal simplicial complex. Most important is the notion of *incidence*. For instance, if two primal edges are on the same primal face, then the corresponding dual faces are incident, that is, they share a common dual edge (which is the dual of the primal common face). As a result of this incidence property, one may easily derive a boundary operator on the dual cell complex and, consequently, a discrete exterior derivative! The reader is invited to verify that this exterior derivative on the dual mesh can be simply written as the opposite of a primal one *transposed*:

$$d_{\text{Dual}}^{n-k} = (-1)^k (d_{\text{Primal}}^{k-1})^t. \tag{4.2}$$

The added negative sign appears as the orientation on the dual is induced from the primal orientation, and must therefore be properly accounted for. Once again, an implementation can overload the definition of this operator d when used on dual forms using this previous equation. In the remainder of our chapter, we will be using d as a contextual operator to keep the notation a simple as possible. Because we have defined a proper exterior derivative on the dual mesh (still satisfying $d \circ d = 0$), this dual cell complex also carries the structure of a chain complex. The structure on the dual complex may be linked to that of the primal complex using the Hodge star (a metric-dependent operator), as we will discuss in Section 5.

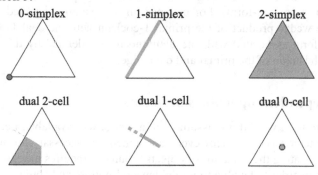

FIGURE 12. A 2-dimensional example of primal and dual mesh elements. On the top row, we see the primal mesh (a triangle) with a representative of each simplicial complex being highlighted. The bottom row shows the corresponding circumcentric dual cells (restricted to the triangle).

4.5.1. Dualization: the ∗ operator. For simplicity, we use circumcentric (or Voronoi) duality to construct the dual cell complex. The circumcenter of a k-simplex is defined as the center of the k-circumsphere, which is the unique k-sphere that has all $k + 1$ vertices of the k-simplex on its surface. In Figure 12, we show examples of circumcentric dual cells of a 2D mesh. The dual 0-cell associated with the triangular face is the circumcenter of the triangle. The dual 1-cell associated with one of the primal edges is the line segment that joins the circumcenter of the triangle to the circumcenter of that edge, while the dual 2-cell associated with a primal vertex is corner wedge made of the convex hull of the circumcenter of the triangle, the two centers of the adjacent edges, and the vertex itself (see Figure 12, bottom left). Thereafter, we will denote as ∗ the operation of *duality*; that is, a primal simplex σ will have its dual called $\ast\sigma$ with the orientation induced by the primal orientation and the manifold orientation. For a formal definition, we refer the reader to [23] for instance. It is also worth noting that other notions of duality such as the barycentric duality may be employed. For further details on dual cell (or "block") decompositions, see [33].

4.5.2. Wedge product. In the continuous setting, the wedge product ∧ is an operation used to construct higher-degree forms from lower-degree ones; it is the antisymmetric part of the tensor product. For example, let α and β be 1-forms on a subset $\mathcal{R} \subset \mathbb{R}^3$, their wedge product $\alpha \wedge \beta$ is a 2-form on $\mathcal{R}$. In this case, one can relate the wedge product to the cross product of vector fields on $\mathcal{R}$. Indeed, if one considers the vector representations of α and β, the vector proxy to $\alpha \wedge \beta$ is the cross product of the two vectors. Similarly, the wedge product of a 1-form γ with the 2-form $\omega = \alpha \wedge \beta$ is a 3-form $\mu = \gamma \wedge \omega$ (also called volume-form) on $\mathcal{R}$ which is analogous to the scalar triple product of three vectors.

A discrete treatment of the wedge operator can be found in [23]. In this work, we only need to introduce the notion of a discrete *primal-dual wedge product*: given a primal k-cochain γ and a dual $(n-k)$-cochain ω, the discrete wedge product $\gamma \wedge \omega$ is an n-form (or a volume-form). For instance, in the example depicted in the inset, the wedge product of the primal 1-cochain with the dual 1-cochain is a 2-form associated with the diamond region defined by the convex hull of the union of the primal and dual edge.

5. Metric-dependent operators on forms

Notice that up to now, we did *not* assume that a metric was available, i.e., we never required anything to be *measured*. However, such a metric is necessary for many purposes. For instance, simulating the behavior of objects around us requires measurements of various parameters in order to be able to model laws of motion, and compare the numerical results of simulations. Consequently, a certain number of operations on forms can only be defined once a metric is known, as we shall see in this section.

5.1. Notion of metric and inner product

A *metric* is, roughly speaking, a non-negative function that describes the "distance" between neighboring points of a given space. For example, the Euclidean metric assigns to any two points in the Euclidean space $\mathbb{R}^3$, say $\mathbf{X} = (x_1, x_2, x_3)$ and $\mathbf{Y} = (y_1, y_2, y_3)$, the number:

$$d(\mathbf{X}, \mathbf{Y}) = \|\mathbf{X} - \mathbf{Y}\| = \sqrt{(x_1 - y_1)^2 + (x_2 - y_2)^2 + (x_3 - y_3)^2}$$

defining the "standard" distance between any two points in $\mathbb{R}^3$. This metric then allows one to measure length, area, and volume. The Euclidean metric can be expressed as the following quadratic form:

$$g^{\text{Euclid}} = \begin{pmatrix} 1 & 0 & 0 \\ 0 & 1 & 0 \\ 0 & 0 & 1 \end{pmatrix}.$$

Indeed, the reader can readily verify that this matrix g satisfies: $d^2(\mathbf{X}, \mathbf{Y}) = (\mathbf{X} - \mathbf{Y})^t g (\mathbf{X} - \mathbf{Y})$. Notice also that this metric *induces an inner product* of vectors. Indeed, for two vectors $\mathbf{u}$ and $\mathbf{v}$, we can use the matrix g to define:

$$\mathbf{u} \cdot \mathbf{v} = \mathbf{u}^t g \, \mathbf{v}.$$

Once again, the reader is invited to verify that this equality does correspond to the traditional dot product when g is the Euclidean metric. Notice that on a non-flat manifold, subtraction of two points is only possible for points infinitesimally close to each other, thus the metric is actually defined point-wise for the tangent space at each point: it does not have to be constant. Finally, notice that a volume form can be induced from a metric by defining $\mu^n = \sqrt{\det(g)}\, dx^1 \wedge \cdots \wedge dx^n$.

5.2. Discrete metric

In the discrete setting presented in this paper, we only need to measure length, area, and volume of the simplices and dual cells (note these different notions of sizes depending on dimension will be denoted "intrinsic volumes" for generality). We therefore do not have a full-blown notion of a metric, only a *discrete metric*. Obviously, if one were to use a finer mesh, more information on the metric would be available: having more values of length, area, and volume in a neighborhood provides a better approximation of the real, continuous metric.

5.3. The differential Hodge star

Let us go back for a minute to the differential case to explain a new concept. Recall that the metric defines an inner product for vectors. This notion also extends to forms: given a metric, one can define the product of two k-forms $\in \Omega^k(\mathcal{M})$ which will measure, in a way, the projection of one onto the other. A formal definition can be found in [1]. Given this inner product denoted $\langle\ ,\ \rangle$, we can introduce an operator $\star$, called the *Hodge star*, that maps a k-form to a complementary $(n-k)$-form:

$$\star : \Omega^k(\mathcal{M}) \to \Omega^{n-k}(\mathcal{M}),$$

and is defined to satisfy the following equality:

$$\alpha \wedge \star\beta = \langle \alpha, \beta \rangle\, \mu^n$$

for any pair of k-forms α and β (recall that μ^n is the volume form induced by the metric g). However, notice that the wedge product is very special here: it is the product of k-form and a $(n-k)$ form, two complementary forms. This fact will drastically simplify the discrete counterpart of the Hodge star, as we now cover.

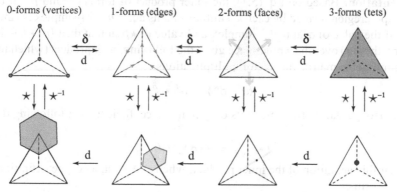

FIGURE 13. On the first line, the 'primal' chain complex is depicted and on the second line we see the dual chain complex (i.e., cells, faces, edges and vertices of the Voronoi cells of each vertex of the primal mesh).

5.4. Discrete Hodge star

In the discrete setting, the Hodge star becomes easier: we only need to define how to go from a *primal* k-cochain to a *dual* $(n - k)$-cochain, and vice-versa. By definition of the dual mesh, k-chains and dual $(n - k)$-chains are represented by vectors of the same dimension. Similarly to the discrete exterior derivative (coboundary) operator, we may use a matrix (this time of size $|\mathcal{K}^k| \times |\mathcal{K}^k|$) to represent the Hodge star. Now the question is: what should the coefficients of this matrix be?

For numerical purposes we want it to be symmetric, positive definite, and sometimes even diagonal for faster computations. One such diagonal Hodge star can be defined with the diagonal elements as the ratio of intrinsic volumes of a k-simplex and its dual $(n - k)$-simplex. In other words, we can define the discrete Hodge star through the following simple rule:

$$\frac{1}{|\sigma^k|} \int_{\sigma^k} \omega = \frac{1}{|\ast \sigma^k|} \int_{\ast \sigma^k} \star \omega. \tag{5.1}$$

Therefore, any primal value of a k-form can be easily *transferred* to the dual mesh through proper scaling—and vice-versa; to be precise, we have:

$$\star_k \star_{n-k} = (-1)^{k(n-k)} \mathrm{Id}, \tag{5.2}$$

which means that $\star$ on the dual mesh is the inverse of the $\star$ on the primal *up to a sign* (the result of antisymmetry of the wedge product, which happens to be positive for any k-form when $n = 3$).

So we must use the inverse of the Hodge star to go from a dual $(n - k)$-cochain to a k-cochain. We will, however, use $\star$ to indistinguishably mean either the star or its inverse, as there is no ambiguity once we know whether the operator is applied to a primal or a dual form: this is also a context-dependent operator.

Implementation. Based on Eq. (5.1), the inner product of forms α^k and β^k at the diamond-shaped region formed by each k-simplex and its dual $(n - k)$-simplex is simply the product of the value of α at that k-simplex and value of $\star \beta$ at that dual $(n - k)$-simplex. Therefore, the sum over the whole space gives the following inner product (which involves only linear algebra matrix and vector multiplications):

$$\langle \alpha^k, \beta^k \rangle = \alpha^t \star \beta. \tag{5.3}$$

where the Hodge star matrix has, as its only nonzero coefficients, the following diagonal terms:

$$(\star_k)_{qq} = |(\ast \sigma)_q| / |(\sigma_q)|.$$

Notice that this definition of the inner product, when $\alpha = \beta$, induces the definition of the norm of k-forms.

Again, there are three different Hodge stars in $\mathbb{R}^3$, one for each simplex dimension. But as we discussed for all the other operators, the dimension of the form on which this operator is applied disambiguates which star is meant. So we will not encumber our notation with unnecessary indices, and will only use the symbol $\star$ for any of the three stars implied.

The development of an accurate, yet fast to compute, Hodge star is still an active research topic. However, this topic is beyond the scope of the current chapter.

5.5. Discrete codifferential operator δ

We already have a linear operator d which maps a k-form to a $(k + 1)$-form, but we do not have a linear operator which maps a k-form to a $(k - 1)$-form. Having defined a discrete Hodge star, one can now create such an *adjoint* operator δ for the discrete exterior derivative d. Here, adjoint is meant with respect to the inner product of forms; that is, this operator δ satisfies:

$$\langle d\alpha, \beta \rangle = \langle \alpha, \delta\beta \rangle \quad \forall \alpha \in \Omega^{k-1}(M), \beta \in \Omega^k(M).$$

For a smooth, compact manifold without boundary, one can prove that the operator $(-1)^{n(k-1)+1} \star d \star$ satisfies the above condition [1]. Let us try to use the same definition in the discrete setting; i.e., we wish to define the discrete δ applied to k-forms by the relation:

$$\delta \equiv (-1)^{n(k-1)+1} \star d \star. \tag{5.4}$$

Beware that we use the notation d to mean the context-dependent exterior derivative. If you apply δ to a primal k-form, then the exterior derivative will be applied to a *dual $(n - k)$-form*, and thus, Equation 4.2 should be used. Once this is well understood, it is quite straightforward to verify the following series of equalities:

$$\langle d\alpha, \beta \rangle \overset{(5.3)}{=} (d\alpha)^t \star \beta = \alpha^t d^t \star \beta$$

$$\overset{(5.2)}{=} \alpha^t (-1)^{(k-1)(n-(k-1))} \star \star d^t \star \beta$$

$$= \alpha^t (-1)^{n(k-1)+1} \star \star (-1)^k d^t \star \beta$$

$$\overset{(5.4)}{=} \langle \alpha, \delta\beta \rangle$$

holds on our discrete manifold. So, indeed, the discrete d and δ are also adjoint, in a similar fashion in the discrete setting as they were in the continuous sense. For this reason, δ is called the *codifferential operator*.

Implementation of the codifferential operator. Thanks to this easily-proven adjointness, the implementation of the discrete codifferential operator is a trivial matter: it is simply the product of three matrices, mimicking exactly the differential definition mentioned in Eq. (5.4).

5.6. Exercise: Laplacian

At this point, the reader is invited to perform a little exercise. Let us first state that the Laplacian Δ of a form is defined as: $\Delta = \delta d + d\delta$. Now, applied to a 0-form, notice that the latter term disappears. Question: in 2D, what *is* the Laplacian of a function f at a vertex i? The answer is actually known: it is the now famous *cotangent formula* [34], since the ratio of primal and dual edge lengths leads to such a trigonometric equality.

6. Interpolation of discrete forms

In Section 3.4, we argued that k-cochains are discretizations of k-forms. This representation of discrete forms on chains, although very convenient in many applications, is not sufficient to fulfill certain demands such as obtaining a point-wise value of the k-form. As a remedy, one can use an interpolation of these chains to the rest of space. For simplicity, these interpolation functions can be taken to be linear (by linear, we mean with respect to the coordinates of the vertices).

6.1. Interpolating 0-forms

It is quite obvious how to linearly interpolate discrete 0-forms (as 0-cochains) to the whole space: we can use the usual vertex-based linear interpolation basis, often referred to as the *hat function* in the finite-element literature. This basis function will be denoted as φ_i for each vertex v_i. By definition, φ_i satisfies:

$$\varphi_i = 1 \quad \text{at } v_i, \qquad \varphi_i = 0 \quad \text{at } v_j \neq v_i$$

while φ_i linearly goes to zero in the one-ring neighborhood of v_i. The reader may be aware that these functions are, within each simplex, *barycentric coordinates*, introduced by Möbius in 1827 as mass points to define a *coordinate-free geometry*.

With these basis functions, one can easily check that if we denote a vertex v_j by σ_j, we have:

$$\int_{v_j} \varphi_{v_i} = \int_{\sigma_j} \varphi_{\sigma_i} = \int_{\sigma_j} \varphi_i = \begin{cases} 1 & \text{if } i = j, \\ 0 & \text{if } i \neq j. \end{cases}$$

Therefore, these interpolating functions represent a basis of 0-cochains, that exactly corresponds to the dual of the natural basis of 0-chains.

6.2. Interpolating 1-forms

We would like to be able to extend the previous interpolation technique to 1-forms now. Fortunately, there is an existing method to do just that: the *Whitney* 1-form (used first in [39]) associated with an edge σ_{ij} between v_i and v_j is defined as:

$$\varphi_{\sigma_{ij}} = \varphi_i \, d\varphi_j - \varphi_j \, d\varphi_i.$$

A direct computation can verify that:

$$\int_{\sigma_{kl}} \varphi_{\sigma_{ij}} = \begin{cases} 1 & \text{if } i = k \text{ and } j = l, \\ -1 & \text{if } i = l \text{ and } j = k, \\ 0 & \text{otherwise.} \end{cases}$$

Indeed, it is easy to see that the integral is 0 when we are not integrating it on edge e_{ij}, because at least one of the vertices (say, i) is not on the edge, thus, $\varphi_i = 0$ and $d\varphi_i = 0$ on the edge. However, along the edge σ_{ij}, we have $\varphi_i + \varphi_j = 1$, therefore:

$$\int_{\sigma_{ij}} \varphi_{\sigma_{ij}} = \int_{\varphi_i=1}^{\varphi_i=0} (\varphi_i d(1 - \varphi_i) - (1 - \varphi_i)d\varphi_i) = \int_{\varphi_i=1}^{\varphi_i=0} (-d\varphi_i) = 1.$$

We thus have defined a correct basis for 1-cochains.

6.3. Interpolating with Whitney k-forms

One can extend these 1-form basis functions to arbitrary k-simplices. In fact, Whitney k-forms are defined similarly:

$$\varphi_{\sigma_{i_0,i_1,\ldots,i_k}} = k! \sum_{j=0,\ldots,k} (-1)^j \varphi_{i_j} d\varphi_{i_0} \wedge \cdots \wedge \widehat{d\varphi_{i_j}} \cdots \wedge d\varphi_{i_k}$$

where $\widehat{d\varphi_{i_p}}$ means that $d\varphi_{i_p}$ is excluded from the product. Notice how this definition exactly matches the case of vertex and edge bases, and extends easily to higher-dimensional simplices.

Remark. If a metric is defined (for instance, the Euclidean metric), we can simply identify $d\varphi$ with $\nabla\varphi$ for the real calculation. This corresponds to the notion of *sharp* ($\sharp$), but we will not develop this point other than for pointing out the following remark: the traditional gradient of a linear function f in 2D, known to be constant per triangle, can indeed be re-written à la Whitney:

$$\nabla f = \sum_i f_i \nabla\varphi_i \overset{\varphi_i+\varphi_j+\varphi_k=1}{=} \sum_{i,j,i\neq j} (f_j - f_i)(\varphi_i \nabla\varphi_j - \varphi_j \nabla\varphi_i).$$

The values $f_j - f_i$ are the edge values associated with the gradient, i.e., the values of the one-form df.

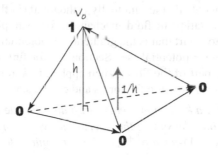

FIGURE 14. $\nabla\varphi$ for the vertex on top.

Basis of forms. The integration of the Whitney form φ_{σ_k} associated with the k-simplex σ_k will be 1 on that particular simplex, and 0 on all others. Indeed, it is a simple exercise to see that the integration of φ_{σ_k} is 0 on a different k-simplex, because there is *at least one* vertex of this simplex v_j that does not belong to σ_k, so its hat function φ_j is valued 0 everywhere on σ_k. Since φ_j or $d\varphi_j$ appears in every term, the integral of φ_{σ_k} is 0. To see that the integral is 1 on the simplex itself, we can use Stokes's theorem (as our discrete forms satisfy it exactly on simplices): first, suppose $k < n$, and pick a $(k+1)$-simplex, such that the k-simplex σ_k is a *face* of it. Since it is 0 on other faces, the integral of the Whitney form is equal to the integral of $d\varphi_{\sigma_k} = (k+1)! d\varphi_{i_0} \wedge \cdots \wedge d\varphi_{i_k}$ on the $(k+1)$-simplex, if we use φ_{i_j} as a local reference frame for the integration, $\int_{\sigma_{k+1}} d\varphi_{i_0} \wedge \cdots \wedge d\varphi_{i_k}$ is simply the volume of a standard simplex, which is $\frac{1}{(k+1)!}$, thus the integral is 1. The case when $k = n$ is essentially the same as $k = n - 1$.

This means that these Whitney forms are forming a basis of their respective form spaces. In a way, these bases are an extension of the finite-element bases defined on nodes, or of the finite-volume elements that are constant per tet.

Note finally that the Whitney forms are not continuous; however, they *are* continuous *along the direction of the k-simplex* (i.e., tangential continuity for 1-forms, and normal continuity for 2-forms); this is the only condition needed to make the integration well defined. In a way, this property is the *least* we can ask them to be. We would lose generality if we were to add any other condition! The interested reader is referred to [4] for a more thorough discussion on these Whitney bases and their relations to the notion of weak form used in the finite-element method.

7. Application to Hodge decomposition

We now go through a first application of the discrete exterior calculus we have defined up to now. As we will see, the discrete case is often much simpler than its continuous counterpart; yet it captures the same properties.

7.1. Introducing the Hodge decomposition

It is convenient in some applications to use the Helmholtz-Hodge decomposition theorem to decompose a given continuous vector field or differential form (defined on a smooth manifold $\mathcal{M}$) into components that are mutually orthogonal (in the $\mathcal{L}^2$ sense), and easier to compute (see [1] for details). In fluid mechanics, for example, the velocity field is generally decomposed into a part that is the gradient of a potential function and a part that is the curl of a stream vector potential (see Section 8.3 for further details), as the latter one is the incompressible part of the flow. When applied to k-forms, this decomposition is known as the *Hodge decomposition* for forms and can be stated as follows:

> Given a manifold $\mathcal{M}$ and a k-form ω^k on $\mathcal{M}$ with appropriate boundary conditions, ω^k can be decomposed into the sum of the exterior derivative of a $(k-1)$-form α^{k-1}, the codifferential of a $(k+1)$-form β^{k+1}, and a harmonic k-form h^k:
>
> $$\omega^k = d\alpha^{k-1} + \delta\beta^{k+1} + h^k.$$

Here, we use the term *harmonic* to mean that h^k satisfies the equation $\Delta h^k = 0$, where Δ is the Laplacian operator defined as $\Delta = d\delta + \delta d$. The proof of this theorem is mathematically involved and requires the use of elliptic operator theory and similar tools, as well as a careful study of the boundary conditions to ensure uniqueness. The discrete analog that we propose has a very simple and straightforward proof as shown below.

7.2. Discrete Hodge decomposition

In the discrete setting, the discrete operators such as the exterior derivative and the co-differential can be expressed using matrix representation. This allows one to easily manipulate these operators using tools from linear algebra. In particular, the discrete version of the Hodge decomposition theorem becomes a simple exercise in linear algebra. Note that we will assume a boundary-less domain for simplicity (the generalization to domains with boundary is conceptually as simple).

Theorem 7.1. *Let $\mathcal{K}$ be a discrete manifold and let $\Omega^k(\mathcal{K})$ be the space of discrete Whitney k-forms on $\mathcal{K}$. Consider the linear operator $d^k : W^k \to W^{k+1}$, such that $d^{k+1} \circ d^k = 0$, and a discrete Hodge star which is represented as a symmetric, positive definite matrix. Furthermore, define the codifferential (the adjoint of the operator d) as done in Section 5.5; namely, let $\delta^{k+1} = (-1)^{n(k-1)+1}(\star^k)^{-1}(d^k)^t \star^{k+1}$. In this case, the following orthogonal decomposition holds for all k:*

$$\Omega^k(\mathcal{K}) = d\,\Omega^{k-1}(\mathcal{K}) \oplus \delta\Omega^{k+1}(\mathcal{K}) \oplus \mathcal{H}^k(\mathcal{K})$$

where $\oplus$ means orthogonal sum, and $\mathcal{H}^k(\mathcal{K})$ is the space of harmonic k-forms on $\mathcal{K}$, that is, $\mathcal{H}^k(\mathcal{K}) = \{h \mid \Delta^k h = 0\}$.

Proof. For notational convenience, we will omit the superscript of the operators when the rank is obvious. We first prove that the three component spaces are orthogonal. Clearly, using the facts that the Laplacian operator Δ is equal to $d\delta + \delta d$ and that d and δ are adjoint operators, one has that for all $h \in \mathcal{H}^k$:

$$\langle \Delta h, h \rangle = 0 \Rightarrow \langle d\delta h, h \rangle + \langle \delta dh, h \rangle = \langle dh, dh \rangle + \langle \delta h, \delta h \rangle = 0$$

$$\Rightarrow dh = 0 \text{ and } \delta h = 0.$$

Also, for all $\alpha \in \Omega^{k-1}(\mathcal{K})$ and $\beta \in \Omega^{k+1}(\mathcal{K})$, one has:

$$\langle d\alpha, \delta\beta \rangle = \langle dd\alpha, \beta \rangle = 0$$

and

$$\langle d\alpha, h \rangle = \langle \alpha, \delta h \rangle = 0 \qquad \langle h, \delta\beta \rangle = \langle dh, \beta \rangle = 0.$$

Now, any k-form that is perpendicular to $d\Omega^{k-1}(\mathcal{K})$ and $\delta\Omega^{k+1}(\mathcal{K})$ must be in $\mathcal{H}^k(\mathcal{K})$, because this means $dh = 0$ and $\delta h = 0$, so $\Delta h = d\delta h + \delta dh = 0$.

Alternatively, we can prove that:

$$\Omega^k(\mathcal{K}) = \Delta\Omega^k(\mathcal{K}) \oplus \mathcal{H}^k(\mathcal{K}).$$

By analogy to the previous argument, it is easy to show that $\Delta\Omega^k$ is orthogonal to $\mathcal{H}^k$. Additionally, the dimension of these two spaces sum up to the dimension of Ω^k, which means the decomposition is complete. $\qquad\square$

Note that the reader can find a similar proof given in Appendix B of [16], where it is used for Kirchhoff's circuit laws. There, Frankel does not mention that we can actually use cochains as the discretization of forms, and his operations using a "metric" of cochains can be interpreted as a Hodge star.

Implementation of the discrete Hodge decomposition. Before we discuss how to numerically implement the discrete Hodge decomposition, we prove a useful result (that has a continuous analog).

Lemma 7.2. *In the discrete setting, one can find exactly one harmonic cochain from each cohomology equivalence class.*

Proof. It can be readily shown that the bases of harmonic cochains and the cohomology groups both have the dimension equal to dim(Ker d^k) − dim(Im d^{k-1}). To this end, recall that a cohomology basis is defined as is Ker(d^k)/Im(d^{k-1}) and has dimension dim(Ker d^k) − dim(Im d^{k-1}). Now, in order to see that the space of harmonic cochains has this same dimension, simply note that: Ker(d^k) = $d\Omega^{k-1} \oplus \mathcal{H}^k$.

Now, the equation $\delta(\omega + df) = 0$ has a solution for each ω in one cohomology equivalence class. We know that the cochains forming different cohomology groups are linearly independent, hence, we conclude that these harmonic cochains span $\mathcal{H}^k$.   □

By virtue of the above lemma, the implementation of the Hodge decomposition is simply recursive in the rank of the form (i.e., cochain). The case of 0-forms is trivial: fix one vertex to a constant, and solve the Poisson equation for 0-forms. Now suppose that we have a decomposition working for $(k-1)$-forms, and we look for the decomposition of k-forms. Our approach is to get the harmonic component h^k first, so that we only need to solve a Poisson equation for the rest:

$$\Delta\omega^k = f^k - h^k. \tag{7.1}$$

One is left with the problem of finding a basis of harmonic forms. Since we are given a Hodge star operator, we will use it to define the metric on the space of cochains. This metric allows us to define a basis for harmonic k-forms (the dimension of this harmonic space is generally small, since it is the k-th Betti number β_k). First, one needs to calculate the cohomology basis $\{P_i\}$ based on the algorithm in Section 4.4.4. Once we have $\{P_i\}$, we solve one special decomposition of $(k-1)$-forms by first computing the forms f_i satisfying

$$\Delta f_i = -\delta P_i. \tag{7.2}$$

Now $H^k = P_i + df_i$ gives us the forms in basis for harmonic k-form space. After normalization, we have the basis to calculate the projection $h^k = HH^t f^k$, where we assemble all H^k into a matrix H. This completes the procedure of calculating the decomposition.

A nonsingular matrix is often preferable when it comes to solve a linear system efficiently; we can change the Laplacian matrix slightly to make the Poisson equation satisfy this requirement. First, we can get an orthonormal basis for the space of harmonic forms (the dimension is β^k). Now for the basis e^j (column vector with j-th element equal to 1, and 0 everywhere else), take the distance of e^j to the harmonic space $|e^j - HH^t e^j|$; notice that this can be done in constant time. Now take out the j-th column and j-th row of Δ if e^j has the smallest distance from harmonic space, and repeat the step for β^k times. We are left with a nonsingular matrix, and the solution to the new linear system is a solution to the original Poisson equation.

8. Other applications

8.1. Form-based proof of Tutte's theorem

The notion of forms as convenient, intrinsic substitutes for vector fields has been used to provide a concise proof of the celebrated *Tutte's embedding theorem*. This important result in graph theory states that if one fixes the boundary of a 3-connected graph (i.e., a typical polygonal mesh) to a convex domain in the plane and ensures that every non-boundary vertex is a *strict convex combination* of its neighbors, then one obtains a planar straight-line embedding of the graph. In other words, this embedding procedure will not result in fold-overs. A significantly shorter alternative to the original proof of this theorem was proposed by Gortler, Gotsman, and Thurston [17], using discrete 1-forms on edges. We now present a sketch of their approach, using a formulation more in line with the terms we used in this paper.

A Tutte embedding assigns to each vertex v_i of a graph G some 2D coordinates $\mathbf{X}(v_i) = (x(v_i), y(v_i))$. By definition, each interior vertex v_i satisfies a linear condition on its coordinates of the form: $\mathbf{X}(v_i) = \sum_{v_j \in \mathcal{N}(i)} w_{ij} \mathbf{X}(v_j)$, where $\mathcal{N}(i)$ is the set of 1-ring neighbors of vertex v_i. These coefficients w_{ij} are all non-negative due to the condition of strict convex combination mentioned above. Now, for a given Tutte embedding, one can construct a 0-form $z(v) = \alpha x(v) + \beta y(v)$ for any pair of positive coefficients α and β. Notice that this 0-form satisfies the same convex combination condition: $z(v_i) = \sum_{v_j \in \mathcal{N}(i)} w_{ij} z(v_j)$. As they are non-negative, one can identify these coefficients w_{ij} with the diagonal Hodge star of primal 1-forms (see Section 7) defined by a particular metric. Therefore, the relationship $0 = \sum_{v_j \in \mathcal{N}(i)} w_{ij}(z(v_j) - z(v_i))$ is equivalent to $d \star dz = 0$. There are two immediate conclusions:

- the 1-form $\omega = dz$ is closed (since it is the exterior derivative of a 0-form), and

- it is also coclosed since $\delta\omega = (\star d \star)dz = \star(d \star dz) = 0$.

To use the previously defined 1-form ω to prove Tutte's theorem, Gortler *et al.* then invoke the usual definition of index of vector fields, *i.e.*, the number of revolutions that the direction of the vector fields does along any small curve around this vertex. This concept is one of the oldest in algebraic topology, initially stated by Poincaré and then developed by Hopf and Morse in the continuous case. Its discrete counterpart was first proposed by Banchoff, and used for instance in [26]. A discrete Poincare-Hopf index theorem also holds, stating that the sum of all indices must be equal to 2 for a genus-0 patch. The final argument uses the link between (co)closed forms and their indices. Indeed, because we found a closed *and* coclosed form ω, it can be easily shown that these two properties induce that the index of each face must be less or equal to zero, as well as the index of each vertex. Because the boundary of the patch is convex, only two vertices on the boundary have index 1. Since all the indices must sum to 2 and each interior index must be less than zero, we can conclude that *each interior index is zero*. Because this argument is valid for every positive pair (α, β), one can easily deduce that each interior face is convex and each vertex is a "wheel"; thus, injectivity can be guaranteed.

This rather elegant proof demonstrates how discrete forms and their obvious links to algebraic topology can be quite powerful in a variety of applications. We also point the interested reader to other papers, such as [30, 19], for which special discrete Hodge stars are defined to satisfy a discrete definition of conformality: there are also very interesting research on this particular topic, once again using the calculus of exterior forms.

8.2. Electromagnetism with forms

Electromagnetism can be formulated very elegantly using differential forms. For a detailed exposition of the geometric structure in E&M, we refer the reader to [4] and [38]. In this approach, the electric field E is represented by a 1-form as the integral of E along a path traced by a test charge q, and is equal to the electromotive force experienced by that charge. The electric displacement L as well as the current density J are represented by 2-forms. The charge distribution ρ is a 3-form. The magnetic field B is represented by a 2-form since it is measured as a flux, whereas the magnetic field intensity H is a 1-form.

With these conventions, Maxwell's equations can be rewritten as follows:

$$\partial_t B + dE = 0, \qquad -\partial_t L + dH = J, \qquad dL = \rho, \qquad (8.1)$$

subject to the *constitutive* equations:

$$L = \epsilon E, \qquad H = \mu B, \qquad (8.2)$$

where ϵ is the permittivity, and μ is the permeability. The constitutive relations (8.2) are very similar to the Hodge star operator that transforms a k-form to an $(n-k)$-form. Here, ϵ operates on the electric field E (1-form) to yield the electric displacement L (2-form) while μ transforms the magnetic field B (2-form) into the magnetic field intensity H (1-form). To this end, one may think of both ϵ and μ as Hodge star operators induced from appropriately chosen metrics. Note that the balance laws in (8.1) are metric-independent.

As the reader can guess, one can readily discretize this representation of the physical quantities $E, L, \ldots$ and the associated system of equations (8.1-8.2) using the tools presented in this chapter. The resulting numerical algorithm preserves *exactly* the geometric structure of the system, see [4].

8.3. Fluids

The geometric structure of fluid mechanics, specifically Euler's equations for inviscid fluids, has been investigated (see [28] and references therein). In this geometric framework, vorticity is represented as a two-form (an area-form) and Euler's equations can be written as vorticity advection. Roughly speaking, vorticity measures the rotation of a fluid parcel; we say the fluid parcel has vorticity when it spins as it moves along its path. Vorticity advection means that the vorticity (as a two-form) moves dynamically as if it is pushed forward by the fluid flow. The integral of the vorticity on a given bounded domain is equal, by Stokes's theorem, to the circulation around the loop enclosing the domain. This quantity, as the loop is advected by the fluid, is conserved in the absence of external forcing, as well as the total energy of the fluid. Inspired by this geometric viewpoint and in light of the present development of discrete exterior calculus, we have proposed a discrete differential approach to fluid mechanics and an integration scheme that satisfy the properties of conservation of circulation, see [12] for further details.

9. Conclusions

In this contribution, we have provided an introduction to discrete differential forms and explained how they can be extremely useful in computational science. A convenient discrete exterior calculus solely based on values stored on a discrete manifold has been derived. In the common 3D case, this calculus for scalar and vector fields can be summarized by the following schematic graph:

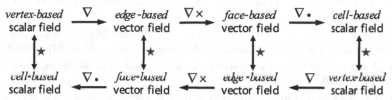

We have also given a discrete version of the Hodge decomposition, useful for a number of computations in various fields. This geometric approach to computations is particularly novel, thus many details need to be explored and proven superior to the current approaches. In order to work towards this goal, more work needs to be done to further demonstrate that this idea of forms as fundamental readily-discretizable elements of differential equations can be successfully used in various other contexts where predictive power is crucial.

Further Reading

Despite a large number of theoretical books, we are aware of only a few books with a truly "applied flavor" in line with this chapter. For applications based on this exterior calculus or other geometric algebras, see [4, 14, 3, 10, 18, 35]. The reader interested in the application of differential forms to E&M is further referred to [38], for applications in fluid mechanics see [28], and in elasticity see [25] and [16]. The reader is also invited to check out current developments of variants of DEC, for instance, in [9, 36, 40, 20].

Acknowledgments

The authors wish to first thank Jerrold E. Marsden for his constant support along the way. We are also indebted to Peter Schröder and Herbert Edelsbrunner for their thorough proofreading of this document. Bugs in an initial version were reported by Fengtao Fan. Finally, we acknowledge the support of Alain Bossavit, Anil Hirani, Melvin Leok, David Cohen-Steiner, Sharif Elcott, Pierre Alliez, and Eitan Grinspun.

References

[1] R. Abraham, J.E. Marsden, and T. Ratiu (eds.), *Manifolds, Tensor Analysis, and Applications*, Applied Mathematical Sciences Vol. 75, Springer, 1988.

[2] Anders Björner and Volkmar Welker, *The homology of "k-equal" manifolds and related partition lattices*, Advances in Math. **110** (1995), 277–313.

[3] A. Bobenko and R. Seiler (eds.), *Discrete integrable geometry and physics*, Clarendon Press, 1999.

[4] Alain Bossavit, *Computational electromagnetism*, Academic Press, Boston, 1998.

[5] William L. Burke, *Applied differential geometry*, Cambridge University Press, 1985.

[6] Sean Carroll, *Spacetime and geometry: An introduction to general relativity*, Pearson Education, 2003.

[7] Élie Cartan, *Les systèmes differentiels exterieurs et leurs applications géometriques*, Hermann, Paris, 1945.

[8] Mathieu Desbrun, Melvin Leok, and Jerrold E. Marsden, *Discrete Poincaré Lemma*, Appl. Num. Math. (2004).

[9] Aristophanes Dimakis and Folkert Müller-Hoissen, *Discrete Differential Calculus, Graphs, Topologies, and Gauge Theory*, Journal of Mathematical Physics **35** (1994), 6703–6735.

[10] Chris Doran and Anthony Lasenby (eds.), *Geometric algebra for physicists*, Cambridge University Press, 2003.

[11] Herbert Edelsbrunner, David Letscher, and Afra Zomorodian, *Topological persistence and simplification*, IEEE Symposium on Foundations of Computer Science, 2000, pp. 454–463.

[12] Sharif Elcott, Yiying Tong, Eva Kanso, Mathieu Desbrun, and Peter Schröder, *Discrete, Circulation-Preserving and Stable Simplicial Fluid*, ACM Trans. on Graphics **26(1)** (2007).

[13] Harley Flanders, *Differential forms and applications to physical sciences*, Dover, 1990.

[14] Harley Flanders (ed.), *Geometric methods for computational electromagnetics*, EMW Publishing, Cambridge Mass., 2001.

[15] Robin Forman, *Bochner's Method for Cell Complexes and Combinatorial Ricci Curvature*, J. of Discrete and Computational Geom. **29** (2003), no. 3, 323–374.

[16] Theodore Frankel, *The geometry of physics*, Second Edition, Cambridge University Press, 2004.

[17] Steven Gortler, Craig Gotsman, and Dylan Thurston, *One-Forms on Meshes and Applications to 3D Mesh Parameterization*, Computer Aided Geometric Design **23** (2006), no. 2, 83–112.

[18] Paul W. Gross and Robert Kotiuga, *Electromagnetic Theory and Computation: A Topological Approach*, Cambridge University Press, 2004.

[19] Xianfeng Gu and Shing-Tung Yau, *Global conformal surface parameterization*, Proceedings of the Eurographics/ACM SIGGRAPH Symposium on Geometry Processing, Eurographics Association, 2003, pp. 127–137.

[20] Jenny Harrison, *Ravello Lecture Notes on Geometric Calculus – Part I*, Tech. report, UC Berkeley, 2005.

[21] Allen Hatcher, *Algebraic Topology*, Cambridge University Press, 2004.

[22] Klaus Hildebrandt and Konrad Polthier, *Anisotropic filtering of non-linear surface features*, Computer Graphics Forum (Grenoble, France) (M.-P. Cani and M. Slater, eds.), vol. 23, September 2004, Proc. Eurographics 2004.

[23] Anil N. Hirani, *Discrete Exterior Calculus*, Ph.D. thesis, Caltech, 2003.

[24] J.M. Hyman and M. Shashkov, *Natural Discretizations for the Divergence, Gradient, and Curl*, International Journal of Computers and Mathematics with Applications **33** (1997).

[25] Eva Kanso, Marino Arroyo, Yiying Tong, Arash Yavari, Jerrold E. Marsden, and Mathieu Desbrun, *On the Geometric Character of Continuum Mechanics*, Z. angew. Math. Phys. **58** (2007).

[26] Francis Lazarus and Anne Verroust, *Level Set Diagrams of Polyhedral Objects*, Proceedings of the 5th ACM Symposium on Solid Modeling and Applications, 1999, pp. 130–140.

[27] David Lovelock and Hanno Rund, *Tensors, differential forms, and variational principles*, Dover Publications, 1993.

[28] Jerrold E. Marsden and Alan Weinstein, *Coadjoint orbits, vortices and Clebsch variables for incompressible fluids*, Physica D **7** (1983), 305–323.

[29] Jerrold E. Marsden and Matthew West, *Discrete Mechanics and Variational Integrators*, Acta Numerica **10** (2001), 357–514.

[30] Christian Mercat, *Discrete Riemann Surfaces and the Ising Model*, Commun. Math. Phys. **218** (2001), no. 1, 177–216.

[31] Mark Meyer, Mathieu Desbrun, Peter Schröder, and Alan H. Barr, *Discrete differential-geometry operators for triangulated 2-manifolds*, Visualization and Mathematics III (H.-C. Hege and K. Polthier, eds.), Springer, Berlin, 2003, pp. 35–57.

[32] Shigeyuki Morita, *Geometry of differential forms*, Translations of Mathematical Monographs, Vol. 201, Am. Math. Soc., 2001.

[33] James R. Munkres, *Elements of Algebraic Topology*, Addison-Wesley, Menlo Park, CA, 1984.

[34] Ulrich Pinkall and Konrad Polthier, *Computing Discrete Minimal Surfaces*, Experimental Mathematics **2** (1993), no. 1, 15–36.

[35] Vasu Ramaswamy and Vadim Shapiro, *Combinatorial Laws for Physically Meaningful Design*, J. of Computing and Information Science in Engineering **4** (2004), no. 1, 3–10.

[36] Urs Schreiber, *On Superstrings in Schrödinger Representation*, Preprint (2003).

[37] Richard W. Sharpe, *Differential geometry: Cartan's generalization of Klein's Erlangen programm*, Springer-Verlag, NY, 1997.

[38] Karl F. Warnick, Richard H. Selfridge, and David V. Arnold, *Teaching Electromagnetic Field Theory Using Differential Forms*, IEEE Trans. on Education **40** (1997), no. 1, 53–68.

[39] H. Whitney, *Geometric integration theory*, Princeton Press, Princeton, 1957.

[40] Roman Zapatrin, *Polyhedral Representations of Discrete Differential Manifolds*, preprint (1996).

Mathieu Desbrun
California Institute of Technology
MS 256–80
Pasadena, CA 91125
USA
e-mail: mathieu@caltech.edu

Eva Kanso
University of Southern California
Dept. of Aerospace and Mechanical Engineering
Los Angeles, CA 90089–1453
USA
e-mail: kanso@usc.edu

Yiying Tong
California Institute of Technology
MS 256–80
Pasadena, CA 91125
USA
e-mail: yiying@caltech.edu

Discrete Differential Geometry, A.I. Bobenko, P. Schröder, J.M. Sullivan and G.M. Ziegler, eds.
Oberwolfach Seminars, Vol. 38, 325–337

A Discrete Model of Thin Shells

Eitan Grinspun

Abstract. We describe a discrete model for the dynamics of thin flexible structures, such as hats, leaves, and aluminum cans, which are characterized by a curved undeformed configuration. Previously such *thin-shell* models required complex continuum mechanics formulations and correspondingly complex algorithms. We show that a simple shell model can be derived geometrically for triangle meshes and implemented quickly by modifying a standard cloth simulator. Our technique convincingly simulates a variety of curved objects with materials ranging from paper to metal, as we demonstrate with several examples including a comparison of a real and simulated falling hat.

Keywords. Discrete differential geometry, physical simulation, thin shell, curvature energy, bending, buckling.

1. Introduction

Thin shells are thin flexible structures with a high ratio ($>$ 100) of width to thickness [7]. While their well-known counterparts, thin *plates*, relax to a *flat* shape when unstressed, thin *shells* are characterized by a *curved undeformed configuration*. Cloth, recently studied in the computer animation literature, may be modeled as a thin plate, since garments are typically constructed from flat textiles. In contrast, thin-walled objects which are naturally curved (e.g., leaves, fingernails), or put into that shape through plastic deformation

FIGURE 1. Frames from the simulation of tumbling thin shell.

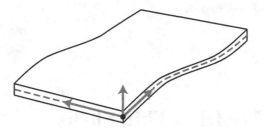

FIGURE 2. The local coordinate frame in a small neighborhood of a thin shell: two axes span the *middle surface*, and the normal *shell director* spans the thickness.

(e.g., hats, cans, carton boxes, pans, car bodies, detergent bottles) are thin shells and cannot be modeled using plate formulations.

Thin shells are remarkably difficult to simulate. Because of their degeneracy in one dimension, shells do not admit to straightforward tessellation and treatment as three-dimensional solids; indeed, the numerics of such approaches become catastrophically ill-conditioned, foiling numerical convergence and accuracy. Robust finite-element methods for thin-shell equations continue to be an active and challenging research area.

This chapter, which is based on [18], develops a simple model for thin shells with applications to computer animation. Our *discrete model* of shells captures the same characteristic behaviors as more complex models, with a surprisingly simple implementation. The model described here was previously demonstrated to be useful in computer animation; its ability to capture the qualitative behavior of thin shells was evaluated in various experiments, including comparisons to real-world shells (see Figure 3).

2. Kinematics

Since it is thin, the geometry of the shell is well described by its *middle surface* (see Figure 2). At any point on the middle surface the local tangent plane and surface normal induce a coordinate frame in which to describe "motion along the surface" and "motion along thickness".

In the discrete setting, the topology of the middle surface is represented by the combinatorics of an oriented 2-manifold simplicial complex, $M = \{v, e, f\}$, where $v = \{v_1, v_2, \ldots\}$, $e = \{e_1, e_2, \ldots\}$, $f = \{f_1, f_2, \ldots\}$ are sets of vertices, edges and faces, respectively. The geometry of the middle surface is given by the discrete *configuration map*, $C : v \mapsto \mathbb{R}^3$, which assigns to every vertex, v_i, a position, $C(v_i)$, in the ambient space. Together M and C correspond to the usual notion of a *triangle mesh* in $\mathbb{R}^3$; in our exposition we assume fixed combinatorics M, and discuss a temporally evolving configuration, C_t where the subscript refers to a specific instant in time.

Restricting our attention to elastic ("memory-less") materials, the physics can be understood in terms of the *undeformed* configuration (the stress-free shape) and the *deformed* configuration (the stressed shape at the current instant in time), C_0 and C_1, respectively. The elastic response of a material depends on the *change in shape* of the elastic body, and on the *constitutive laws* that describe the restoring forces associated to this change in shape. The former is a purely geometric quantity.

What is the change in shape between C_0 and C_1? Since rigid motions (translations and rotations) do not affect shape, the answer must be invariant under composition of C_0 (likewise C_1) with any rigid-body transformation. A simple theorem is that any reasonable measure of change in shape, or *generalized strain*, may be written as a function of only the edge lengths and dihedral angles of C_0 and C_1. The proof lies in showing that the configuration can be completely recovered from the edge lengths and dihedral angles, up to an unknown (but here inconsequential) rigid-body transformation. We will also expect our measure of strain to be zero when shape has not changed, and nonzero whenever shape has changed. In particular, strain should "see" *any local change in shape*.

Among the simplest forms of generalized strain that satisfy these requirements are two expressions that are evaluated at a specific edge e_i. Comparing C_0 to C_1, let $s^e(e_i)$ be the difference in length of edge e_i, and let $s^\theta(e_i)$ be the difference in dihedral angle at e_i.

While these are perhaps the simplest possible measures of generalized strain, other more complex formulas can offer attendant advantages. Recent research in discrete shell models has focused on functions evaluated over mesh faces which aggregate in one term the configuration of all the incident edge lengths and dihedral angles [13]. Nevertheless, our goal here is to develop the *simplest* discrete model of thin shells capturing their qualitative elastic behavior.

3. Constitutive model

Having defined the *geometry* of thin shells, we turn our attention to the governing *physical* equations. The *stored elastic energy* of a thin shell is at the heart of the equations which govern its response to elastic deformations. The stored energy, $W[C_0, C_1]$, should be a function of the local strain, integrated over the middle surface.

We choose the simplest expression for energy that is consistent with Hookean mechanics. In 1676 Robert Hooke stated

The power [*sic*] of any springy body is in the same proportion with the extension.

This statement was the birth of modern elasticity, which states that a first order approximation for the response of a material is a force proportional to strain, and, consequently (by the definition of work as force over distance), that the first approximation of stored energy is quadratic in strain. We propose an energy with two kinds of terms, measuring stretching and bending modes, respectively:

$$W_M[C_0, C_1] = \sum_{e_i \in M} k_i^e \cdot \left(s^e(e_i)\right)^2 + \sum_{e_k \in M} k_k^\theta \cdot \left(s^\theta(e_k)\right)^2.$$

This expression has several desirable properties. First, it is positive whenever the shapes of C_0 and C_1 differ, and zero otherwise. Second, evaluations over subsets of M satisfy the usual inclusion/exclusion principle: for $A, B \subset M$, $W_M = W_A + W_B - W_{A \cap B}$, which is consistent with continuum formulations in which energy is defined as an integral of energy density over the middle surface. Third, because strain is invariant under rigid-body transformations of the undeformed and deformed configurations, Noether's theorem guarantees that the resulting dynamics will conserve linear and angular momentum.

We offer the following interpretations for the above membrane and bending terms:
Membrane. Elastic surfaces resist stretching (local change in length). While some materials such as rubber sheets may undergo significant deformations in the stretching or shearing (*membrane*) modes, we focus on inextensible shells which are characterized by nearly *isometric* deformations, i.e., possibly significant deformations in bending but unnoticeable deformation in the membrane modes. Most membrane models for triangle meshes satisfy this small-membrane-strain assumption with choice of suitably large membrane stiffness coefficient, k_i^e.

Rewriting the membrane term in the following form permits an alternative interpretation:

$$W^e(e_i) = k^e \left(|e_i| - |\bar{e}_i|\right)^2 = k^e |\bar{e}_i|^2 \left(\frac{|e_i|}{|\bar{e}_i|} - 1\right)^2$$

where $|e_i|$ is the length of edge i, quantities with a bar (such as $\bar{e}_i$) refer to the undeformed configuration C_0 and remaining quantities are with respect to C_1; note that we have dropped the subscript on k_i^e indicating a uniform material stiffness over the domain of interest. This is a unitless strain measurement, squared, and then integrated over the area of the local neighborhood, and multiplied by the material-dependent parameters. Observe that under regular refinement of a triangle mesh, the local area indeed scales as $|\bar{e}_j|^2$, which has units of area. The units of the material parameters are energy per unit area, i.e., surface energy density. In engineering models of shells, the material parameter is given as a volume energy density, and the energy is integrated over shell thickness yielding a surface energy density. Efficient implementations of this formula precompute the quantities $k^e |\bar{e}_i|^2$, which depend only on the undeformed configuration.

Bending. Consider the proposed discrete bending energy in relation to its continuous analogues. Models in mechanics are typically based on tensors, and in particular shell models use the difference of the second fundamental forms in the deformed and undeformed configurations (pulling back the deformed tensor onto the undeformed configuration). These treatments derive tensorial expressions over smooth manifolds, and, as a final step, discretize the energy functional (or directly the forces) to carry out the numerics. Our approach instead builds up a discrete energy functional by starting from a discrete shape operator.

The *shape operator* is the derivative of the Gauss map[1]: geometrically, it measures the local curvature at a point on a smooth surface. Our bending energy is an extrinsic measure of the difference between the shape operator evaluated on the deformed and

---

[1] This is the map from the surface to the unit sphere, mapping each surface point to its unit surface normal.

FIGURE 3. Real footage vs. Simulation: top, a real hat is dropped on a table; bottom, our shell simulation captures the bending of the brim. Notice that volumetric-elasticity, plate, or cloth simulations could not capture this behavior, while earlier work on shell simulation required significant implementation and expertise.

undeformed surfaces. We express this difference as the *squared difference of mean curvature*:

$$[\mathrm{Tr}(\varphi^*S) - \mathrm{Tr}(\bar{S})]^2 = 4(H \circ \varphi - \bar{H})^2 , \qquad (3.1)$$

where $\bar{S}$ and S are the shape operators evaluated over the undeformed and deformed configurations, respectively; likewise, $\bar{H}$ and H are the mean curvatures; φ^*S is the pullback of S onto $\bar{\Omega}$, and we use $\mathrm{Tr}(\varphi^*S) = \varphi^* \mathrm{Tr}(S) = \mathrm{Tr}(S) \circ \varphi = H \circ \varphi$ for a diffeomorphism φ. This measure is *extrinsic*: it sees only changes in the *embedding* of the surface in $\mathbb{R}^3$. Integrating (3.1) over the reference domain we find the continuous flexural energy $\int_{\bar{\Omega}} 4(H \circ \varphi - \bar{H})^2 d\bar{A}$. Next, we discretize this integral over the *piecewise linear mesh* that represents the shell.

We derive the discrete, integral *mean-curvature squared* operator as follows. We first partition the undeformed surface into a disjoint union of diamond-shaped tiles, $\bar{T}$, associated to each mesh edge, e, as indicated on the side figure. Following [23], one can use the barycenter of each triangle to define these regions—or, alternatively, the circumcenters. Over such a diamond, the mean curvature integral is $\int_{\bar{T}} \bar{H} d\bar{A} = \bar{\theta}|\bar{e}|$ (for a proof see [10]). A similar argument leads to: $\int_{\bar{T}}(H \circ \varphi - \bar{H})d\bar{A} = (\theta - \bar{\theta})|\bar{e}|$. Using the notion of area-averaged value from [23], we deduce that

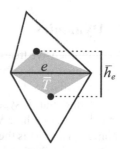

$(H \circ \varphi - \bar{H})|_{\bar{T}} = (\theta - \bar{\theta})/\bar{h}_e$, where $\bar{h}_e$ is the span of the undeformed tile, which is one sixth of the sum of the heights of the two triangles sharing $\bar{e}$. For a sufficiently fine, non-degenerate tessellation approximating a smooth surface, the average over a tile (converging pointwise to its continuous counterpart) *squared* is equal to the squared average, leading to: $\int_{\bar{T}}(H \circ \varphi - \bar{H})^2 d\bar{A} = (\theta - \bar{\theta})^2|\bar{e}|/\bar{h}_e$.

Comparing an argument presented in [23], we observe that there may be numerical advantages in using circumcenters instead of barycenters for the definition of the diamond tiles (except in triangles with obtuse angles). This affects the definition of $\bar{h}_e$ and of the lumped mass below. Since we only need to compute these values for the undeformed shape, the implementation and performance *only of initialization code* would be affected. Bobenko notes that when circumcenters are used, this formulation of discrete shells coincides (for flat undeformed configurations) with the derivation of the discrete Willmore energy based on circle packing [4].

As we have just seen, we can express our *discrete flexural energy* as a summation over mesh edges,

$$W^\theta(e_k) = K^\theta \left(\theta_k - \bar{\theta}_k \right)^2 \frac{|\bar{e}_k|}{\bar{h}_k}, \tag{3.2}$$

where the term for edge e_k is where θ_k and $\bar{\theta}_k$ are corresponding complements of the dihedral angle of edge e_k measured in the deformed and undeformed configuration, respectively, K^θ is the material bending stiffness, and $\bar{h}_k$ is a third of the average of the heights of the two triangles incident to the edge e_k (see the appendix for another possible definition of $\bar{h}_k$). Note that the unit of K^θ is energy (not surface energy density). This formulation is consistent with the physical scaling laws of thin shells: if the (deformed and undeformed) geometry of a thin shell is uniformly scaled by λ along each axis, then surface area scales as λ^2 as does the total membrane energy, *however* the total bending (curvature squared) energy is *invariant* under uniform scaling.

Following the reasoning for (3.1), we could have formed a second energy term taking the determinant instead of the trace of S. This would lead to a difference of Gaussian curvatures, but this is always zero under isometric deformations (pure bending). This is not surprising, as Gaussian curvature is an *intrinsic* quantity, i.e., it is independent of the embedding of the two-dimensional surface into its ambient three-dimensional space. In contrast, flexural energy measures *extrinsic* deformations.

4. Dynamics

A comprehensive treatment of the temporal evolution of a thin shell is beyond the scope of this chapter; we briefly summarize the basic components required to simulate the motion of thin shells.

Our dynamic system is governed by the ordinary differential equation of motion $\ddot{\mathbf{x}} = -\mathbf{M}^{-1} \nabla W(\mathbf{x})$ where $\mathbf{x}$ is the vector of unknown DOFs (i.e., the vertices of the deformed geometry) and $\mathbf{M}$ is the mass matrix. We use the conventional simplifying hypothesis that the mass distribution is lumped at vertices: the matrix $\mathbf{M}$ is then diagonal, and the mass assigned to a vertex is a third of the total area of the incident triangles, scaled by the area mass density.

Newmark time-stepping. We adopt the Newmark scheme [24] for ODE integration,

$$\mathbf{x}_{i+1} = \mathbf{x}_i + \Delta t_i \dot{\mathbf{x}}_i + \Delta t_i^2 \left((1/2 - \beta) \ddot{\mathbf{x}}_i + \beta \ddot{\mathbf{x}}_{i+1} \right),$$

$$\dot{\mathbf{x}}_{i+1} = \dot{\mathbf{x}}_i + \Delta t_i \left((1 - \gamma) \ddot{\mathbf{x}}_i + \gamma \ddot{\mathbf{x}}_{i+1} \right),$$

where Δt_i is the duration of the ith timestep, $\dot{\mathbf{x}}_i$ and $\ddot{\mathbf{x}}_i$ are configuration velocity and acceleration at the beginning of the ith timestep, respectively, and β and γ are adjustable parameters linked to the accuracy and stability of the time scheme. The numerical advantages of the Newmark scheme are discussed in [30]. Newmark is either an explicit ($\beta = 0$) or implicit ($\beta > 0$) integrator: we used $\beta = 1/4$ for final production, and $\beta = 0$ to aid in debugging. Newmark gives control over numerical damping via its second parameter γ. We obtained the best results by minimizing numerical damping ($\gamma = 1/2$).

Dissipation. We model dissipation due to flexural oscillations by introducing a damping force proportional to $(\dot{\theta} - \bar{\dot{\theta}})\nabla_x\theta$, where $\nabla_x\theta$ is the gradient of the bending angle with respect to the mesh position. For elastic deformations, $\bar{\dot{\theta}} = 0$; for plastoelastic materials, $\bar{\dot{\theta}}$ is in general nonzero. This approach may be viewed as a generalization of Rayleigh damping forces based on the strain rate tensor [2].

Discussion. This discrete flexural energy (3.2) generalizes established formulations for (*flat*) plates both continuous and discrete: (a) Ge and coworkers presented a geometric argument that the stored energy of a continuous inextensible plate has the form $\int_{\bar{\Omega}} c_H H^2 + c_K K dA$ for material-specific coefficients c_H and c_K [12]; (b) Haumann used a discrete hinge energy [20], similarly Baraff and Witkin used a discrete constraint-based energy [2], of the form $W_B(\mathbf{x}) = \sum_{\bar{e}} \theta_e^2$. Our approach generalizes both (a) and (b), and produces convincing simulations beyond the regime of thin plate and cloth models (see Section 5).

The proposed discrete model has three salient features: (a) the energy is invariant under rigid-body transformation of both the undeformed and the deformed shape: our system conserves linear and angular momenta; (b) the piecewise nature of our geometry description is fully captured by the purely intrinsic membrane terms, and the purely extrinsic bending term; most importantly, (c) it is *simple* to implement.

5. Results

We exercised our implementation on various problems, including fixed beams, falling hats, and pinned paper. Computation time, on a 2GHz Pentium 4 CPU, ranged from 0.25s–3.0s per frame; timings are based on a research implementation that relies on automatic differentiation techniques.

Beams. We pinned to a wall one end of a V-beam, and released it under gravity. Figure 4 demonstrates the effect of varying flexural stiffness on oscillation amplitude and frequency. The flexural energy coefficient has a high dynamic range; extreme values (from pure-membrane to near-rigid) remain numerically and physically well behaved. Observe that increasing flexural stiffness augments structural rigidity. Compare the behavior of beams: the non-flat cross section of the V-beam contributes to structural rigidity. This difference is most pronounced in the operating regime of low flexural stiffness (but high membrane stiffness). Here the material does not inherently resist bending, but a V-shaped cross section effectively converts a bending deformation into a stretching deformation. One can mimic this experiment by holding a simple paper strip by its end; repeat after folding a V-shaped cross section.

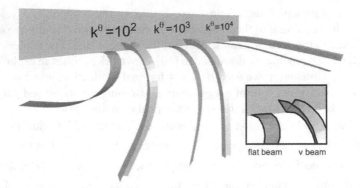

FIGURE 4. Three pairs of flat and V-beams with increasing flexural stiffness K^θ (left to right) of 100, 1000, and 10000; the non-flat cross section of the V-beam contributes to structural rigidity.

Elastic hats. We dropped both real and virtual hats and compared (see Figure 3): the deformation is qualitatively the same, during impact, compression, and rebound. Adjusting the damping parameter, we capture or damp away the brim's vibrations. Adjusting the flexural stiffness, we can make a hat made of hard rubber or textile: a nearly rigid hat or a floppy hat).

Plastoelasticity. As discussed in the early work of Kergosien and coworkers, a compelling simulation of paper would require a mechanical shell model [22]. Using our simple shell model, we can easily simulate a sheet of paper that is rolled, then creased, then pinned (see Figure 5). Here the physics require *plastic* as well as elastic deformations. We begin with a flat surface, and gradually increase the undeformed angles, $\bar{\theta}_e$. Notice: modifying the *undeformed* configuration effects a *plastic* deformation. The kinematics of changing $\bar{\theta}_e$ span only physically realizable bending, i.e., *inextensible* plastic deformations. In contrast, directly modifying $\bar{x}$ could introduce plastic deformations with unwanted membrane modes. We introduced elastic effects by applying three pin constraints to the *deformed* configuration. Observe the half-crease on the left side. The (plastically deformed) left and (untouched) right halves have incompatible undeformed shapes. In this situation, there does not exist a globally consistent undeformed (strain-free) state.

Recent extensions. More recently, we demonstrated that simple, discrete models of thin shells can also produce striking examples of shattering glass (see Figure 6) [13], and paper origami (see Figure 7) [5].

Implementation. An attractive practical aspect of the proposed model is that it may be easily incorporated into working code of a standard cloth or thin-plate simulator such as those commonly used by the computer graphics community [2]. One must replace the bending energy with (3.2). From an implementation point of view, this involves minimal work. For example, consider that [2] already required all the computations relating to θ_e. These and other implementation details were outlined in [18].

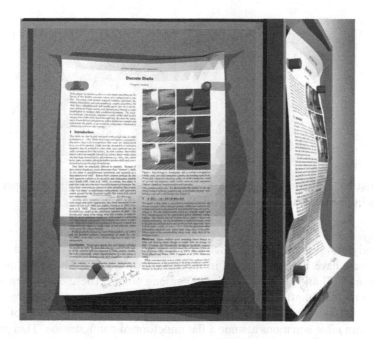

FIGURE 5. Modeling a curled, creased, and pinned sheet of paper: by altering dihedral angles of the reference configuration, we effect plastic deformation. While the rendering is texture-mapped we kept flat-shaded triangles to show the underlying mesh structure.

FIGURE 6. A measure of discrete strain is used to fracture a thin shell in this simulation of a shattering lightbulb.

6. Further reading

A comprehensive survey of this expansive body of literature is far beyond the scope of this chapter; as a starting point see [1, 9] and references therein. Here we highlight only a few results from the graphics and engineering literature.

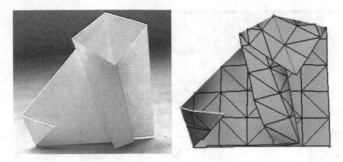

FIGURE 7. Virtual origami: user-guided simulated folding of a paper
sheet produces a classical origami dog.

Recently, novel numerical treatments of shells, significantly more robust than earlier approaches, have been introduced in mechanics by Cirak et al. [8] and in graphics
by Green et al. [15] and Grinspun et al. [19] among others. These continuum-based approaches use the Kirchhoff–Love constitutive equations, whose energy captures curvature
effects in curved coordinate frames; consequently, they model a rich variety of materials.
In contrast, thin *plate* equations assume a flat undeformed configuration. Thin plate models are commonly used for cloth and garment simulations and have seen successful numerical treatment in the computer graphics literature (see [21] and references therein). Thin
plates have also been useful for variational geometric modeling [6, 16, 29] and intuitive
direct manipulation of surfaces [25, 26]. In graphics, researchers have used two kinds of
approaches to modeling plates: finite elements and mass-spring networks. In the latter re-
sistance to bending is effected by springs connected to opposite corners of adjacent mesh
faces. Unfortunately, this simple approach does not carry over to curved undeformed configurations: the diagonal springs are insensitive to the sign of the dihedral angles between
faces.

In this chapter we have developed a very simple discrete model of thin shells. One
price that must be paid for this simplicity is that, while we have taken care to ensure
the correct scaling factors for each energy term, for an arbitrary triangle mesh we cannot
guarantee the convergence of this model to its continuum equivalent. In [17] we present
experimental results comparing the convergence of the discrete shell and other discrete
curvature operators.

In [3, 27, 11], we consider nearly inextensible thin shells. Under the assumption
of *isometric surface deformations*, we show that the smooth thin-plate bendingy energy
is quadratic in surface positions, and the smooth thin-shell bending energy is cubic in
surface positions. To carry these obesrvations over to the discrete setting, we build on an
axiomatic treatment of discrete Laplace operators (further developed in [28]), deriving
a family of *discrete isometric bending models* for triangulated surfaces in 3-space. The
resulting family of thin-plate (resp. shell) discrete bending models has simple linear (resp.
quadratic) energy gradients and constant (resp. linear) energy Hessians. The simplicity
of the energy gradients and Hessians enables fast time-integration of cloth [3] and shell

dynamics [11] and near-interactive rates for Willmore smoothing of large meshes [27]. In a simulation environment, methods that assume isometric deformations must be coupled with an efficient technique for ensuring material inextensability. In [14] we present an efficient algorithm, based on constrained Lagrangian mechanics, for projecting a given surface deformation onto the manifold of isometric deformations.

Acknowledgments

The work described here is the fruit of a collaboration with Mathieu Desbrun, Anil Hirani, and Peter Schröder [18]. References to recent work on modeling thin shells refer to an active collaboration with Miklós Bergou, Akash Garg, Yotam Gingold, Rony Goldenthal, David Harmon, Saurabh Mathur, Jason Reisman, Adrian Secord, Max Wardetzky, Zoë Wood, and Denis Zorin. The author is indebted to Alexander Bobenko, Jerry Marsden, and Anastasios Vayonakis for insightful discussions. Pierre Alliez, Ilja Friedel, Charles Han, Jeff Han, Harper Langston, Anthony Santoro and Steven Schkolne were pivotal in the production of the images shown here. The author is grateful for the generous support of the National Science Foundation (MSPA Award No. IIS–05–28402, CSR Award No. CNS–06–14770, CAREER Award No. CCF–06–43268).

References

[1] Douglas Arnold, *Questions on shell theory*, Workshop on Elastic Shells: Modeling, Analysis, and Computation, 2000.

[2] David Baraff and Andrew Witkin, *Large steps in cloth simulation*, Proceedings of SIGGRAPH, 1998, pp. 43–54.

[3] Miklós Bergou, Max Wardetzky, David Harmon, Denis Zorin, and Eitan Grinspun, *A quadratic bending model for inextensible surfaces*, Eurographics Symposium on Geometry Processing, June 2006, pp. 227–230.

[4] Alexander I. Bobenko, *A conformal energy for simplicial surfaces*, Combinatorial and Computational Geometry, vol. 52, MSRI Publications, August 2005, pp. 133–143.

[5] R. Burgoon, E. Grinspun, and Z. Wood, *Discrete shells origami*, Proceedings of CATA (Seattle, WA, USA), 2006, pp. 180–187.

[6] George Celniker and Dave Gossard, *Deformable curve and surface finite elements for freeform shape design*, Computer Graphics (Proceedings of SIGGRAPH 91) **25** (1991), no. 4, 257–266.

[7] Philippe Ciarlet, *Mathematical Elasticity. Vol. III*, Studies in Mathematics and its Applications, vol. 29, North Holland, Amsterdam, May 2000.

[8] F. Cirak, M. Ortiz, and P. Schröder, *Subdivision surfaces: A new paradigm for thin-shell finite-element analysis*, Internat. J. Numer. Methods Engrg. **47** (2000), no. 12, 2039–2072.

[9] F. Cirak, M.J. Scott, E.K. Antonsson, M. Ortiz, and P. Schröder, *Integrated modeling, finite-element analysis, and engineering design for thin-shell structures using subdivision*, Computer Aided Design **34** (2002), no. 2, 137–148.

[10] David Cohen-Steiner and Jean-Marie Morvan, *Restricted delaunay triangulations and normal cycle*, Proceedings of the 19th Annual Symposium on Computational Geometry, 2003, pp. 312–321.

[11] Akash Garg, Eitan Grinspun, Max Wardetzky, and Denis Zorin, *Cubic shells*, ACM/Eurographics Symposium on Computer Animation, Eurographics Association, 2007, pp. 91–98.

[12] Z. Ge, H.P. Kruse, and J.E. Marsden, *The limits of hamiltonian structures in three-dimensional elasticity, shells, and rods*, Journal of Nonlinear Science **6** (1996), 19–57.

[13] Yotam Gingold, Adrian Secord, Jefferson Y. Han, Eitan Grinspun, and Denis Zorin, *Poster: A discrete model for inelastic deformation of thin shells*, ACM/Eurographics Symposium on Computer Animation '04 (Grenoble, France), 2004.

[14] Rony Goldenthal, David Harmon, Raanan Fattal, Michel Bercovier, and Eitan Grinspun, *Efficient simulation of inextensible cloth*, ACM Trans. Graph. **26** (2007), no. 3, 49.

[15] Seth Green, George Turkiyyah, and Duane Storti, *Subdivision-based multilevel methods for large scale engineering simulation of thin shells*, Proceedings of ACM Solid Modeling, 2002, pp. 265–272.

[16] G. Greiner, *Variational design and fairing of spline surfaces*, Computer Graphics Forum **13** (1994), no. 3, 143–154.

[17] Eitan Grinspun, Yotam Gingold, Jason Reisman, and Denis Zorin, *Computing discrete shape operators on general meshes*, Computer Graphics Forum **25** (2006), no. 3, 547–556.

[18] Eitan Grinspun, Anil Hirani, Mathieu Desbrun, and Peter Schröder, *Discrete shells*, ACM/Eurographics Symposium on Computer Animation, 2003, pp. 62–67.

[19] Eitan Grinspun, Petr Krysl, and Peter Schröder, *CHARMS: a simple framework for adaptive simulation*, ACM Transactions on Graphics **21** (2002), no. 3, 281–290.

[20] R. Haumann, *Modeling the physical behavior of flexible objects*, Topics in Physically-based Modeling, *Eds. Barr, Barrel, Haumann, Kass, Platt, Terzopoulos, and Witkin*, SIGGRAPH Course Notes, vol. 16, ACM SIGGRAPH, 1987.

[21] Donald H. House and David E. Breen (eds.), *Cloth modeling and animation*, A.K. Peters, 2000.

[22] Y.L. Kergosien, H. Gotoda, and T.L. Kunii, *Bending and creasing virtual paper*, IEEE Computer Graphics and Applications (1994), 40–48.

[23] Mark Meyer, Mathieu Desbrun, Peter Schröder, and Alan H. Barr, *Discrete differential-geometry operators for triangulated 2-manifolds*, Visualization and mathematics III, Math. Vis., Springer, Berlin, 2003, pp. 35–57.

[24] N.M. Newmark, *A method of computation for structural dynamics*, ASCE J. of the Engineering Mechanics Division **85** (1959), no. EM 3, 67–94.

[25] Hong Qin and Demetri Terzopoulos, *D-NURBS: A physics-based framework for geometric design*, IEEE Transactions on Visualization and Computer Graphics **2** (1996), no. 1, 85–96.

[26] Demetri Terzopoulos and Hong Qin, *Dynamic nurbs with geometric constraints for interactive sculpting*, ACM Transactions on Graphics **13** (1994), no. 2, 103–136.

[27] Max Wardetzky, Miklós Bergou, David Harmon, Denis Zorin, and Eitan Grinspun, *Discrete Quadratic Curvature Energies*, Computer Aided Geometric Design **24** (2007), no. 8–9, 499–518.

[28] Max Wardetzky, Saurabh Mathur, Felix Kälberer, and Eitan Grinspun, *Discrete Laplace operators: No free lunch*, Eurographics Symposium on Geometry Processing, Jul 2007, pp. 33–37.

[29] William Welch and Andrew Witkin, *Variational surface modeling*, Computer Graphics (Proceedings of SIGGRAPH 92) **26** (1992), no. 2, 157–166.

[30] M. West, C. Kane, J.E. Marsden, and M. Ortiz, *Variational integrators, the Newmark scheme, and dissipative systems*, International Conference on Differential Equations 1999 (Berlin), World Scientific, 2000, pp. 1009–1011.

Eitan Grinspun
Department of Computer Science
Columbia University
500 W 120th Street
New York, NY, 10027
USA
e-mail: eitan@cs.columbia.edu

Index

Oberwolfach Seminars (OWS)

The workshops organized by the *Mathematisches Forschungsinstitut Oberwolfach* are intended to introduce students and young mathematicians to current fields of research. By means of these well-organized seminars, also scientists from other fields will be introduced to new mathematical ideas. The publication of these workshops in the series *Oberwolfach Seminars* (formerly *DMV seminar*) makes the material available to an even larger audience.

BIRKHÄUSER

OWS 38: Bobenko, A.I. / Schröder, P. / Sullivan, J.M. / Ziegler, G.M. (Eds.), Discrete Differential Geometry (2008). ISBN 978-3-7643-8620-7

OWS 37: Galdi, G.P. / Rannacher, R. / Robertson, A.M. / Turek, S., Hemodynamical Flows (2008). ISBN 978-3-7643-7805-9

This book surveys results on the physical and mathematical modeling as well as the numerical simulation of hemodynamical flows, i.e., of fluid and structural mechanical processes occurring in the human blood circuit. The topics treated are continuum-mechanical description, choice of suitable liquid and wall models, mathematical analysis of coupled models, numerical methods for flow simulation, parameter identification and model calibration, fluid-solid interaction, mathematical analysis of piping systems, particle transport in channels and pipes, artificial boundary conditions, and many more. Hemodynamics is an area of active current research, and this book provides an entry into the field for graduate students and researchers.

OWS 36: Cuntz, J. / Meyer, R. / Rosenberg, J.M., Topological and Bivariant K-theory (2007). ISBN 978-3-7643-8398-5

Topological K-theory is one of the most important invariants for noncommutative algebras. Bott periodicity, homotopy invariance, and various long exact sequences distinguish it from algebraic K-theory. We describe a bivariant K-theory for bornological algebras, which provides a vast generalization of topological K-theory. In addition, we discuss other approaches to bivariant K-theories for operator algebras. As applications, we study K-theory of crossed products, the Baum-Connes assembly map, twisted K-theory with some of its applications, and some variants of the Atiyah-Singer Index Theorem.

OWS 35: Itenberg, I. / Mikhalkin, G. / Shustin, E., Tropical Algebraic Geometry (2007). ISBN 978-3-7643-8309-1

Tropical geometry is algebraic geometry over the semifield of tropical numbers, i.e., the real numbers and negative infinity enhanced with the (max,+)-arithmetics. Geometrically, tropical varieties are much simpler than their classical counterparts. Yet they carry information about complex and real varieties.
These notes present an introduction to tropical geometry and contain some applications of this rapidly developing and attractive subject. It consists of three chapters which complete each other and give a possibility for non-specialists to make the first steps in the subject which is not yet well represented in the literature. The intended audience is graduate, post-graduate, and Ph.D. students as well as established researchers in mathematics.

OWS 34: Lieb, E.H. / Seiringer, R. / Solovej, J.P. / Yngvason, J., The Mathematics of the Bose Gas and its Condensation (2005). ISBN 978-3-7643-7336-8

OWS 33: Kreck, M. / Lück, W., The Novikov Conjecture: Geometry and Algebra (2004). ISBN 978-3-7643-7141-8

DMV 32: Bolthausen, E. / Sznitman, A.-S., Ten Lectures on Random Media (2002). ISBN 978-3-7643-6703-9

DMV 31: Huckleberry, A. / Wurzbacher, T. (Eds.), Infinite Dimensional Kähler Manifolds (2001). ISBN 978-3-7643-6602-5

DMV 30: Scholz, E. (Ed.), Hermann Weyl's *Raum—Zeit—Materie* and a General Introduction to His Scientific Work (2001). ISBN 978-3-7643-6476-2

DMV 29: Kalai, G. / Ziegler, G.M. (Eds.), Polytopes — Combinatorics and Computation (2000). ISBN 978-3-7643-6351-2

DMV 28: Cercignani, C. / Sattinger, D., Scaling Limits and Models in Physical Processes (1998). ISBN 978-3-7643-5985-0